The Gravitational Million–Body Problem

The globular star clusters of the Milky Way contain hundreds of thousands of stars held together by gravitational interactions, and date from the time when the Milky Way was forming. This text describes the theory astronomers need for studying globular star clusters. The gravitational million-body problem is an idealised model for understanding the dynamics of a cluster with a million stars. After introducing the million-body problem from various view-points, the book systematically develops the tools needed for studying the million-body problems in nature, and introduces the most important theoretical models. Including a comprehensive treatment of few-body interactions, and developing an intuitive but quantitative understanding of the three-body problem, the book introduces numerical methods, relevant software, and current problems. Suitable for graduate students and researchers in astrophysics and astronomy, this text also has important applications in the fields of theoretical physics, computational science and mathematics.

Douglas Heggie is Professor of Mathematical Astronomy, and teaches applied mathematics at the University of Edinburgh, Scotland. He has devoted almost his entire research career to stellar dynamics, especially star clusters and the three-body problem. He has been President of Commission 37 of the International Astronomical Union, and is a fellow of the Royal Society of Edinburgh and of the Royal Astronomical Society.

Piet Hut is Professor of Astrophysics at the Institute for Advanced Study in Princeton. He obtained a PhD in astrophysics at the University of Amsterdam, and in 1986 he published with Joshua Barnes the tree code algorithm which soon became the method of choice for large-scale simulations in collisionless stellar dynamics. His latest research project involves the creation of a virtual reality simulator to fly through the full four-dimensional space–time history of a simulated globular cluster.

The Gravitational Million-Body Problem

Douglas Heggie
School of Mathematics,
University of Edinburgh,
Edinburgh EH9 3JZ, Scotland

Piet Hut
Institute for Advanced Study,
Princeton, NJ 08540, U.S.A.

PUBLISHED BY THE PRESS SYNDICATE OF THE UNIVERSITY OF CAMBRIDGE
The Pitt Building, Trumpington Street, Cambridge, United Kingdom

CAMBRIDGE UNIVERSITY PRESS
The Edinburgh Building, Cambridge CB2 2RU, UK
40 West 20th Street, New York, NY 10011-4211, USA
477 Williamstown Road, Port Melbourne, VIC 3207, Australia
Ruiz de Alarcón 13, 28014 Madrid, Spain
Dock House, The Waterfront, Cape Town 8001, South Africa

http://www.cambridge.org

First published 2003

Printed in the United Kingdom at the University Press, Cambridge

Typefaces Times 10.25/13.5 pt and Joanna   *System* LaTeX 2_ε [TB]

A catalogue record for this book is available from the British Library

Library of Congress Cataloguing in Publication data

Heggie, D. C.
The gravitational million-body problem / Douglas Heggie, Piet Hut.
    p.   cm.
Includes bibliographical references and index.
ISBN 0 521 77303 2 – ISBN 0 521 77486 1 (pbk.)
1. Few-body problem.   2. Stars – Globular clusters.   3. Gravitation.
I. Hut, Piet, 1952–   II. Title.
QB362.F47 H44   2002
523.8′55 – dc21   2002067410

ISBN 0 521 77303 2 hardback
ISBN 0 521 77486 1 paperback

For Eiko, Linda, Caroline and Alison

Contents

Preface

Since this book is aimed at a broad audience within the physical sciences, we expect most of our readers not to be experts in either astrophysics or mathematics. For those readers, the title of this book may seem puzzling at least. Why should they be interested in the *gravitational* attraction between bodies? What is so special about a *million*-body problem, rather than a billion or a trillion[1] bodies? What kind of *bodies* do we have in mind? And finally, what is the *problem* with this whole topic?

In physics, many complex systems can be modelled as an aggregate of a large number of relatively simple entities with relatively simple interactions between them. It is one of the most fascinating aspects of physics that an enormous richness can be found in the collective phenomena that emerge out of the interplay of the much simpler building blocks. Smoke rings and turbulence in air, for example, are complex manifestations of a system of air molecules with relatively simple interactions, strongly repulsive at small scales and weakly attractive at larger scales. From the spectrum of avalanches in sand piles to the instabilities in plasmas of more than a million degrees in labs to study nuclear fusion, we deal with one or a few constituents with simple prescribed forces. What is special about *gravitational* interactions is the fact that gravity is the only force that is mutually attractive. Unlike a handful of protons and electrons, where like charges repel and opposite charges attract, a handful of stars shows attraction between every pair of stars. As a result, a star cluster holds itself together: there is no need for a container (as with a gas or plasma in a lab) or a table (as with a sand pile). And for astronomers, an extra reason to study star clusters is simply: because they are there.

---

[1] By trillion we mean 10^{12}.

Now why are we interested in a *million* stars? Usually in physics we analyse a system of interacting components in two limits: one in which the number of components is one, or a few; and one in which the number of components tends to infinity. And at first sight, a million and a billion stars both seem to be 'close enough' to infinity to allow a common treatment. However, there is a large difference in behaviour between the two types of systems. In a typical rich star cluster, with a million stars, each star feels enough of the granularity of the gravitational field of the other stars that the consequent perturbations lead to a total loss of memory of the initial conditions of its orbit, within the lifetime of the Universe. In contrast, in a typical galaxy, with a number of stars between a billion and a trillion, an individual star for all intents and purposes feels only the smooth average background field. This is the reason that galaxies still retain much prettier shapes than the shapeless (indeed globular) clusters for which orbital memories have been wiped out. In other words, in this part of astronomy, when we count 'one, two, a few, many', a few means a million.

The term *body* is just an archaic term for a material object, be it a molecule or a stone or a star. In many cases in astronomy, when we study the motion of a group of stars we can completely neglect the finite size of the stars, and we may as well treat them as if they were simple mass points, with a mass but without an extension. When stars come closer, we may want to approximate them as finite-size bodies, with a simple description of their mass distribution. Only when stars physically collide, as they sometimes do in the dense inner regions of star clusters, do we have to remind ourselves how complex these celestial bodies really are,[2] in order to say something about the transformations that stars undergo when they merge.

Finally, why do we call all this a *problem*? In mathematics, more than in physics, we talk about specific topics as problems, as long as they have not (yet) been solved. For example, the famous four-colour problem, the question of how many different colours we need to introduce to colour a map such that neighbouring countries can always be given different colours, remained a real problem until the 1970s. Partly with the help of computers, this problem was solved (the answer was: four colours suffice), and then the four-colour problem turned into the four-colour theorem. In the days of Newton, at the beginning of mathematical physics, the same term 'problem' was used to describe the challenge to find the motions of two, three, or more bodies under the influence of their mutual gravitational attractions: hence the two-body problem, the three-body problem, etc. Unlike the four-colour problem, though, we don't expect to ever 'solve' the gravitational N-body problem, for arbitrary N – and even for $N = 2$, where we *do* have analytic solutions, we still follow tradition in calling this the two-body *problem*. So apart from this quaint piece of history, we could have called this book 'the gravitational million-body system'.

----

[2] Addressing an astronomer who had just remarked that 'after all, a star is a pretty simple thing', R.O. Redman pointed out, that, 'at a distance of 10 parsecs *you'd* look pretty simple!'

Why would someone want to study the gravitational million-body problem? There are at least four quite different motivations, centring on the fields of astronomy, theoretical physics, computational physics, and mathematics.

Let us first take the point of view of *astronomy*. Throughout the last hundred years or more, astronomers have been studying an important class of objects called *globular star clusters*, which are roughly spherical collections of stars, each much smaller than a galaxy but much bigger than the solar system. The number of stars they contain varies from one object to another, but a million is the right order of magnitude. Astronomers study globular clusters for many reasons, of which we can touch on only a few. It is thought that they were among the first recognisable structures that were born in galaxies like ours, and their age is a vital constraint on that of the Universe. All the stars in each cluster were born with different masses at roughly the same time, and their present stellar population gives a snapshot of the results of about ten billion years of stellar evolution. They are exceptionally rich in some of the more exotic kinds of star which astronomers now study: binary X-ray sources, millisecond pulsars, etc. Finally, the stars inside a cluster are sufficiently densely packed that they can and do collide with each other, and so clusters provide us with a sort of laboratory where we may hope to understand the more dramatic effects of the dense stellar environment in the nuclei of galaxies. Thus the gravitational million-body problem has an important place in modern astronomy.

Now let us consider the point of view of *theoretical physics*. Whichever way a physicist approaches them, gravitational problems pose particular difficulties. Classical thermodynamics is poorly developed for such problems, because of the long-range nature of the gravitational force. Kinetic theory requires ad hoc approximations, for the same reasons. The methods of plasma physics fail because gravitational forces, being attractive, are unshielded. Gravitational systems are intractable with many traditional methods because the natural equilibria are not spatially uniform. The gravitational many-body problem also has a wider significance in physics, because of its historical roots. The science of mechanics became firmly rooted, in large part, because of the success of Newton's programme for the study of planetary motion. The further development of this theory led to the development of Lagrangian methods, and then, in the hands of Poincaré, to the foundation of the study of chaos. Much of the development of physics since the time of Newton has rested on foundations set in place with the aid of the gravitational many-body problem, which became the model for how a successful physical theory should look. Finally, the most difficult models to predict and to interpret their behaviour are the ones where we are dealing with neither a very large nor a very small number of particles. When there are huge numbers of particles, as in a gas, there are successful approximate methods, like fluid mechanics, which greatly simplify the problems. When there are very few particles, especially one or two, the problem is either completely soluble or can be understood approximately. When there are an intermediate number (a million or less in our case), elements of both extremes are in play simultaneously, thereby thwarting either type of approximation.

This brings us naturally to *computational physics*. Just as the theoretical physicist does, the computational physicist immediately grasps the fact that the gravitational million-body problem poses formidable problems whichever way it is approached. It is not even clear whether the solution can be obtained reliably at all. Even if we lay such fundamental issues aside, running an *N*-body simulation with $N \gtrsim 10^4$ is exceptionally time-consuming for even the fastest generally available computers. This in turn has led to the development of special computers (the GRAPE family), whose sole task is to solve this problem very quickly. Around 1995 these were the fastest computers in the world, a place they recaptured in 2001. This approach has become a model of the way in which future work in computational science in other areas may be carried out economically and quickly, given the right flair and ingenuity. Providing the hardware is only half the problem though; developing the software for running *N*-body simulations is equally challenging. In our particular case of globular cluster simulations, we are confronted with length scales spanning the range from kilometres to parsecs (a factor of more than 10^{13}), and with time scales spanning the range from milliseconds to the lifetime of the universe (a factor of more than 10^{20}). As a result, stellar dynamicists have developed special integration methods that are not encountered in any textbooks on differential equations.

And finally we turn to *mathematics*. When Newton laid the foundations of classical physics in the *Principia*, among the mathematical tools he deployed was the infinitesimal calculus, which he had invented for the purpose. This illustrates the rich potential for the invention of new mathematics which results from intense scrutiny of gravitational problems, and other issues in dynamics. Newton set a precedent which has repeated itself several times since. The work of Poincaré was developed in the context of celestial mechanics, and the famous theorem of Kolmogorov, Arnold and Moser, foreshadowed as it was by the work of C.L. Siegel in his book *Vorlesungen über Himmelsmechanik*,[3] owes much to this discipline, and the emphasis which it helped to place on Hamiltonian problems. And yet when one turns to the *million*-body problem, as opposed to the *few*-body problems of the solar system, the flow of information has been from mathematics to stellar dynamics, rather than the other way around. A number of techniques introduced by mathematicians for the solution of quite abstract problems have turned out to be just what was needed to improve the numerical solution of our problem. This flow of ideas has been sufficiently influential that the astronomer's understanding of the problem would not have advanced so far without it. One of our purposes in writing this book, perhaps a far-fetched one, is the hope that it may help reverse the flow, and stimulate the birth of new mathematics. At the very least, we hope that mathematicians will enjoy learning how their work has been put to use. As one of us has enjoyed a place in the tolerant community of mathematicians for many years, it is a way of giving something back.

---

[3] *Lectures on Celestial Mechanics.*

We believe that the gravitational million-body problem has a seminal role to play in all four areas: astronomy, physics, mathematics and computational science. But while the above remarks have sometimes emphasised the difficulties it poses, we will have failed in one of our aims if, after reading the book, the reader from any of these disciplines is not impressed by how much of the problem can be understood, and how much it has to offer. In other words, one of our goals is to convey some of the beauty and simplicity underlying classical dynamics, as illustrated through the gravitational many-body problem. This is a book that is meant to be read by a variety of scientists who share a curiosity for the roots of physics, the recent fruits that have sprouted directly from those old roots, and its interconnections with neighbouring sciences.

That having been said, it is mostly within the astronomical community that this subject has been developed. And it has been another firm intention in writing the book that it may serve the role of a *graduate textbook* on the theory of stellar dynamics of dense stellar systems. Wherever possible, therefore, our statements and results are developed from first principles. We include some exercises and problems, and hints for their solution.

In writing the book we were aware that the reader may or may not be an astronomer, and may or may not be interested in learning the details of the theory. We realised also that the variety of topics which make up the gravitational million-body problem are best treated at a variety of levels, and that the interconnectedness of these topics forces interconnections between the chapters. Some of the topics we address might even strike some readers as being faintly whimsical. Recognising, therefore, that not all chapters will be equally accessible or interesting to all readers, we start each part of the book (each of which consists of several connected chapters) with an outline of its contents. We hope that readers will thus find their own optimal route between the contents page and the index.

For much the same variety of reasons we have relegated some material to boxes. Often these contain details of some derivation or discussion, and a reader in a hurry could avoid them. Sometimes, however, they contain background which might help a reader over some difficulty. Sometimes they collect some useful results which might act as a reference resource. Sometimes they explore interesting or amusing by-ways that would otherwise interrupt the text. Perhaps, therefore, they should not be passed over too readily.

We hope the index will both help and intrigue the reader, but the quirky and personal selection of names there is not meant as a comprehensive answer to the question of who did what. (Last time we looked, we could not even find our own names there.) Even in the extensive list of references to published work we have not attempted completeness. While this risks antagonising the reader for the omission of his or her own most cherished paper, for which we apologise, it could be just as hazardous if we tried to attribute every advance to its originator. Our main reason for giving references is to give the reader an entry into the research literature. Nowadays it is quite easy to move both forwards and backwards from a suitable entry point, using such invaluable bibliographical resources as the Astronomy Abstract Service of the NASA Astrophysics Data System.

And now we have a confession to make. Studying problems in theoretical physics, the solved as well as the unsolved, is interesting and enjoyable all by itself. Finding new and unexpected insights, and extending the world's body of knowledge about these problems, is an added pleasure. Sharing these interests with others is still more rewarding. But we have other reasons for our interest in the gravitational N-body problem, and among these is its status as one of the *oldest* unsolved problems in the exact sciences, and one with an exceptionally distinguished pedigree. Some of the great names of the subject have been mentioned in this preface, but we would add some other much admired names of the past, including those of Jeans, Chandrasekhar and Spitzer. With equal pleasure we call to mind the many contemporaries with whom we have worked and continue to work on these problems. The community is world-wide, but it is not a large one, it is rather close-knit, and works together and openly in the best tradition of the 'community of scholars'. We hope that the common enjoyment of this joint enterprise surfaces from time to time throughout this book.

Acknowledgements

We have a number of these colleagues to thank with particular regard to the help they have given with parts of the book. Large sections have been read and commented on by Haldan Cohn, Herwig Dejonghe, Mirek Giersz, Jeremy Goodman, Jarrod Hurley, Jun Makino, Simon Portegies Zwart, Rainer Spurzem and Peter Teuben, and we thank them warmly for their comments. If there are problems with the book, it is proba-bly because we did not *always* take their advice. Simon Portegies Zwart and Chris Eilbeck also gave considerable help with Appendix A, and we also thank Maurizio Salaris and Max Ruffert for their input. For permission to reproduce illustrations we would like to thank O. Gnedin, J. Goodman, L. Hernquist, K. Johnston, S. Kulkarni, C. Lada, J. Makino, S. McMillan, J. McVean, E. Milone, F. Rasio, the American Astronomical Society, Blackwell Publishing (publishers of *Monthly Notices of the Royal Astronomical Society*), and other individuals and institutions named beside individual figures. We are also most grateful to staff at Cambridge University Press, especially A. Black (who has since moved on), J. Garget, S. Mitton, J. Aldhouse and J. Clegg, for their advice, encouragement and patience, and to our copy editor, F. Nex. DCH is deeply grateful to the Institute for Advanced Study, Princeton, for its generous hospitality over several visits during which much of the planning and drafting of the book was carried out.

Edinburgh and Princeton
2002

Part I

Introductions

Newton's equations for the gravitational N-body problem are the starting point for all four chapters in Part I, but each time seen in a different light. To the astrophysicist (Chapter 1) they represent an accurate model for the dynamical aspects of systems of stars, which is the subject known as *stellar dynamics*. We distinguish this from *celestial mechanics*, and sketch the distinction between the two main flavours of stellar dynamics. This book is largely devoted to what is often (but maybe mis- leadingly) called *collisional* stellar dynamics. This does not refer to actual physical collisions, though these can happen, but to the dominant role of *gravitational* encoun- ters of pairs of stars. In dense stellar systems their role is a major one. In *collisionless* stellar dynamics, by contrast, motions are dominated by the average gravitational force exerted by great numbers of stars.

We lay particular emphasis on the stellar systems known as *globular star clusters*. We survey the gross features of their dynamics, and also the reasons for their impor- tance within the wider field of astrophysics. Though understanding the million-body problem is not among the most urgent problems in astrophysics, through globular clusters it has close connections with several areas which are. Another practical topic we deal with here is that of *units*, which may be elementary, but is one area where the numbers can easily get out of hand.

Chapter 2 looks at the N-body equations from the point of view of theoretical physicists. To them, the gravitational many-body problem is just one of many inter- esting N-body problems. In some ways it is simple (being scale-free), and in other ways complicated (long and short ranges have comparable importance, for example, unlike in plasmas). Theoretical physicists understand the two-body problem, just like astrophysicists do, but for large N their usual methods don't apply very well.

Statistical mechanics is one casualty. The study of limiting cases is stymied by the fact that there are no parameters to alter, except N.

The enormous range of length scales (and, consequently, time scales) is also one reason why the N-body equations are a severe challenge to the computational physicist (Chapter 3). There are good reasons why the faster methods used in other (e.g. collisionless) problems are deprecated in this field. But the penalty is that computational progress has been slow: we are only now approaching the capability of directly modelling even a small globular cluster. Though software advances are continually playing a role, what has made this possible is the development of special-purpose computers tailored precisely for this problem.

Mathematicians (Chapter 4) seek a *qualitative* understanding of the N-body equations. They are concerned with the non-obvious question of what one even means by a 'solution'. They pay particular attention to the (collision) singularities of the equations, and ask how solutions behave in their vicinity, even if they occur only at complex values of the time. The mathematician asks whether the equations can exhibit singularities of other kinds. Historically, the fertile equations of the three-body problem have bred a variety of topics which remain mainstream issues in mathematics, foremost among them being questions of integrability.

1

Astrophysics Introduction

When Newton studied dynamics and calculus he was motivated, at least in part, by a desire to understand the movements of the planets. The mathematical model which he reached may be described by the equations

$$\ddot{\mathbf{r}}_i = -G \sum_{j=1, j \neq i}^{j=N} m_j \frac{\mathbf{r}_i - \mathbf{r}_j}{|\mathbf{r}_i - \mathbf{r}_j|^3}, \tag{1.1}$$

where $\mathbf{r}_j$ is the position vector of the jth body at time t, m_j is its mass, G is a constant, and a dot denotes differentiation with respect to t. Now, over 300 years later, our motivation has a very different emphasis. The study of the solar system has lost none of its importance or fascination, but it now competes with the study of star clusters, galaxies, and even the structure of the universe as a whole. What is remarkable is that Newton's model is as central to this extended field of study as it was in his own time.

Stellar dynamics

It may seem surprising that the single, simple set of equations (1.1) forms a good first approximation for modelling many astrophysical systems, such as the solar system, star clusters, whole galaxies as well as clusters of galaxies (Fig. 1.1). The reason is that gravity, being an attractive long-range force, dominates everything else in the Universe. The only other long-range force, electromagnetism, is generally not important on very large scales, since positive and negative charges tend to screen

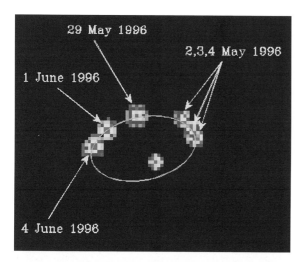

Fig. 1.1. Two extreme astrophysical many-body problems. *Top* A binary star, ζ^1 Ursae Majoris, imaged interferometrically (*Credit: J. Benson et al., NPOI Group, USNO, NRL*). *Bottom* The galaxy M87. This is not a straightforward *N*-body problem, as much of the mass is not stellar. Many of the spots of light in the fringes of the galaxy are individual globular clusters (*Credit: NOAO/AURA/NSF*).

each other. Short-range forces, such as gas pressure, are usually only important on small scales, such as in the interiors of stars. On large scales, comparable to the size of a galaxy, pressure is rarely important. On intermediate scales we have giant gas clouds in which both pressure and magnetic fields can play a role.

This dominance of gravity on cosmic scales is a fortunate feature of our Universe. It implies that it is relatively simple to perform detailed computer simulations of many astronomical systems. Mastery of the much more complicated physics of other specialisations in astronomy, such as plasma astrophysics, radiative transfer, or nuclear astrophysics, is often not immediately necessary. Nor, in applications to star clusters, is it usually necessary to use the more refined theory of general relativity. The velocities are too low, except in a few situations involving degenerate stars (neutron stars). Even the motions of stars round the black hole at the centre of our

Galaxy (Ghez *et al.* 2000), though it takes place on a human time scale, can be understood with classical dynamics.

Stellar dynamics can be defined as studying the consequences of Eq. (1.1) in astrophysical contexts. Traditionally, these equations were discovered by studying the motions of the Moon and planets, and for the next few centuries they were applied mainly to planetary dynamics. Before the advent of electronic computers, most effort went into developing analytical approximations to the nearly regular motion of the planets. This field, known as celestial mechanics, had an important influence on developments both in physics and mathematics. An example is the study of chaos, which first arose as an annoying complexity barring attempts to make long-time predictions in celestial mechanics.

The solar system, however, is a very regular system. All planets move in orbits close to the ecliptic, and all revolve in the same direction. The orbits are well-separated, and consequently no close encounters take place. When we look around us on larger scales, that of star clusters and galaxies, then no such regularity applies. In our Galaxy, most stars move in the galactic disk, revolving in the same sense as the Sun, but close encounters are not excluded. In many globular star clusters, there is not even a preferred direction of rotation, and all stars move as they please in any direction. Does this mean that analytic approximations are not very useful for such systems?

The answer is: 'it depends on what you would like to know'. In general, the less regular a situation is, the larger the need for a computer simulation. For example, during the collision of two galaxies each star in each of the galaxies is so strongly perturbed that it becomes very difficult to predict the overall outcome with pen and paper. This is indeed an area of research which had to wait for computers to even get started, in the early seventies (Toomre & Toomre 1972; but see the reference to Holmberg in Chapter 3). On the other hand, when we want to understand the conditions in a relatively isolated galaxy, such as our own Milky Way, then there is some scope for pen-and-paper work. For example, a rich variety of analytic as well as semi-numerical models has been constructed for a range of galaxies. However, even in this case we often have to switch to numerical simulations if we want to obtain more precise results.

Two flavours of stellar dynamics

The field of stellar dynamics can be divided into two subfields, traditionally called collisional and collisionless stellar dynamics. The word 'collision' is a bit misleading here, since it is used to describe a close encounter between two or more stars, not a physical collision. After all, physical collisions are excluded in principle as soon as we have made the approximation of point particles, as we will do throughout most of the present book, with the exception of the last few chapters, where we will take up the question of real stellar traffic accidents.

Collisional stellar dynamics is concerned about the long-term effects of close (as well as not-so-close) stellar encounters. The evolution of a star cluster is governed by the slow diffusion of 'heat' through the system from the inside towards the edge. The heat transport occurs through the frequent interactions of pairs of stars, in a way similar to the heat conduction in the air in a room, which is caused by collisions between pairs of gas molecules. The main difference here is that individual stars in a star cluster have mean free paths that are much longer than the size of the system. In other words, little heat exchange takes place during a system-crossing time scale.

Collisionless stellar dynamics is the subfield of stellar dynamics in which the heat flow due to pairwise interactions of stars is neglected. For small systems this approximation is appropriate when we consider the evolution of the star system on a time scale which does not exceed the crossing time by a very large amount. For a system with very many particles, such as a galaxy, the collisionless approximation is generally valid even on time scales comparable to the age of the Universe (which is comparable to the age of most galaxies).

Life in a collisional stellar system

So far, we have talked about the astrophysical many-body problem in a rather generalised way. We have dealt with stars or planets moving in the gravitational field of their mutually attractive forces, and have ignored their internal structure, but we have not focused on any particular kind of system to which these idealisations apply. Fortunately, there are systems in nature which approach the idealisations of collisional stellar dynamics to a remarkable degree. These are the globular clusters, among the oldest components of our galactic system, and going their own way in wide orbits around the Galaxy (see Webbink 1988). They are nearly spherical because they don't rotate much by astronomical standards (see Merritt *et al.* 1997, Davoust & Prugniel 1990). They are largely isolated from perturbing influences of the galactic disk, and therefore form ideal laboratories for stellar dynamics. Typically they have about a million stars each (Table 1.1), with a range of individual stellar masses (see Paresce & De Marchi 2000).

Other galaxies also have globular clusters (Fig. 1.1, bottom; Ashman & Zepf 1998). In some galaxies they are not the oldest but some of the *youngest* stellar components (see Lançon & Boily 2000). For instance, the cluster NGC 1866, which lies in the Large Magellanic Cloud, has a mass of about $10^5 M_\odot$ but is only about 10^8 yrs old (Fischer *et al.* 1992, Testa *et al.* 1999). Even in our own galaxy, the most massive young clusters are found in the apparently hostile environment of the Galactic Centre (Figer *et al.* 1999), and they are as massive as a modest old globular cluster (though they will certainly not last as long, see Kim *et al.* 2000, Portegies Zwart *et al.* 2001a).

Table 1.1. *Basic facts about the globular clusters of the Galaxy*

Number known	147
Median distance from Galactic Centre	9.3kpc
Median absolute V magnitude	−7.27
Median concentration	1.50
Median core relaxation time	3.39×10^8yr
Median relaxation time at the half-mass radius	1.17×10^9yr
Median core radius	1.32pc
Median half-mass radius	3.08pc
Median tidal radius	34.5pc
Median mass	$8.1 \times 10^4 M_\odot$
Median line-of-sight velocity dispersion	5.50km/s

The data are based on tables in Harris (1996; for updated online versions see physun.physics.mcmaster.ca/Globular.html), Mandushev *et al.* (1991) and Pryor & Meylan (1993), and different data are based on somewhat different samples. The terms *concentration*, *relaxation time*, *core* and *tidal radius* are defined elsewhere in the book. Units are discussed in Box 1.1.

The realisation that the formation of globular clusters is still going on in the Universe around us has taken hold only relatively recently, and has helped to rejuvenate the study of these stellar systems. For the most part, though, we concentrate on the old and better observed systems in our own Galaxy.

Box 1.1. Units

Consider a stellar system whose total mass and energy are M and E, respectively. Then we may choose M for the unit of mass, and in simple cases a common and useful unit of length is the *virial radius* $R_v = -\frac{GM^2}{4E}$. (For example, for a binary star this is roughly the diameter of the relative orbit.) Then, if we want units in which the constant of gravitation is unity, the unit of time is required to be $U_t = R_v^{3/2}/\sqrt{GM}$. Another common unit of time is the *crossing time*, which is defined to be $2\sqrt{2}U_t$.

Now in the evaluation of these expressions R_v and M will often be in units favoured by astrophysicists (e.g. parsecs and solar masses, respectively), while G is most familiar in one of the standard physical systems, such as SI. The evaluation of U_t then requires conversion to a common system of units. At this point, because of the enormous distances and time scales which prevail in astronomy, the formulae bristle with enormous exponents, until the final result is re-expressed in astrophysically sensible units, and the numbers become reasonable once again. Indeed, it is one of the minor pleasures and surprises of a theorist's life, and a clue that he is on the right lines, when sensible numbers emerge at the end of a long and tiresome calculation studded with huge powers of ten. Sir Harold Jeffreys once remarked that 'incorrect geophysical hypotheses usually fail by extremely large margins' (Jeffreys 1929). The same is true in astrophysics.

To avoid this unpleasantness, it is much simpler to calculate entirely within the astrophysical system, *provided that you know the value of G in these units*. The following remarks will therefore save much time. Consider units such that masses are measured in $M_\odot$, speeds are measured in km/s, and distances are measured in parsecs (pc). Then $G = 1/232$ approximately. Also, the corresponding unit of time is 10^6 yr approximately.

In integrating the N-body equations numerically, still other units are used. Part of the reason for this is to avoid carrying around the values of physical constants, such as G, that would be needed in any physical system, partly to avoid the huge and unreadable numbers that would result from an inappropriate choice of physical units, but mainly so as not to obscure the intrinsic scalability of many simple N-body calculations.

In simulations it is natural to choose units such that $G = 1$, and, following Hénon (1971), the units of length and mass are conventionally chosen so that the total mass of the system is $M = 1$ and the total energy is $E = -1/4$. These choices might seem bizarre, but it then follows that, if the system is in virial equilibrium (Chapter 9), the potential energy is $W = -1/2$, and the virial radius is then $R_v = 1$. These units are referred to as N-body units, and are in widespread, but not quite universal, use. They become rather ambiguous when external fields or many binary stars are included in the computations.

In order to get a feel for the physical presence of a globular cluster, imagine that we live in the very core of a dense globular cluster (Fig. 1.2). The density of stars there can easily be as high as $10^6 M_\odot/\text{pc}^3$. This is a factor 10^6 higher than in our own neighbourhood, in the part of the galactic disk where the Sun happens to reside. Therefore, we can get an impression of the night sky by bringing each star that we normally see at night closer to us by a factor 10^2. Each star would thus become brighter by a factor 10^4, which corresponds to a difference of ten magnitudes (in the astronomical system where a factor ten corresponds to $\Delta m = 2.5$).

The brightest stars would thus appear at magnitudes at or above $m \sim -10$, comparable to that of the full Moon. They would be too bright to look at directly, because their size would be so much smaller than that of the Moon (they would still look point-like). It would be easy to read books by the light of the night sky. This is what the huge stellar density in these systems means from the 'human' point of view. Later on we shall consider what effect it has on the stars.

A recipe for a star cluster

Looking at Fig. 1.2, we cannot immediately tell how such an assembly of stars has got there, or what its fate will be. One might even guess, as Eddington (1926a) once did, that the stars cannot 'escape the fate of ultimately condensing into one confused mass'. Nowadays, though many details are unclear, we would argue that the clusters

Fig. 1.2. The central region of the globular cluster M15. Image by courtesy of P. Guhathakurta, UC Santa Cruz.

have survived for almost as long as the universe itself, and that they are only now slowly dying off as their stars escape. Here we paint a few details on this picture, showing what we think are the stages through which a cluster would have to pass in order to arrive at something resembling those we see around us.

To make a star cluster, in practice, one should start out with a large gas cloud, which under the right conditions undergoes internal collapse on several length and time scales, to produce a large number of stars. This is a process which is being modelled with increasing sophistication (see, for example, the beautiful movie at www.ukaff.ac.uk/movies.shtml, or Klessen & Burkert 2001). Nevertheless, the physical processes related to star formation are very complex, and at present only partly understood. Therefore, let us perform a thought experiment in which we will cheat a bit. Let us take a bucket full of ready-made single stars, and sprinkle them into a limited region of space, in order to watch just the stellar dynamics between the stars at play.

The simplest way to place the stars in space is to start from rest. As soon as the stars are allowed to start moving, the whole system will begin to collapse under the mutual gravitational attraction of all the stars. If the stars are sprinkled in at random, the contraction will not proceed very far, since the individual star–star interactions will cause deflections from a purely radial infall. The stars will pass at some distance from the centre, each one at a somewhat different time, and then move out again. As a consequence, the whole system will breath a few times: globally shrinking and expanding. However, phase coherence will be lost very soon, and after a few breathing motions, the stars will be pretty well mixed. Some stars will be lost, spilled into the surrounding space, never to return, but most stars will remain bound to the system.

After those first few crossing times, not much will happen for quite a while. During a typical crossing time, a typical star will move through the system on a rosette-shaped orbit, while almost conserving its energy and angular momentum. On a longer time scale, though, the cumulative effect of many distant encounters, and occasional closer encounters, will affect the orbit. If we wait long enough, the memory of the original energy and angular momentum of the star will be lost. This time scale is called the relaxation time scale.

Even if the system was set off in a non-spherical distribution, such as that of a pancake, say, the original order will be erased after a few relaxation time scales, and the system will become spherical. There will also be a steady trickle of stars that escape from the system. In addition, energy will be transported from the centre to the outer regions, and as a result the inner regions tend to contract. Before long, the centre will gain in density, by many orders of magnitude, a phenomenon called core collapse.

After core collapse, double stars will be formed, which provide a source of energy for the system, empowering the ongoing evaporation of the system, until the whole star cluster is dissolved. While the star cluster slowly evaporates, the system will undergo core oscillations, with vast swings in central density, if the number of stars is large enough, well over 10^4. These so-called gravothermal oscillations are the result of the fact that the inner regions, having much shorter relaxation times than the rest of the system, grow impatient in the post-collapse phase, and start collapsing all over again. Interestingly, these oscillations exhibit mathematical chaos, one more example of the gravitational many-body problem providing a stage for some of the most modern forms of applied mathematics.

This general picture, of a star cluster being formed, coming to rest, undergoing a slow form of core 'collapse', followed by evaporation, will be spun out in much greater detail throughout the rest of the book. But before we take a closer look at the specific million-body problem, let us see how much the picture can change if we treat the stars as stars and not as point masses.

Dynamics of dense stellar systems

Most stars do not interact much with their environment, after they leave the cradle of the interstellar cloud they are born in. Some stars are born single, and stay that way throughout their life, although they may have acquired a planetary system during the late stages of their formation. However, most stars are members of a double star or an even more complex multiple system (triples, quadruples, etc.). In such a system, when two stars are sufficiently close, all kind of interesting interactions may take place, including the transfer of mass from one star to another, and even the spiral-in and eventual merging of two or more stars.

Although binary-star evolution is much more complex than single-star evolution, it can generally still be studied in isolation from its wider environment. After all, the typical separation between stars in the solar neighbourhood (itself typical for our Galaxy) is some hundred million times larger than the diameters of individual stars. The exception to this rule occurs in unusually dense stellar systems. Examples are star clusters, both in the disk of the galaxy and outside (the globular clusters), and the nuclei of galaxies.

Especially in the cores of globular clusters and in the nuclei of galaxies, the densities of stars can easily exceed that of the solar neighbourhood by a factor of a million or more (Fig. 1.2). At any given time, such a system is still dilute enough to make physical collisions unlikely, even during many crossing times, thus allowing point mass dynamics to provide useful first-order approximations. However, when viewed over a time scale of billions of years, such collisions become unavoidable in many of these systems.

In recent years much progress has been made in the study of physical collisions, both theoretically in computer simulations, and observationally by looking for 'star wrecks' as tell-tale signs of violent encounters. For example, observations with the Hubble Space Telescope have shown us the presence of so-called blue stragglers, right down in the centre of the most crowded star clusters. Blue stragglers are unusual types of stars that are at least compatible with being the products of stellar collision. Earlier, millisecond pulsars and X-ray binaries already have hinted to us more indirectly about the sagas of their formation and subsequent interactions.

In order to interpret this wealth of observational information, vigorous attempts are being made by several groups to make theoretical models for the evolution of dense stellar systems. The first step is to determine the long-time behaviour of a large system of point masses (a million or so), a classical problem that is still far from solved, and that has given rise to fascinating new insights, even over the last ten years, including the surprising discovery of the presence of mathematical chaos in the late stages of its evolution. It is this step that is the main theme for the present book.

The second step is to integrate our understanding of the dynamics of a system of point masses with the extra complexity introduced by the non-point-behaviour, in the form of stellar evolution, physical collisions between stars, mass loss, etc. This integration gives rise to an extremely complex picture of the ecology of star clusters, as we will briefly discuss in the last part of this book.

Why study the dynamics of globular clusters?

The dynamical study of globular clusters has had a peculiar history, which stretches over the entire twentieth century and beyond. Even up to the 1960s, however, many astrophysicists regarded it as a quiet backwater, even though it had attracted the attention of the best theorists of each generation: Jeans, Eddington, Chandrasekhar, Spitzer, Hénon, Lynden-Bell, Ostriker, ... Perhaps they were initially attracted to it because of what Lyman Spitzer called its 'appealing but deceptive simplicity'.

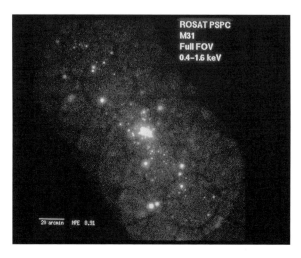

Fig. 1.3. A recent X-ray image of the Andromeda galaxy, M31. More than 5% of the sources lie in globular clusters belonging to Andromeda. Original colour image by courtesy of the ROSAT Mission (www.xray.mpe.mpg.de/) and the Max-Planck-Institut für extraterrestrische Physik (www.mpe.mpg.de/).

This dynamical study of the clusters went in parallel with the astrophysical study of the stars inside them, as these were seen to be excellent test beds for checking the theory of stellar evolution, the chemical history of the Galaxy, the age of the Universe, and so on. In fact there was hardly any major problem of astronomy – from star formation to cosmology – in which globular star clusters did not have something important to say, and this remains true to the present day.

Until the early 1970s these parallel tracks in the study of globular clusters – the dynamical and the astrophysical – proceeded entirely independently. The two communities had nothing in common and did not even need to talk to each other. By the 1990s all this had changed. What happened?

More than anything else, it was the discovery of variable X-ray sources in globular clusters in the 1970s which brought down the barriers between the two communities (Fig. 1.3). Though their origin is still not quite clear, it was soon realised that dynamical processes were important. Dynamicists also realised that here was a new process which could have a profound influence on their understanding of dynamical evolution. Suddenly the dynamics of globular clusters had rejoined the mainstream of astrophysics, and in the 1970s there was not much that seemed more glamorous than X-ray astronomy.

The integration of the dynamical and stellar studies of globular clusters has progressed steadily in the intervening decades, and there is even a new breed of 'cross-over' astrophysicists who are at home in both camps. This grand unification of cluster studies has been strengthened by a stream of exciting discoveries: binary stars (through the rapid explosion of data on radial velocities, and other techniques), millisecond pulsars, white dwarfs, blue stragglers in dense clusters, core collapse, and so on. They are ideal targets for searching for extrasolar planets, or should be (Gilliland *et al.* 2000).

The sheer visual beauty of globular clusters, and the richness of the problems they present, must not lead us to lose sight of their place in the bigger picture. In particular, they allow us to study the behaviour of stars in dense environments, and will prove to be an important stepping stone for the understanding of the even denser stellar systems in galactic nuclei – surely one of the major problems of astrophysics for the twenty-first century. Nor are we here restricting attention to *active* galactic nuclei. For instance, the nucleus of the quiescent nearby galaxy M33 can really be regarded as a gigantic globular cluster sitting at the centre of its parent galaxy, and may well have exhibited the kind of core collapse studied in the context of globular clusters (Hernquist *et al.* 1991).

Sidney van den Bergh (1980) once expressed this philosophy very nicely: 'One of the dangers of living on a beautiful isle, such as Vancouver Island, is that it is very easy to become insular. By the same token it is all too easy for us, working in various exciting areas of cluster studies, to forget about the broader impact that such work might have on other areas of astronomy. The justification of our work on clusters is not just that it is fun but also that such investigations often help to illuminate other branches of science.'

Problems

(1) Compute the virial radius (Box 1.1) of a binary consisting of two stars of mass m_1 and m_2 in a circular relative orbit of radius a.

(2) Compute the speed of a star in a circular orbit at the median tidal radius (Table 1.1) around a cluster of median mass. Do this in astrophysical, physical and N-body units.

(3) Using the fact that the magnitude, m, and luminosity, L, of a source at a distance d are related by

$$m = -2.5 \log_{10} \frac{L}{L_\odot} + 5 \log_{10} \frac{d}{10\mathrm{pc}} + M_\odot,$$

where $L_\odot$ and $M_\odot$ are the luminosity and absolute magnitude of the Sun, find the mean surface brightness (within the tidal radius) of the median globular cluster (using data from Table 1.1). Express your result in magnitudes per square arc second.

2

Theoretical Physics Introduction

A single coupling constant

The gravitational N-body problem can be defined as the challenge to understand the motion of N point masses, acted upon by their mutual gravitational forces (Eq. (1.1)). From the physical point of view, a fundamental feature of these equations is the presence of only one *coupling constant*: the constant of gravitation, $G = 6.67 \times 10^{-8}$ cm^3 g^{-1} s^{-2} (see Seife 2000 for recent measurements). It is even possible to remove this altogether by making a choice of units in which $G = 1$. Matters would be more complicated if there existed some length scale at which the gravitational interaction departed from the inverse square dependence on distance. Despite continuing efforts, no such behaviour has been found (Schwarzschild 2000).

The fact that a self-gravitating system of point masses is governed by a law with only one coupling constant (or none, after scaling) has important consequences. In contrast to most macroscopic systems, there is no decoupling of scales. We do not have at our disposal separate dials that can be set in order to study the behaviour of local and global aspects separately. As a consequence, the only real freedom we have, when modelling a self-gravitating system of point masses, is our choice of the value of the dimensionless number N, the number of particles in the system.

As we will see, the value of N determines a large number of seemingly independent characteristics of the system: its granularity and thereby its speed of internal heat transport and evolution; the size of the central region of highest density after the system settles down in an asymptotic state; the nature of the oscillations that may occur in this central region; and to a surprisingly weak extent the rate of exponential divergence of nearby trajectories in the system. The significance of the value

of N is underlined by the fact that N often makes its appearance in the very names
of the problems we study, e.g. the famous 'three-body problem', and in the titles of
books like this one.

$$N = 2, 3$$

The three-body problem is so famous precisely because it is one of the oldest problems
that *cannot* be solved. In contrast, the two-body problem is one of the oldest *solved*
problems. It was Newton's great triumph that he was able to explain why planets move
in elliptical orbits around the Sun, as had been discovered earlier by the brilliant
Kepler, using observational data by Tycho Brahe. The significance of Newton's
solution can hardly be overestimated, since the result was a breaching of the seemingly
separated realms of the temporal, human, sublunar world and the eternal world beyond
the Moon's orb.

Though Newton had shown how to solve the two-body problem, the presence
of even an infinitesimal third body (the so-called 'restricted' three-body problem)
is insoluble in the usual sense.[1] We should not forget, however, that Newton was
able to solve some three-body problems approximately. He knew, for instance, that
it is better to treat a planet as revolving around the barycentre of the inner solar
system than about the Sun itself. If we do not insist on a precise solution, as with
the two-body problem, in other words, if we look at it from the point of view of
physics rather than mathematics, as Newton did, many aspects of the behaviour of
the three-body problem are not so hard to understand. Indeed, it is one of the quiet
triumphs of recent decades that our intuitive understanding of three-body motions
has developed to the extent it has. Most physics undergraduates are still exposed to
the two-body problem, and can easily develop a feel for the motion in ellipses and
hyperbolae, especially when the scattering problems of atomic physics are encountered. We think that it is possible to arrive at a similar feel for the richer behaviour
of three-body systems, and one of our aims in writing this book is to demonstrate
this.

The *mathematical* development of the three-body problem continued in the hands
of Euler, Lagrange, Laplace and many others, who systematised the methods of approximate solution. These led to the remarkable and successful development of the
theory of planetary motion; the successes of the recovery of Ceres and the discovery
of Neptune; and the familiar but no less remarkable ability of astronomers to predict
the motions of the planets, satellites, asteroids, and comets. Equally, they led to profound insights in the emerging fields of dynamical systems and chaos (Chapter 4),
which in turn have illuminated our understanding of N-body problems. But they also

---

[1] As an aside, it is amusing to see the progress of physics reflected in our (in)ability to solve
N-body problems. In Newtonian gravity we cannot solve the three-body problem. In general
relativity we cannot solve the two-body problem. In quantum electrodynamics we cannot solve
the one-body problem. And in quantum chromodynamics we cannot even solve the zero-body
problem, the vacuum.

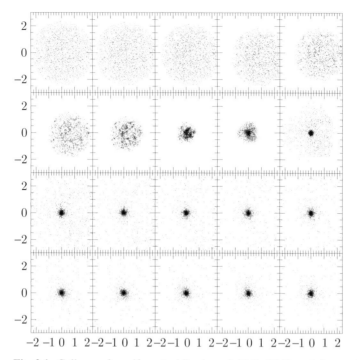

Fig. 2.1. Collapse of a uniform 'cold' sphere. Initially 2048 particles are distributed randomly within a sphere. In units such that $G = 1$, the initial radius is 2.4, the total mass is 1, and successive frames are taken at intervals of $2\sqrt{2}/5$ (cf. Box 1.1). (The successive frames are displayed left to right. Frame 1 top left, frame 20 bottom right.) Late in the collapse (frames 7 and 8) the distribution of particles has become very irregular. In frame 9 the particles which arrived first have begun to re-expand. By the last row of frames the remaining central condensation has settled nearly into dynamical equilibrium.

led away from the sort of approach we need if our goal is to find quantitative answers to questions about the million-body problem.

$$N \to \infty$$

Throughout classical physics there are several important analogues to the million-body problem, and it is from this background, and not so much from celestial mechanics, that the most fruitful approaches have come. At first sight the closest analogy is with plasma physics, where inverse square laws, as in Eq. (1.1), also arise, but the analogy has not proved as fruitful as one might think.

In the first place plasmas are often nearly neutral, and for this reason it makes sense to conceive of plasmas which are nearly uniform, nearly at rest and of large spatial extent. By contrast, it is impossible to conceive of an infinite uniform gravitational medium in equilibrium (Fig. 2.1). Newton himself glimpsed this problem, and solved it by noting that the stars had been placed at immense distances from each other, lest they should, 'by their gravity, fall on each other'. This was a satisfactory solution if one supposed that the system of fixed stars were young enough. For stellar systems the modern solution is that the random motions or circulation of the stars maintain *dynamic* equilibrium, just as the motions of the planets prevent their falling into the Sun. Even so, we shall see that the contraction of stellar systems in dynamic *quasi*-equilibrium is a real and fascinating issue, though it is a much gentler process than the headlong rush suggested by the phrase 'gravitational collapse'. In problems of cosmology, Newton's dilemma is avoided by an overall expansion or contraction of the entire Universe, though a general relativistic treatment is required.

The second basic feature of an infinite plasma is that several important effects are localised within a *Debye length*, beyond which the rearrangement of charges effectively screens off the influence of an individual charge (see, for example, Sturrock 1994). If one naïvely computes the Debye length for a stellar system, it is found usually to be of the order of the size of the system itself (Problem 2). Thus it is difficult to treat gravitational interactions as being localised within a stellar system, though it is a difficulty that stellar dynamicists habitually ignore. In any event, all these considerations help to explain why many of the well known properties of plasmas have little relevance to stellar systems. And yet there is one phrase that stellar dynamicists use all the time which betrays their debt to plasma physics: the words 'Coulomb logarithm', which occur in the theory of relaxation (Chapter 14).

Thermodynamics

The plasma analogy exploits the form of the force equation but not the size of N. Physicists routinely study problems with great numbers of particles using concepts of thermodynamics. How fruitful is such an approach to the million-body problem?

From a formal point of view, unfortunately, the field of thermodynamics excludes a description of self-gravitating systems. The reason is that the existence of gravitational long-range forces violates the notion of an asymptotic thermodynamic limit in which physical quantities are either intensive or extensive. Gravity exhibits what is known in particle physics as an infra-red divergence. This means that the effect of long-distance interactions cannot be neglected, even though gravitational forces fall off with the inverse square of the distance (Problem 3).

Take a large box containing a homogeneous swarm of stars. Now enlarge the box, keeping the density and temperature of the star distribution constant. The total mass M of the stars will then scale with the size R of the box as $M \propto R^3$, and the total kinetic energy E_{kin} will simply scale with the mass: $E_{kin} \propto M$. The total potential energy E_{pot}, however, will grow faster: $E_{pot} \propto M^2/R \propto M^{5/3}$. Unlike intensive thermodynamic variables that stay constant when we enlarge the system, and unlike extensive variables that grow linearly with the mass of the system, E_{pot} is a superextensive variable, growing faster than linear.

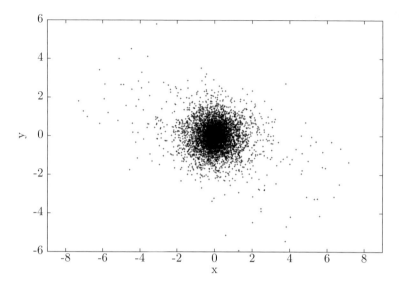

Fig. 2.2. Escape from an *N*-body system. This computer model of a star cluster shows a broad stream of escapers emerging at the left and right. The escapers are channelled into these streams because of external forces, and the streams are curved because of Coriolis forces.

As a consequence, the specific gravitational potential energy of the system, the total potential energy of the system divided by the particle number *N*, grows without bounds when we increase *N*. This causes various problems. For example, the kinetic energy of a stable self-gravitating system is directly related to the gravitational potential energy through the so-called virial theorem (Chapter 9). Therefore, we have to make a choice when enlarging a self-gravitating system. Either we increase the temperature steadily while increasing *N*, in order to increase the specific kinetic energy enough to satisfy the virial theorem and guarantee stability. Or we keep the temperature constant, and quickly lose stability when enlarging our system. In the latter case, the system will 'curdle': it will fall apart in more and more subclumps, and the original homogeneity will be lost quickly.

In conclusion, there is no way that we can reach an asymptotic thermodynamic limit, with the system size becoming arbitrarily large while holding the intensive variables fixed.[2] Therefore, the traditional road to equilibrium thermodynamics is blocked. There are no arbitrarily large homogeneous distributions of stars. As soon as the Universe became old and cold enough to let matter condense out of the original fire ball into 'islands' in the form of galaxy clusters and galaxies, the original homogeneity was lost. And each individual clump of self-gravitating material, be it a galaxy or a star cluster, is ultimately unstable against evaporation, and will fall apart into a

---

[2] A consistent theory is possible if we let $R \to \infty$ with N/R fixed (de Vega & S'anchez 2000), though what relevance this might have to the million-body problem is unclear.

bunch of escaping particles (Fig. 2.2). Most of these escapers will be single, some will escape as stable pairs, and a few will even manage to form stable triples or higher-number multiples of particles.

The presence of these few-body systems is a robust feature of N-body systems, as we shall see throughout this book. From our present perspective, it is an indication of another problem: a short-range ('ultra-violet') divergence. The usual Boltzmann factor used in calculations of canonical ensemble averages, i.e. $\exp(-E/kT)$, gives divergent results in the limit when two particles approach each other within a small distance r. This factor then contains a term $\exp\left(\dfrac{Gm^2}{rkT}\right)$.

A lack of handles

Even though we cannot use thermodynamics in a formal way, when dealing with a star cluster, we can still describe the motion of the stars in a way that is analogous to the treatment of the motion of molecules in a gas studied in a laboratory. One important difference is that a swarm of stars forms an open system, while a body of gas in a lab has to be contained. Typical textbook experiments in thermodynamics show the gas to reside inside a cylinder, with a movable piston that allows the experimenter to change the volume of the gas. In a star cluster, there are no cylinder and piston. Instead, the stars are confined by their collective gravitational field.

The structural simplicity of a star cluster thus allows far less experimentation than is the case for a body of gas in a lab situation. Whether in thought experiments, computer simulations, or in actual table-top experiments, the macroscopic parameters of a laboratory gas can be changed freely, independent of the microscopic parameters governing the attraction and repulsion between individual molecules. Temperature, density, and size of the system all can be varied at will. In contrast, once the number of particles in a self-gravitating system has been chosen, we are left with no degree of freedom at all, apart from trivial scalings in the choice of units of length, time, and mass.

The fact that there are no dials that can be turned in a self-gravitating experiment, apart from the choice of the total number of stars, is directly related to the ultra-violet and infra-red divergences of classical gravity. Having a simple shape for the gravitational potential energy well, with an energy inversely proportional to distance, leaves no room for preferred length scales. In contrast, molecular interactions show far more complicated forces, typically strongly repulsive at shorter distances and weakly attractive at larger separations between the molecules. This change in behaviour automatically specifies particular length scales, for example the distance at which repulsion changes into attraction. In contrast, gravity is attractive everywhere, at least in the classical Newtonian approximation.

Towards an understanding of the million-body problem

It is now being realised that the gravitational N-body problem is just one of a growing list of known systems with long-range interactions where the non-extensivity of energy looks like an obstacle (Cipriani & Pettini 2001). (In other contexts these are known as 'small systems', to indicate that their spatial extent is comparable with the range of the relevant interaction.) It turns out that non-extensivity is only an issue if we insist on treating these problems with the traditional tools of canonical ensembles (corresponding to systems immersed in a heat bath). This hardly seems natural in the stellar dynamical context. It is the *microcanonical* ensemble which corresponds best to the isolated N-body systems which have been studied so much in stellar dynamics. Indeed it is known that the different ensembles one studies in thermodynamics, and which are usually regarded as equivalent for many purposes, have very different properties for self-gravitating systems (see, for example, Youngkins & Miller 2000).

The *formal* inability to apply traditional thermodynamic concepts, then, does not seriously hinder us from thinking and working with them. Gravity is just as important in the theory of stellar structure and evolution, where thermodynamics is just as much the stock-in-trade as in any other area of gas dynamics. Indeed, it was precisely the search for extrema of the entropy of stellar systems that led V.A. Antonov (Antonov 1962) to one of the most profound insights into the behaviour of stellar systems – gravothermal stability.

In addition to fundamental approaches like this, there are several other routes by which the behaviour of N-body systems are explored (Chapter 9). One can construct toy models, one can borrow models from other areas (such as the theory of stellar evolution, or the kinetic theory of gases), and these can be studied analytically or by numerical methods. Finally, simulations may be based, with the minimum of simplifying assumptions, on the numerical integration of Eqs. (1.1). As a result of all these types of approaches, over the last few decades we have developed a relatively deep and accurate understanding of the many subtle aspects of the evolution of a system of gravitating point masses, the topic of this book. Indeed the study of the gravitational N-body problem must rank as one of those mature areas of research where the variety of approaches enrich each other like a community of craftsmen.

Let us highlight a few points of interest for a general physics point of view.

Perhaps the most fascinating aspect of self-gravitating systems is their instability,[3] exemplified by the fact that these systems have a negative heat capacity (Chapter 5). Thermodynamic purists would shudder, but the idea is intuitively simple to grasp and very powerful. Removing heat from a system means reducing its kinetic energy

---

[3] We exclude $N = 2$, and certain kinds of stable larger systems, e.g. hierarchical triples (Chapter 25).

of random motions. If this is done to a self-gravitating system, its constituents fall towards each other a little, and in doing so actually pick up more kinetic energy than they lost in the first place. When this concept is applied to *part* of a stellar system (as it easily can be in thought experiments) an instability can result. Imagine what might happen to a block of material with a negative specific heat, as heat flows from a hot spot: the more heat it lost, the hotter it would get, and it would quickly burst into flame.

Next, the *N*-body problem is a fascinating example of a system for which no useful equilibrium exists. Particles can and do escape (Fig. 2.2). A Maxwellian distribution of velocities would always allow particles of arbitrarily high speed, but in self-gravitating systems there is a finite escape speed. Therefore the notion of 'local thermodynamic equilibrium', which is such a powerful idea in many areas, has limited usefulness. Even if we prevent particles from escaping (in a thought experiment), thermodynamic equilibrium is possible, but may be unstable (because of the negative specific heat). There is certainly no equilibrium in the sense of solutions of Eqs. (1.1) in which all particles are at rest. Even the concept of *dynamic* equilibrium (if we ignore for a moment the escape of particles) brings with it the difficulty of showing why such equilibria are stable. This in turn depends on an understanding of the global modes of oscillation of a system in dynamic equilibrium, and for stellar systems like globular clusters this area is still in its infancy, despite a great deal of work by many experts. One of the obstacles, of course, is the severe spatial inhomogeneity of gravitating systems.

The *N*-body problem exemplifies some of the most perplexing issues in statistical mechanics. Even though the equations of motion are reversible, particles escape and are never captured. A stellar system with a collapsing core (Chapter 18) will collapse, even if the velocities are reversed. In every reasonable sense it is a chaotic system, which rapidly forgets its initial conditions, and collisions can almost always be treated with Boltzmann's classic 'Stosszahlansatz'. The fact that there are exceptions, e.g. the mutual interaction of stellar orbits in the dominating field of a central black hole (Rauch & Tremaine 1996), is an avenue (so far unexplored) for investigating the foundations and limitations of the stochastic treatment of collisions.

Problems

(1)   How would Newton have solved the problem of the collapse of the system of fixed stars if he had had access to a GRAPE?[4]

(2)   In plasma physics the Debye length is defined to be

$$\lambda_D = \sqrt{\frac{kT_e}{4\pi e^2 N_e}}.$$

---

[4] instead of an Apple.

Translate this into the language of stellar dynamics, bearing in mind that the rms speed of electrons is related to the electron temperature by

$$v_{Te} = \sqrt{\frac{3kT_e}{m_e}}.$$

If a stellar system is in dynamic equilibrium then it approximately obeys the *virial relation* $2T + W = 0$, where T is now the *total kinetic energy* and W the total potential energy (Chapter 9). By making suitable estimates for the density and other parameters of the system, show that its radius is comparable with the Debye length.

(3) Newton's Theorems on the gravitational force due to a uniform spherical shell imply that the force inside vanishes, and the force outside is the same as that due to an equal point mass at the centre of the shell. Hence show that the force at a point due to an infinite uniform medium can take any value we please.

(4) Using Newton's Theorems (Problem 3) show that the acceleration of a point at a distance r from the centre of a uniform sphere of finite radius a and total mass M is

$$\ddot{r} = -\frac{GMr}{a^3}.$$

Deduce that the sphere collapses homologously, and that collapse is complete at time

$$t = \sqrt{\frac{\pi^2 a_0^3}{8GM}},$$

where a_0 is the initial value. Compute this for the system displayed in Fig. 2.1.

3

Computational Physics Introduction

Following the evolution of a star cluster is among the most computer-intensive and delicate problems in science, let alone stellar dynamics. The main challenges are to deal with the extreme discrepancy of length and time scales, the need to resolve the very small deviations from thermal equilibrium that drive the evolution of the system, and the sheer number of computations involved. Though numerical algorithms of many kinds are used, this is not an exercise in numerical analysis: the choice of algorithm and accuracy are dictated by the need to simulate the physics faithfully rather than to solve the equations of motion as exactly as possible.

Length/time scale problem

Simultaneous close encounters between three or more stars have to be modelled accurately, since they determine the exchange of energy and angular momentum between internal and external degrees of freedom (Chapter 23). Especially the energy flow is important, since the generation of energy by double stars provides the heat input needed to drive the evolution of the whole system, at least in its later stages (Chapter 27). Unfortunately, the size of the stars is a factor 10^9 smaller than the size of a typical star cluster. If neutron stars are taken into account, the problem is worse, and we have a factor of 10^{14} instead, for the discrepancy in length scales.

The time scales involved are even worse, a close passage between two stars taking place on a time scale of hours for normal stars, milliseconds for neutron stars (Table 3.1). In contrast, the time scale on which star clusters evolve can be as long

Table 3.1. *Time scales of the million-body problem*

Time scale	Stellar dynamics	Stellar evolution	Human evolution
Seconds	White dwarf collision	Formation of neutron star	Heartbeat
Years	Hard/soft binary period[1]	Long-period variable	Malt whisky
Myrs	Crossing time[2]	Shortest stellar lifetime	Human evolution
Gyrs	Relaxation time[3]	Lifetime of the Sun	Life on Earth

[1] Chapter 19; [2] Chapter 1; [3] Chapter 14

as the age of the Universe, of order ten billion years, giving a discrepancy of time scales of a factor 10^{14} for normal stars, and 10^{20} for neutron stars.

Sophisticated algorithms have been developed over the years to deal with these problems, using individual time step schemes, local coordinate patches, and even the introduction of mappings into four dimensions in order to regularise the 3d Kepler problem (through a Hopf map to a 4d harmonic oscillator, see Chapter 15). While these algorithms have been crucial to make the problem tractable, they are still very time-consuming.

Near-equilibrium problem

In the central regions of a star cluster, the two-body relaxation time scale, which determines the rate at which energy can be conducted through the system, can be far shorter than the time scale of evolution for the system as a whole, by several orders of magnitude. For example, in globular clusters, density contrasts between the centre and the half-mass radius can easily be as large as 10^4, which implies a similar discrepancy in relaxation time scales.

As a consequence, thermal equilibrium is maintained to a very high degree. Since it is precisely the deviation from thermal equilibrium that drives the evolution of the system (Chapter 2), it is extremely difficult to cut corners in the calculation of close encounters. If any systematic type of error slipped in here, even at the level of, say, 10^{-6}, the result could easily invalidate the whole calculation. It is for this reason that none of the recently developed fast methods for approximate force calculations has been adopted in this area, e.g. tree codes, P^3M codes, etc. (Barnes & Hut 1986, Greengard 1990, Efstathiou *et al.* 1985). All such methods gain speed at the expense of relatively large errors in the force computation. The result is that the N-body simulations of stellar dynamics can boast far fewer particles than in, say, cosmological simulations. This is a pity, because N is often used as a crude 'figure of merit' in the art of simulation, whereas what really matters is the value of the science that comes out.

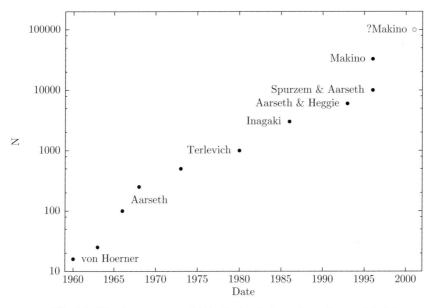

Fig. 3.1. The slow progress of N-body simulations of star clusters. Models computed well into the late evolution are plotted against publication date. For a human perspective, see Aarseth (1999) and von Hoerner (2001).

Computational requirements

The cpu cost of a direct N-body calculation scales $\propto N^3$, where the inter-particle forces contribute two powers in N (Problem 1) and the increased time scale for heat conduction contributes the third factor of N. For this reason the progress of N-body simulations of star clusters has been painfully slow, from the earliest published work of S. von Hoerner in 1960 (Fig. 3.1).

Almost none of this progress has been made by large general-purpose supercomputers. The use of parallel computers, such as Cray-T3E, has had less impact on this area than on many others, because of the communication bottleneck. The force on each particle necessarily depends on the position of every other, and therefore it is not usually efficient to parallelise the force calculations (which are the main bottleneck in serial codes). Another way of exploiting parallelism is to advance many particles simultaneously (Spurzem & Baumgardt 2001). While this works well in simple cases, the enormous range of time scales can ruin the efficiency of this approach also: individual time steps were introduced precisely so that it should not be necessary to advance all particles with the tiny time step required for one close binary!

Currently, with routine calculations, it is only feasible to model the evolution of a globular cluster containing a few thousand stars, since this requires some 10^{15} floating point calculations, equivalent to 10 Gflops-day, or several months to a year on a typical workstation. Therefore, a calculation with half a million stars, resembling a typical globular star cluster, will require $\sim$10 Pflops-day.

In contrast, the memory requirements are and will remain very modest. All that is needed is to keep track of $N = 5 \times 10^5$ particles, each with a mass, position, velocity, and a few higher derivatives for higher-order integration algorithms. Adding a few extra diagnostics per particle still will keep the total number of words per particle to about 25 or so. With 200 bytes per particle, the total memory requirement will be a mere 100 Mbytes.

Output requirements will not be severe either. A snapshot of the positions and velocities of all particles will only take 10 Mbytes. With, say, 10^5 snapshot outputs for a run, the total run worth 10 Pflops-day will result in an output of only 1 Tbyte.

Special-purpose hardware

While general-purpose supercomputers have not yet made much impact in this field, from time to time it has attracted attention as a possible application of *special*-purpose hardware (Fukushige *et al.* 1999). The earliest idea along these lines was put into practice by Holmberg as long ago as 1941 (Holmberg 1941; see also Tremaine 1981). He arranged a set of light bulbs like the stars in a stellar system, and used photometers to determine the illumination at each. Since light also obeys an inverse square law, this provided an analogue estimate of the gravitational field.

The next step in astronomy took place not in stellar dynamics but in celestial mechanics, with the building and development of the Digital Orrery (Applegate *et al.* 1985). For several years it performed ground-breaking calculations on the stability of the solar system, including the discovery of chaos in the motion of Pluto (Sussman & Wisdom 1988), until eventually being regretfully laid to rest in the Smithsonian Museum.

A significant step toward the modelling of globular star clusters was made in 1995 with the completion of a special-purpose piece of hardware, the GRAPE-4, by an ingenious group of astrophysicists at Tokyo University (Makino & Taiji 1998). GRAPE, short for GRAvity PipE, is the name of a family of pipeline processors that contain chips specially designed to calculate the Newtonian gravitational force between particles. A GRAPE processor operates in cooperation with a general-purpose host computer, typically a normal workstation. Just as a floating point accelerator can improve the floating point speed of a personal computer, without any need to modify the software on that computer, so the GRAPE chips act as a form of Newtonian accelerator (Box 3.1).

The force integration and particle pushing are all done on the host computer, and only the inter-particle force calculations are done on the GRAPE. Since the latter require a computer power that scales with N^2, while the former only require power $\propto N$, load balance can always be achieved by choosing N values large enough.

Box 3.1. GRAPE design

The design of a typical GRAPE chip (Fig. 3.2) reflects the N-body equations (Eq. (1.1)). The position and mass of an attracting particle (index j) are read from memory, the difference $\mathbf{r} = \mathbf{r}_i - \mathbf{r}_j$ is computed, then r^2, then r^3, and finally one term on the right of the equations of motion. Contributions from all attracting particles are summed. One reason for the efficiency of GRAPE is the fact that, because of the 'pipelined' design, one contribution is computed for each clock cycle. On a conventional computer each arithmetic operation usually requires several clock cycles, and each contribution requires about 30 such operations.

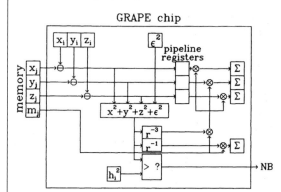

Fig. 3.2. Schematic of a GRAPE-3 chip (from Okumura *et al.* 1993).

This and other versions of the GRAPE chips perform a few other intensive calculations at the same time. Also, several chips or pipelines are installed, along with control hardware and memory for the particle data, on one board, rather like the mother board of a conventional PC. This board communicates with the host computer via a cable and an interface board, such as a PCI card. In larger installations, several GRAPE boards are arranged in a tree with suitable communications interfaces to the single host computer.

In order to make use of an installed GRAPE, sections of a simulation program are replaced by calls to software libraries which have been written by the GRAPE team in Tokyo. For example, computation of forces (and potential) on all n particles on the now-obsolete GRAPE-3 with eight chips was computed as follows, using the library function g3frc:

```
        do 119 i=1,n,8
          ii = 8
          if (i+ii .gt. n+1) ii = n- i + 1
          call g3frc(x(1,i),awork,pwork,ii)
          do 1198 j = 1,ii
            do 1197 k = 1,3
            f(k,i+j-1)=awork(k,j)
1197        continue
            pot(i+j-1) = pwork(j) + mass(i+j-1)*epsinv
1198      continue
119   continue
```

The particle data are loaded beforehand with similar instructions.

Fig. 3.3. The GRAPE-6 at the University of Tokyo, with J. Makino (right). With permission.

For example, the complete GRAPE-4 configuration, with a speed of more than 1 Tflops, could be efficiently driven by a workstation of 100 Mflops. Although such a workstation operates at a speed that is lower than that of the GRAPE by a factor of 10^4, load balance could be achieved for particle numbers of order $N \sim 5 \times 10^5$. In practice, even with this hardware, routine calculations did not greatly exceed a particle number of about 10^4, since much larger simulations could not be completed in less than a few months, and it has been found scientifically more productive to compute large numbers of relatively modest simulations rather than to break records. (Note, by the way, that the extreme parallelism of the GRAPE does not allow the most efficient scalar algorithm to be implemented.) In addition, cut-down versions of this computer can be and have been used for simulations in a wide range of other problems in astrophysics (Hut & Makino 1999) and, indeed, in other fields of science, such as plasma physics, molecular dynamics, the study of turbulence, and even protein folding.

There are several reasons for GRAPE's success. In the first place it was developed quickly, always keeping ahead of general-purpose computers. Secondly, the mathematical model of inverse square laws is quite fixed, and can be 'hard-wired'. Thirdly, the GRAPE group ensured that the devices could be made available to potential users throughout the world, and this maximised the scientific returns.

The introduction of special-purpose hardware has been a truly revolutionary advance, and not just in speed. Before GRAPE and its predecessor, the Digital Orrery, the hardware used by theorists was bought off the shelf from a computer dealer. By contrast, observers have always built their own hardware (or have had it built to their own specification), even back to the time of Galileo. From this perspective, GRAPE represents a remarkable culture shift in the way theorists can do science. In retrospect it is not surprising that it is in the area of dynamical astronomy that this has happened, as it is here that the governing equations and the underlying physical model are most stable. And we are not yet at the end of the road: GRAPE-5 is already at work (Kawai *et al.* 2000) and, as we write, GRAPE-6 is coming on stream (Fig. 3.3). It is about 100 times faster than GRAPE-4.

Software environments

Generating data is only half the job in any simulation. The other half of the work of a computational theorist parallels that of an observer, and lies in the job of data reduction. As in the observational case, here too a good set of tools is essential. And not only that: unless the tools can be used in a flexible and coherent software environment, their usefulness will still be limited.

Three requirements are central in handling the data flow from a full-scale star cluster simulation: modularity, flexibility, and compatibility. For example, to set up a major simulation, it is very useful to have a set of model building tools that are sufficiently modular, so that they can be combined in many different ways. If the data representation is flexible enough, it will be possible to add new physical variables whose use may not have been foreseen at the time that the software package was first developed. And in order for those new variables not to interfere with existing programs, compatibility is a vital issue as well.

The two main specially constructed environments in use are called *NEMO* and *Starlab* (Box 3.2). Both consist of large collections of software and tools satisfying the above requirements. In addition a great deal of *N*-body work is carried out in the UNIX-type environments used universally by computational scientists. The principal codes used in this way are the suite of *N*-body programmes written by S.J. Aarseth (1985). An extremely simplified *N*-body code is provided in Appendix A.

Box 3.2. **Starlab**

Starlab is a software package for simulating the evolution of dense stellar systems and analysing the resultant data. It is a collection of loosely coupled programs ('tools') linked at the level of the UNIX operating system. The tools share a common data structure and can be combined in arbitrarily complex ways to study the dynamics of star clusters and galactic nuclei.

Starlab features the following basic modules:

- Three- and four-body automated scattering packages, constructed around a time-symmetrised Hermite integration scheme.
- A collection of initialisation and analysis routines for use with general N-body systems.
- A general Kepler package for manipulation of two-body orbits.
- N-body integrators incorporating both second-order leapfrog and fourth-order Hermite integration algorithms.
- *Kira*, a general N-body integrator incorporating recursive coordinate transformations, allowing uniform treatment of hierarchical systems of arbitrary complexity within a general N-body framework.

In addition, Starlab enables the use of stellar evolution packages such as *SeBa*, which models the evolution of any star or binary from arbitrary starting conditions.

A novel aspect of Starlab is its very flexible external data representation, which guarantees that tools can be combined in arbitrary ways, without loss of data or internally-generated comments. Thus, two tools connected by UNIX pipes may operate on different portions of the same data set, even though neither understands the data structures, or even the physical variables, used by the other. Unknown data are simply passed through unchanged to the next tool in the chain.

Individual Starlab modules may be linked in the 'traditional' way, as function calls to C++ (the language in which most of the package is written), C, or FORTRAN routines, or at a much higher level, as individual programs connected by UNIX pipes. The former linkage is more efficient, and allows finer control of the package's capabilities; however, the latter provides a quick and compact way of running test simulations and managing production runs. The combination affords great flexibility to Starlab, allowing it to be used by both the novice and the expert programmer with equal ease.

To some extent, Starlab is modelled on NEMO, a stellar dynamics software environment developed during the 1980s at the Institute for Advanced Study, Princeton, in large part by Josh Barnes, with input from Peter Teuben and Piet Hut (and subsequently maintained and extended by Peter Teuben). Starlab differs from NEMO mainly in its use of UNIX pipes, rather than temporary files, its use of tree structures rather than arrays to represent N-body systems, and its guarantee of data conservation – data which are not understood by a given module are simply passed on to the next rather than filtered out and lost. The original version of Starlab was written by Piet Hut in 1989, while on sabbatical at Tokyo University. From 1993 onwards, Steve McMillan has extended Starlab, with help from Piet Hut, Jun Makino, Simon Portegies Zwart and Peter Teuben.

Visualisation tools, such as *partiview*, have been added by Stuart Levy. NEMO, Starlab and partiview are all available at the web site www.manybody.org

Recently, the concept of a 'virtual observatory' has been the topic of several workshops and conferences. The idea is to connect the major observational archives, from radio to optical to X-ray observations, to make available a 'digital sky' online. Archives of large-scale simulations, such as those provided by Starlab, will be connected with those virtual observatories as well, facilitating comparisons between observations and simulations (Teuben *et al.* 2001).

Problems

(1)   Use either *N*-body code in Appendix A to investigate how the cpu time depends on the number of particles. Try to explain the dependence you find.

(2)   Code the *N*-body equations (Eq. (1.1)) using a Runge–Kutta solver, either one specially prepared for the purpose, or one drawn from any available numerical library, such as Press *et al.* (1992). Try to ensure that the accuracy (judged, for example, by energy conservation) is comparable with that in Appendix A. Compare the timing with that of Problem 1, and explain the difference.

(3)   Code the *N*-body equations in a symbolic computation package, such as Maple or Mathematica. Compare the timing with that in Problem 2, again with comparable accuracy, and explain the difference.

4

Mathematical Introduction

For the mathematician, the gravitational N-body problem is the problem of understanding *by pure thought* the solutions of the set of differential equations

$$\ddot{\mathbf{r}}_i = -G \sum_{j=1, j \neq i}^{j=N} m_j \frac{\mathbf{r}_i - \mathbf{r}_j}{|\mathbf{r}_i - \mathbf{r}_j|^3}, \tag{4.1}$$

where $\mathbf{r}_j$ is the position vector of the jth body at time t, m_j is its mass, G is a constant, and a dot denotes differentiation with respect to t. Superficially, what distinguishes the work of a mathematician from that of, say, an astrophysicist, is its organisation into theorems, lemmas, and so on, but that is simply a matter of style. There are few formal theorems in Poincaré's *Les Méthodes Nouvelles de la Mécanique Céleste* (Poincaré 1892–9), but it is a rich vein of ideas. At a deeper level, the work of the mathematician aims at a greater level of rigour.

Apparently it was Herman (1710; see Volk 1975) who first solved the two-body problem using Eq. (4.1) (in component form). Since then this manner of expressing the problem has proved remarkably resilient: much the same form of equation for the general case can be found over 200 years later in the book by Moser (1973). Though Eqs. (4.1) are usually referred to as 'Newtonian', there is nothing like them in any of Newton's published works or writings. Instead, his expositions are dressed in the language of geometry or infinitesimals. Curiously, the modern language of geometry has taken an increasingly important role in recent decades: Box 4.1 shows a recent statement of the two-body problem, in terms of a manifold M and its canonical symplectic structure. Normally, however, we prefer to work with Eq. (4.1).

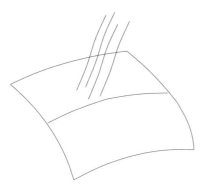

Fig. 4.1. If an orbit in three dimensions intersects a two-dimensional surface, neighbouring orbits do. If it intersects a one-dimensional curve, neighbouring orbits do not (in general).

Box 4.1. The modern two-body problem

In relatively modern language the two-body problem may be defined to be (Abraham & Marsden 1978) the 'system (M, H^μ, m, μ), where:

(i) $M = T^*W$ with canonical symplectic structure,

$$W = \mathbf{R}^3 \times \mathbf{R}^3 \backslash \mathbf{\Delta}, \quad \mathbf{\Delta} = \big\{ (\mathbf{q}, \mathbf{q}) | \mathbf{q} \in R^3 \big\};$$

(ii) $m \in M$ (initial conditions);

(iii) $\mu \in \mathbf{R}, \mu > 0$ (mass ratio);

(iv) $H^\mu \in F(M)$ given by

$$H^\mu(\mathbf{q}, \mathbf{q}', \mathbf{p}, \mathbf{p}') = \frac{\|\mathbf{p}\|^2}{2\mu} + \frac{\|\mathbf{p}'\|^2}{2} - \frac{1}{\|\mathbf{q} - \mathbf{q}'\|},$$

where $\mathbf{q}, \mathbf{q}' \in \mathbf{R}^3, \mathbf{p}, \mathbf{p}' \in (\mathbf{R}^3)^*$ and $\|\ \|$ denotes the Euclidean norm in $\mathbf{R}^{3*}$.

Our opening remark begs the question of what is meant by a 'solution'. The very *existence* of a solution, at least locally, is assured by the usual undergraduate theorem in a course on ordinary differential equations (e.g. Arnold 1978b). Globally, the obvious pitfalls are the numerous surfaces (in phase space) where singularities of the differential equations occur, i.e. two-body singularities (a 'hypersurface' S_{ij} where $\mathbf{r}_i = \mathbf{r}_j$ for some pair i, j), three-body singularities (where two two-body surfaces S_{ij} and S_{ik} intersect), and so on.

Simply by counting dimensions it is easy to see that any given orbit in phase space is very unlikely to encounter one of these singularities. The argument may be illustrated in a three-dimensional context. A single condition such as $x^2 + y^2 + z^2 = 1$ determines a two-dimensional surface. An orbit is one-dimensional, and if a given orbit intersects the surface then neighbouring orbits do (Fig. 4.1). On the other hand, if we have two conditions to satisfy, each yields a surface, and both conditions are satisfied only on their intersection, which is a curve (one-dimensional). It is still possible for the orbit to intersect this curve, but neighbouring orbits do not, in general.

This argument shows that curves which intersect two surfaces simultaneously are rare in three dimensions, or 'of measure zero'.

Returning to the N-body problem, it is easy to show that orbits intersecting any of the surfaces S_{ij} are rare, and so two-body and higher-order collisions are rare in the same sense. The situation changes dramatically, however, if we allow the position vectors $\mathbf{r}_i$ to depend on a *complex* time variable t (Box 4.2). Why should one do this? One answer is that it is an important setting in which to discuss the analytical properties of the solutions, not only because of its pure-mathematical significance, but for 'practical' reasons also. For example, the numerical treatment of the equations of motion requires special care in the vicinity of singularities (cf. Chapter 22), and in the N-body problem these are usually to be found in the complex t-plane. Another application (Chapter 21) is in certain problems of three-body scattering, when a third body temporarily approaches a short-period binary star; it turns out that the change in its energy is determined by the point in the complex plane where the intruder collides with the binary.

We have treated the two-body singularities as though they are to be avoided at all costs. In fact they are quite innocuous. For example, suppose a collision occurs

---

Box 4.2. Singularities of the two-body problem

For the planar two-body problem, complex singularities are easily located using the exact solution. A negative-energy solution of the equation

$$\ddot{\mathbf{r}} = -\mathbf{r}/r^3 \tag{4.2}$$

may be represented as

$$\mathbf{r} = (a(\cos E - e), b \sin E), \tag{4.3}$$

where a, b and e are constants such that $b = a\sqrt{1 - e^2}$, and E is determined by Kepler's equation; this is

$$n(t - t_0) = E - e \sin E, \tag{4.4}$$

where n and t_0 are more constants, with $n^2 a^3 = 1$. From Eq. (4.3) it follows that $r = a(1 - e \cos E)$, and so a two-body collision occurs where $\cos E = 1/e$, i.e. where $E = \pm i \cosh^{-1}(1/e) + 2\pi m$, in which m is an integer. By Eq. (4.4), this corresponds to the complex times $t = t_0 + \{2\pi m \pm i[\cosh^{-1}(1/e) - \sqrt{1 - e^2}]\}/n$.

Now let $e = 1$ and $t_0 = 0$ in Eqs. (4.3) and (4.4). Then a collision occurs at $t = 0$, and the analytical solution may be written as $\mathbf{r} = (x, 0)$, where

$$x = a(\cos E - 1) \tag{4.5}$$

and $nt = E - \sin E$. In the vicinity of the collision at $t = 0$ we have $nt = E^3/6 + O(E^5)$ and $x = -aE^2/2 + O(E^4)$. It is not hard to see from the first of these equations that E is expressible as a series in odd powers of $t^{1/3}$, and then the second equation shows that x may be developed as a series in powers of $t^{2/3}$.

in a two-body problem when $t = 0$, and we try to study this problem by numerical integration of the equations of motion, starting at some negative value of t during the approach to the collision. As t approaches 0 from below, it will be found that the numerical integration requires ever smaller time steps as we approach the singularity at $t = 0$, and there is nothing that can be done to get past it. Our local theorems about the existence of solutions also take us no further. When we examine the same problem analytically, however, it turns out that the coordinates of the two bodies are expressible, in the run-up to the collision, as power series in the variable $t^{2/3}$ (Box 4.2). Since this series contains *non-integer* powers of t we recover the fact that x is not an analytic function at $t = 0$. But we also observe that the series can be expressed in powers of $(t^2)^{1/3}$, which may be equally well evaluated when t is positive as when t is negative. Furthermore, it turns out that this observation is equally valid when we consider two-body collisions which are perturbed by other bodies, where the argument used in Box 4.2 (which is based on the exact solution) no longer holds. In any event, it turns out that this approach shows that there is an *analytic continuation* of the solution beyond the two-body singularity.

Some additional steps are needed before this idea can be turned to practical use, and we shall have something to say about this important technique later on (Chapter 15). In the meantime we may say that there is a choice of independent variable (the quantity $t^{2/3}$ in the above discussion) which allows us to represent the solution of the N-body problem as a function which is analytic on the real axis, even if there are two-body collisions. We may say that the collision singularities have been *regularised*.

There is an artificial but mathematically important class of N-body problems in which these techniques are essential. Two-body collisions occur naturally if one studies the *collinear* N-body problem, but when these collisions have been neutralised, other, higher-order singularities come into view. The most obvious of these is the *triple-collision* singularity, in which the coordinates of three bodies tend to coincidence as some value of t is approached. One way of studying this problem is touched on in Chapter 21, and it shows that these singularities cannot usually be handled by the technique of analytic continuation which works so well for two-body collisions. The problem is that the exponents which occur in the corresponding power series are not usually simple rational numbers, such as the power of $2/3$ which occurs in the case of two-body encounters. Instead, the exponents may involve irrational numbers such as $\sqrt{13}$ (Chapter 21), even in the simplest case of equal masses, and a series involving such powers of t cannot be evaluated (with a real result) for both positive and negative values of t. Nevertheless the study of these singularities has progressed to remarkable lengths, and these investigations are not without their consequences for the astrophysical applications of the N-body problem.

The collinear N-body problem exhibits also other classes of singularities, in which no more than two bodies collide at one time, but the collisions occur more and more frequently as a certain time approaches (see Marchal 1990). In order to give an impression of how this can happen, we have to anticipate some results of Chapter 21.

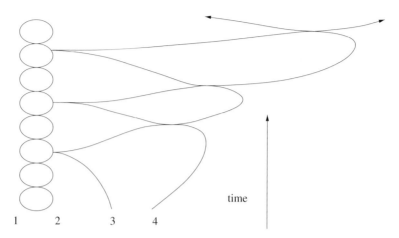

Fig. 4.2. Design of a four-body machine for accelerating a particle to infinite speed in a finite time.

Consider first the notion of a binary. In the collinear problem this is a pair of stars exhibiting a relative motion like that in Eq. (4.5): they periodically bounce off each other (as we assume that the relevant two-body encounters have been regularised). Just after one such bounce the right-hand component moves to the right at high speed. Now suppose a third body of low mass approaches from the right and collides with the right-hand component (Fig. 4.2). After this collision the third body recedes at high speed, its energy having been gained at the expense of the binary, which becomes 'tighter'. Suppose finally there is a fourth body, to the right of the third and moving off to the right. The third body, moving very fast, catches up with it and, being of relatively low mass, bounces back towards the binary. With sufficient care its next encounter can be arranged to occur, once again, just after a collision between the binary components. (In Chapter 20 we shall see in a little more detail how careful choice of initial conditions can lead, in an analogous problem, to an orbit with desired properties.) Now we have the design of a powerful four-body machine which, it may be shown, can accelerate the middle body to arbitrarily high speeds within a finite time.

When we return to a more reasonable number of dimensions it is not hard to avoid triple collisions in the three-body problem. All that is needed is to endow the system with non-zero angular momentum in its barycentric frame. The essential reason is that, if all three particles could be confined (however briefly) into a sphere of radius r, energy conservation shows that their speeds would scale as $r^{-1/2}$ and so their angular momentum would scale as $r^{1/2}$. Thus confinement within an arbitrarily small volume is inconsistent with non-zero angular momentum.

Even though collisions are usually avoided in three dimensions, singularities analogous to the one shown in Fig. 4.2, though without collisions, are still possible, at least for the five-body problem. The story of how this remarkable result was obtained (by J. Xia) is beautifully told in Diacu & Holmes (1996); see also Saari & Xia

(1995). Essentially, Xia's example consists of two Sitnikov machines (see Chapter 20) coupled end-to-end, with cunningly contrived initial conditions.

Though these examples might seem like mathematical playthings, they bear some resemblance to a curious idea with possible implications (admittedly, in the very long run) for mankind. As the Sun expands and heats up it may be possible (in principle) to keep the Earth cool by making its orbit expand. This is done by repeated two-body encounters involving the Earth and an asteroid, and Jupiter and the asteroid (Korycansky *et al.* 2001).

Examples like Xia's are highly contrived and rare. In the N-body problem there will usually be no singularities on the real time-axis. Even if we regularise two-body collisions, however, there will usually be plenty of singularities in the rest of the complex plane. These prevent us from being able to express the solution of the N-body problem as power series in t (or the appropriate regularising variable) which converge for all times. However, there is an amazingly simple transformation of the independent variable (Box 4.3) which does allow us (in principle) to write

---

Box 4.3. A series solution of the three-body problem

For some solution of the three-body problem let us suppose that the singularity which is closest to the real t-axis is at a distance $h > 0$ (i.e. its imaginary part is $\pm ih$). Then the solution of the N-body equations is analytic (without singularities) throughout a strip of width $2h$ (Fig. 4.3). Now the complex transformation

$$t = \frac{2h}{\pi} \log \frac{1 + \tau}{1 - \tau}$$

maps the unit disk $\tau \leq 1$ into the strip just mentioned. For example, when t is real the transformation is $\tau = \tanh(\pi t/(4h))$, and this varies from $\tau = -1$ to $\tau = 1$ as t increases from $-\infty$ to ∞. Since our solution of the N-body problem is analytic in the stated strip in the t plane, it follows that, if τ is used as independent variable, the solution of the N-body problem has no singularities in the unit disk. Therefore, by elementary theorems in the theory of analytic functions, the solution can be expanded as a power series in τ for all τ in the unit disk. In principle, therefore, a solution of the N-body problem may be represented as a convergent power series for all time. See Saari (1990) and Barrow-Green (1996).

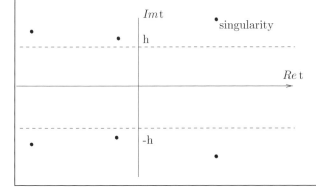

Fig. 4.3. A solution of the three-body problem is analytic in a strip $|\text{Im } t| < h$, which maps to the unit disk in the complex τ plane.

the solution as a series which converges for all times; it is not a power series in t, however. Unfortunately the solution in this form has never been put to practical use.

Since a solution of the three-body problem may be expressed as a convergent series, it is surprising to recall that this problem is often quoted as one of the famous unsolved problems of applied mathematics. Clearly, the issue hinges on what is meant by a 'solution'. Though the series expression is a solution of sorts, it would be very hard to extract from this series any information about the qualitative behaviour of the motion of the N bodies (or even any quantitative results). Nor is it very useful for numerical calculations as the rate of convergence is even more painfully slow than an N-body simulation. One usually expects much more from a satisfactory solution of a dynamical problem.

The best known class of soluble problems in dynamics are the so-called 'integrable' problems. We shall take this to mean a problem in which sufficiently many constants of the motion can be found so that the solution may be written down in terms of 'quadratures', i.e. an integral of a function of a single variable. How one finds these integrals is another matter, and usually boils down to identifying a symmetry of the problem at hand. For example, the motion of a particle in any spherical potential may be integrated using the fact that the angular momentum and energy are constant, and the existence of these integrals results from the invariance of the potential under rotations and time-translation.

The question now arises whether the N-body problem is of this type. For $N = 2$ the answer is affirmative, and indeed it may be reduced to the problem of motion in a spherical potential (see Chapter 7). In fact, in this case the quadratures can be carried out analytically. Even for $N = 3$, however, insufficiently many integrals are known, and the search for other integrals has even led to theorems proving their non-existence under certain conditions (Whittaker 1927, Moser 1973). Chapter 20 describes a particular kind of three-body problem where this question can be settled rather directly. The existence of integrals was one of the questions which motivated Poincaré to study the three-body problem and, in the process, to uncover many of the foundations of current research in Hamiltonian dynamical systems.

What one can do with the known integrals, however, is to reduce the order of the problem, i.e. the dimensionality of the phase space. The integrals associated with the motion of the centre of mass (or barycentre), for instance, reduce the order by six, but this is not much compared with $6N$. With all such tricks even the three-body problem can be reduced only to order seven (though one more order can be removed by transformation of the independent variable). Another way of expressing this is to say that a three-body system moves on a seven-dimensional subspace of phase space. By studying the topology of this subspace one can sometimes reach rigorous conclusions on the stability of three-body systems (the 'c^2h' criterion, see Marchal 1990). But this topological problem also has a pure-mathematical life of its own which recently led to a complete census of all possible types of topology in this context (see Diacu 2000).

The topics raised in this chapter, and even the title of Poincaré's book, illustrate the remarkably close links that have existed between mathematics and dynamical astronomy ever since the time of Newton. It is significant, however, that most of this has occurred within the subject that is even now termed 'celestial mechanics'. Loosely speaking, this is the mathematical study of few-body problems, usually with one dominant mass, such as those found within the solar system. We think that there is equally fertile ground for such cross-fertilisation between mathematics and the N-body problem of *stellar dynamics*, which is the subject of this book. That this is less well-developed than in the area of celestial mechanics can be traced to the fact that astronomy as a whole has become largely a part of physics. And yet it will be found from certain sections of this book that tools which have been developed by mathematicians for their own inscrutable reasons have turned out to have important applications in stellar dynamics, often several decades afterwards.

Problems

(1) Show that the set of initial conditions of the N-body problem which lead to a collision between a specific pair of particles has codimension two (i.e. two less than the dimension of the set of all initial conditions).

(2) Verify that $x = -q(1 - \sigma^2)$, $y = 2q\sigma$ is a solution of Eq. (4.2), provided that $q \neq 0$ and

$$\sigma + \frac{1}{3}\sigma^3 = \frac{t}{\sqrt{2q^3}}.$$

Write down the appropriate collision solution, corresponding to $q = 0$, and show that the orbit varies smoothly with q as one passes through the collision orbit. Determine the geometric nature of the orbits. By treating the x,y plane as a complex z-plane and applying the Levi Civita transformation $\zeta = \sqrt{z}$, determine the geometric nature of the transformed orbits, and verify that they vary smoothly through the collision orbit.

Repeat this problem with the solution given in Eq. (4.3), by varying e and keeping a fixed.

(3) Two particles of mass m move along the x-axis, and are located at $x_1 = \sin^2(E/2)$ and $x_2 = -x_1$ at time $t = E - \sin E$. Verify that their motion satisfies the two-body equations if $G = 1$ and $m = 1/2$.

A third particle moves on the y-axis, and is massless. Show that its equation of motion may be written as the system

$$\dot{y} = v$$
$$\dot{v} = -\frac{y}{\left(y^2 + x_1^2\right)^{3/2}}.$$

Show that there are three possible values of the constant c such that this system has the solution $y = cx_1$.

Change the independent variable in the system to E and code the equations numerically. By taking initial conditions close to one of the non-zero special solutions found previously, show that it is possible for the particle to be ejected with very high speed. (For a comparable problem with equal masses see Szebehely 1974.)

Part II

The Continuum Limit: $N \rightarrow \infty$

Even though this is a book about dense stellar systems (i.e. what is often called 'collisional' stellar dynamics, though no physical collisions need take place), it rests on a foundation of 'collisionless' stellar dynamics, and the relevant aspects are surveyed in these five chapters. In addition, we outline the various ways in which the effects of gravitational encounters can be incorporated, though the details are deferred to later sections of the book.

Chapter 5 begins with a discussion of the main aspects of the thermodynamic behaviour of N-body systems: how a stellar system responds to being put in contact with a 'heat bath', for instance. In fact, stellar systems tend to cool down if heat is added; paradoxical though this might seem, it helps us to understand even the motion of an Earth satellite. A toy model helps to explain what is happening.

Chapter 6 introduces the basic tools used for describing large numbers of gravitating particles: phase space, the distribution function f, the gravitational potential, and the equation governing the evolution of f (the 'collisionless Boltzmann equation'). We outline some of its solutions, and aspects of the manner in which they evolve, especially *phase mixing*. We also look at the development of *Jeans' instability*.

For our purpose the most important distribution functions are those exhibiting spherical spatial symmetry. Therefore Chapter 7 is devoted to the motion of stars in spherical potentials, including constants of the motion and their link with symmetry. An important example which we shall exploit later in other contexts is the *Kepler problem*. We outline the theory for more general potentials as well. Finally, we explore briefly the nature of nearly radial motions (touching on their importance for the *stability* of stellar systems) and those in time-dependent (but still spherical) potentials.

Chapter 8 builds on Chapter 6, by examining those solutions of the collisionless Boltzmann equation which are of particular relevance to the million-body problem. These are the *isothermal model*, and *King's* and *Plummer's* models. We take the opportunity of explaining why it is that the last of these (which is actually one of the family of *polytropic* models) has such a simple form. All these models have a few parameters, which may be adjusted in order to attempt to fit a model to observations, but the chapter closes with a brief account of *non-parametric* methods of fitting theory to observations.

Chapter 9 continues to develop the dynamical theme of Chapter 6, by showing how the evolution of a large N-body system may be modelled approximately. At the crudest level it may be characterised by the mass and energy of the system, especially if we exploit the *virial theorem*. At the next (or, perhaps, the first) level of sophistication we can treat the system like a fluid (a 'star gas'), where the stars are regarded as the atoms of the fluid. This introduces us to the Jeans equations of stellar dynamics, but we also depart from the collisionless emphasis of these chapters by mentioning the manner in which close gravitational encounters between pairs of stars are incorporated (by analogy with collisions in a gas). In the same way we show how they modify the collisionless Boltzmann equation to produce the *Fokker–Planck equation* (in the sense in which this term is used in stellar dynamics). To complete the story we remind ourselves of the most fundamental model: the N-body equations.

5

Paradoxical Thermodynamics

A strange black box

Compared to laboratory situations, a self-gravitating star cluster is a very strange object. Imagine that you were handed a star cluster in a closed box, so that you could only measure the temperature at the surface of the box. Imagine also that you could change the conditions of the star cluster from the outside in two ways: (1) you could put the box inside a larger box with a different temperature, as an effective heat bath, in order to change the temperature inside; (2) you could change the size of the box, compressing or expanding its volume.

So far, there is nothing unusual, and we might still pretend that we are about to carry out a textbook thermodynamics experiment. But when we dip our box into a heat bath, something strange may occur: depending on the exact conditions inside the box, the box may exhibit a most bizarre behaviour. When placed in a colder environment, the box may actually heat up, without limit. The only way to cool the box back to its original temperature would be to place it temporarily inside an even hotter environment – but not for too long, otherwise it will cool to below its original temperature.

This contrary tendency of self-gravitating systems corresponds to the fact that such systems can exhibit a negative heat capacity. We will return to this mysterious character later on, when we analyse its effects in detail, both on macroscopic scales, governing the evolution of a star cluster as a whole, and on a microscopic scale, when we deal with few-body systems.

Reversing the pedals

A simple way to gain some familiarity with the negativity of the heat capacity of a star cluster is to consider a satellite in orbit around the Earth. When the satellite undergoes some friction from the atmosphere, it will tend to lose some energy, and as a consequence it will lose some height. Dropping to a lower orbit, however, brings the satellite closer to the Earth, where the gravitational attraction is larger. This means that a higher velocity is required in order to stay in a circular orbit, balancing the centrifugal forces of inertia with the centripetal forces of gravity (Box 5.1).

The immediate effect, then, of a loss of energy is a speeding up of the motion of the satellite. Net energy is lost, of course, but even more potential energy is lost, and therefore the kinetic part of the energy is actually increased. It would be as if

---

Box 5.1. How air drag speeds up a satellite

Instead of moving on a circular orbit, the satellite describes a tightly wound spiral (Fig. 5.1). Therefore its velocity has a tiny radial inward component. Two forces do work on the satellite: air drag, which acts to oppose the motion, and gravity, which has a tiny component in the direction of motion. Which wins?

In fact the rate of change of kinetic energy is

$$\dot{T} = -\dot{r}\frac{GMm}{r^2} - vF, \tag{5.1}$$

where m and M are the masses of the satellite and the Earth, respectively, r is the distance between them, v is the speed of the satellite, and F is the magnitude of the drag force. In nearly circular motion we have approximately $T = \dfrac{GMm}{2r}$, where we have used the approximate equation of motion $\dfrac{v^2}{r} = \dfrac{GM}{r^2}$, and so $\dot{T} = -\dfrac{GMm}{2r^2}\dot{r}$. Hence it is clear that the first (gravitational) term on the right side of Eq. (5.1) is twice the size of the second (drag) term. In other words gravity wins and the satellite speeds up.

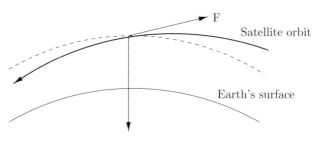

Fig. 5.1. Motion of a satellite on a nearly circular orbit under the action of a small drag force.

stepping on the brake in a car caused the car to speed up, and pushing the accelerator slowed the car down (in a gravitational car, those pedals presumably would be called the push and the decelerator, respectively).

Note the symmetry between the operations. Pushing the brake and seeing the car speed up nonetheless could be accomplished when going down a steep mountain road, with a brake that is not powerful enough. In that case, however, pushing the accelerator would have a qualitatively similar effect, of increasing the velocity (by an even larger amount), and doing nothing would also lead to a rising speed. In the case of a satellite, even a slight acceleration would lead to a higher orbit with a consequently lower speed.

This property of orbits is also of importance in understanding the role of resonances in the maintenance of spiral patterns in galaxies. In the words of Lynden-Bell and Kalnajs (1972), 'stars act like donkeys slowing down when pulled forwards and speeding up when held back'. (Transport analogies seem to be favoured in discussions of this behaviour.)

This paradoxical behaviour does not always occur for centripetal forces like gravity. Consider, for instance, the artificial case of *two-dimensional* gravity. This corresponds to the attraction of infinite parallel rods, which varies as r^{-1}, like the force between two line charges in electrostatics. A system of gravitating rods does not exhibit a negative specific heat (Padmanabhan 1990, and Problem 2).

Never at rest

The situation is actually more complicated. Doing nothing, while perhaps difficult for human beings, is altogether excluded for a self-gravitating star system. Even if we just leave our box with stars alone, and occasionally measure its temperature, we will find, after a long time, that the temperature will begin to creep up. Something is going on inside, and all the signs are pointing towards a perpetuum mobile!

What happens is that, if we wait long enough, a simultaneous close three-body encounter will produce a tightly bound pair (Eq. (21.17)). It can be shown (Chapter 19) that from that moment on, the probability is overwhelmingly large that this pair will grow tighter and tighter, on average, giving off more and more energy. This energy is converted to kinetic energy of all the other particles, including the kinetic energy of the centre-of-mass motion of the bound pair (see Fig. 5.2). So energy is still conserved.

This is an example of a negative heat capacity showing up on a microscopic level, in a self-gravitating system of particles (Chapter 19). It is also an example of what is known in particle physics as an ultra-violet divergence. Since the individual particles are mass points with no spatial extension, they can come arbitrarily close, and therefore their negative gravitational binding energy can become arbitrarily large. Just one pair of particles can therefore provide an unlimited amount of positive energy to the rest of the system.

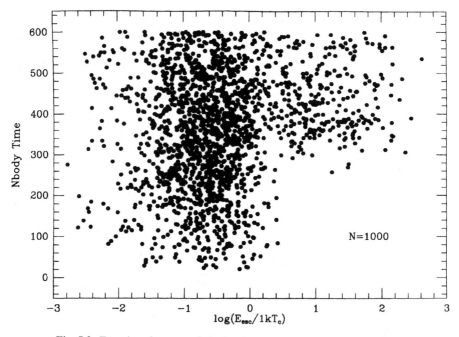

Fig. 5.2. Energies of escapers in isolated *N*-body systems (after Giersz & Heggie 1994b). In the first half of the evolution there are no binaries; escapers are caused by two-body encounters (Chapter 16) and have low energy. The escape rate increases as the core of the system collapses (Chapter 18). Eventually the core is so dense that binaries form. In their evolution they eject stars with much higher energies (Chapter 23). If the system were enclosed in a box, as in the text, they would be perceived through the increased temperature of the walls. Energy is measured in units of 2/3 of the mean kinetic energy per particle. Results from many simulations are collected in the one figure.

In this thought experiment, after closing the lid of the box, we would notice that the walls of the box got hotter, without bound. Even if we slowly extracted heat from the box, its temperature would keep rising. The tightly bound pair of particles in our experiment will have an orbital motion that is much faster than that of the single particles that it encounters. In an attempt to reach equipartition, the bound particles will try to convey some of their rapid motion to the single particles, speeding the latter up in the process. The particles in the bound pair themselves, however, while attempting to slow down, will find themselves falling to an even tighter orbit about each other. A shorter distance in the gravitational two-body problem implies a higher orbital velocity, so the net effect is that *both* the single particles and the bound particles will speed up as a result of their interactions.

A toy model

Imagine a spherical enclosure of radius *R* with a fixed attracting mass *M* at the centre. Another body, of unit mass, freely moves around inside, and bounces elastically

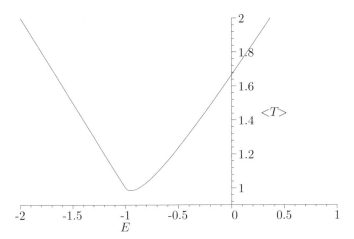

Fig. 5.3. Mean kinetic energy $\langle T \rangle$ against energy E for the toy model discussed in the text. Both energies are scaled by GM/R.

whenever it strikes the inner wall of the sphere. When its energy is low enough, in fact when $E < -GM/R$, it can never reach the sphere, and then its mean kinetic energy is $\langle T \rangle = -E$ (Problem 3). When its energy is high enough to allow collisions with the sphere, the calculation of the mean kinetic energy is a good deal harder.[1] At very high energies, however, it is clear that the particle speeds past the central mass almost without deflection, and its mean kinetic energy is only a little bigger than at the boundary, i.e. $\langle T \rangle \gtrsim E + GM/R$. Using $\langle T \rangle$ as a surrogate for temperature, we see that, above some minimum temperature, a system of a given temperature can be in either a high energy or a low energy state (Fig. 5.3). Furthermore, the high energy state has positive heat capacity $dE/d\langle T \rangle$, and the low energy state has negative heat capacity.

Now suppose we have many slightly interacting particles inside the enclosure, one at low energy and high temperature (i.e. high up on the left branch of the curve in Fig. 5.3), and N others at high energy and low temperature (i.e. lower down on the right branch). Because the slopes of the two branches are comparable (though of opposite sign), it is easy to see that a flow of heat from the hot particle to the remainder heats all the particles, but the inner body is heated more, roughly in the ratio of N:1. Thus the temperature difference is exacerbated, and the energy of the single body decreases still further. The outer bodies strike the wall with increasing energy, which is registered as an increase in the temperature which the system presents to the outside world.

This is a toy model not just of paradoxical thermodynamic behaviour, but of the million-body problem itself. If you turn Fig. 5.3 upside down it resembles part of Fig. 17.2, which is concerned with the stability of self-gravitating systems.

---

[1] See Problem 6.3. The calculations are simpler in a related model discussed by Padmanabhan (1990), but he takes a more advanced viewpoint thermodynamically. Our model was suggested by a reading of this paper, which contains several other highly illuminating toy models and discussions of the thermodynamic behaviour of self-gravitating systems.

Problems

(1)   You are the pilot of a spacecraft about to dock with a satellite which moves ahead of you on the same circular orbit. In which direction should the thrusters of your spacecraft be fired in order to catch up with your objective: forward or backward?

(2)   Using the theory in Box 5.1, show that $\dot{T} = vF$ and $\dot{r} = -\dfrac{2rF}{mv}$. If the attractive force of the Earth were to vary as r^{-n}, show that the results would become $\dot{r} = -\dfrac{2rF}{mv(3-n)}$ and $\dot{T} = vF\dfrac{n-1}{3-n}$. Deduce that paradoxical behaviour only occurs if $1 < n < 3$.

(*Harder*) What happens if (a) $n = 3$, (b) $n > 3$, (c) $n = 1$, (d) $n < 1$?

(3)   A particle of unit mass moves in the gravitational field of a fixed mass M according to the equation of motion

$$\ddot{\mathbf{r}} = -\frac{GM\mathbf{r}}{r^3}.$$

If the energy is $E = v^2/2 - GM/r$, show that $r < R$ provided that $E < -GM/R$. Show that $\dfrac{d^2}{dt^2}r^2 = 2T + 2E$, where T is the kinetic energy. Assuming that the motion is periodic, deduce that $\langle T \rangle = -E$. (This is a version of the virial theorem (Chapter 9) for the Kepler problem.)

6

Statistical Mechanics

In order to progress from qualitative arguments and toy models it is necessary to set up apparatus for describing a gravitational N-body system. There are several ways in which this can be done.

One common approach is to employ the N position vectors $\mathbf{r}_i$ and the N velocity vectors $\mathbf{v}_i$ of the stars at some time. Each of these vectors has three components, and so the entire system can be described by a single $6N$-dimensional vector, i.e. a single point in a $6N$-dimensional space Γ. This is a useful description, because it is sufficient to specify uniquely the entire subsequent evolution of the system, as the equations of motion are of second order; they describe the motion of this point through Γ. Implicitly, therefore, this is the description adopted in N-body methods, even though it is more natural to think of N particles moving in a six-dimensional phase space.

This description in a $6N$-dimensional space can be turned into a statistical one if we imagine a collection of stellar systems, each described by a distinct point in Γ. If their distribution is described by a probability density function f, the evolution of f is determined by the equations of motion, and indeed is equivalent to them. This description is almost never used in stellar dynamics.

Another way of describing a stellar system is to represent each star by a single point in a six-dimensional space with coordinates $\mathbf{r}$ and $\mathbf{v}$. Thus a single system is represented by N such points. We call this space 'phase space', though it is rather more usual to define phase space as a space with coordinates and *momenta*, rather than velocities. We shall see, however, that, in situations where this description is most useful, the basic dynamics is independent of mass. This information, i.e. the positions in phase space of all N stars, given at some time t, is sufficient to determine the entire subsequent evolution.

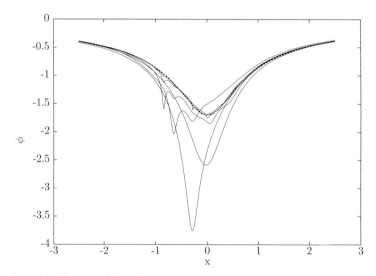

Fig. 6.1. The potential well of a Plummer model (Chapter 8) for various numbers of particles N. The potential ϕ is plotted for points along the x-axis, for $N = 1$ (the deepest), 4, 16, 64, 256 and 1024, and for the exact Plummer model (corresponding to infinite N; this is the heavy dashed curve). The curve for each successive N was obtained by adding the appropriate number of particles to the previous system, i.e. the various systems are not independent realisations of a Plummer model. This figure illustrates the fact that the potential in a phase-space description (the heavy dashed curve) may differ substantially from that in an N-body description.

We can also use phase space for a statistical description. Either we imagine that there are sufficiently many stars in the system that their distribution can be adequately represented by a smooth function, or else we can imagine the evolution of a collection of systems. In either case we suppose that there is a function $f(\mathbf{r}, \mathbf{v}, t)$ which can be interpreted either as a probability density or, in more physical terms, the density of stars in phase space, just as the density of a fluid describes the distribution of atoms in ordinary space. It is often called the 'one-particle' distribution, to distinguish it from the distribution in Γ. There are several ways in which f can be normalised, depending on its interpretation as a number density, a mass density or a probability density. Usually, however, we adopt the convention that $\int f d^3\mathbf{r} d^3\mathbf{v} = N$. This implies that $\int f d^3\mathbf{v} = n$, the number density in space.

The most important fact about f, however, is not these matters of convention but its main limitation: it is insufficient for an exact description of the dynamics (see also Fig. 6.1). It cannot tell us what happens when two stars come very close together, even though a close encounter will drastically alter their orbits. In fact, phase space is a useful description only when such events are sufficiently rare or insignificant: if the dynamics of each star is principally determined by the large mass of relatively distant stars, then the use of phase space is a useful simplification. Put another way, this is the point at which we make a choice between 'collisional' and 'collisionless' stellar dynamics. It is not that we are avoiding *physical* collisions; rather we are neglecting the special effects of gravitational two-body encounters. Fortunately, this assumption is justified in practice in many situations of interest (see Chapter 14).

For the time being, then, we suppose that the dynamics of the stars can be studied if we know only their distribution f. If all N stars have the same mass m then their space density is

$$\rho(\mathbf{r}, t) = m \int f(\mathbf{r}, \mathbf{v}, t) d^3 \mathbf{v}. \tag{6.1}$$

If stars have various masses, then we introduce an m-dependence into f and modify this integral accordingly. In any event, from ρ we determine the potential $\phi(\mathbf{r}, t)$ from Poisson's equation

$$\nabla^2 \phi = 4\pi G \rho, \tag{6.2}$$

where G is the constant of gravitation, and then the equation of motion of an individual star is

$$\ddot{\mathbf{r}} = -\nabla \phi. \tag{6.3}$$

This illustrates two things: (i) that the motion of the star is independent of its mass, and (ii) the motion of two neighbouring stars will be virtually identical, and independent of their neighbour's location. The first of these remarks will play a significant role in Chapter 10; how to improve on the second approximation is the subject of Chapter 14.

In order to complete the approximate statistical description of a stellar system, we now show how the equation of motion, Eq. (6.3), determines the evolution of f. The analogy with fluids is again helpful. If a fluid of density ρ moves in three-space with velocity $\mathbf{v}$, then conservation of mass leads to the 'continuity equation' $\partial \rho / \partial t + \nabla.(\rho \mathbf{v}) = 0$. In our case the space is six-dimensional, with 'coordinates' given by the six-vector $(\mathbf{r}, \mathbf{v})$, the corresponding 'velocity' is the six-vector $(\mathbf{v}, -\nabla \phi)$, and f takes the place of density. Thus the corresponding equation is

$$\partial f / \partial t + \nabla_{\mathbf{r}}.(f\mathbf{v}) - \nabla_{\mathbf{v}}.(f\nabla\phi) = 0, \tag{6.4}$$

where

$$\nabla_{\mathbf{r}}.(f\mathbf{v}) = \sum_{i=1}^{3} \frac{\partial}{\partial r_i}(f v_i),$$

and $\nabla_{\mathbf{v}}$ is similarly defined. Because $\mathbf{r}$ and $\mathbf{v}$ are independent, and ϕ does not depend on $\mathbf{v}$, Eq. (6.4) simplifies to

$$\partial f / \partial t + \mathbf{v}.\nabla_{\mathbf{r}} f - \nabla\phi.\nabla_{\mathbf{v}} f = 0. \tag{6.5}$$

This equation is rather fundamental. It is often called the 'collisionless Boltzmann equation', astrophysicists being rather reluctant to use the term 'Vlasov equation', as is more common in plasma physics.[1] Another way of writing it is in the form

$$\frac{\partial f}{\partial t} + [H, f] = 0, \tag{6.6}$$

where $H = v^2/2 + \phi$ is the Hamiltonian for the motion of a particle in the potential field ϕ and $[H, f]$ denotes the Poisson bracket of H and f (see, e.g., Goldstein 1980). This is very useful for problems where Hamiltonian methods are appropriate, and in that context Eq. (6.6) is called *Liouville's Theorem* (see Goldstein 1980). To write the collisionless Boltzmann equation in curvilinear coordinates, however, it is better to write Eq. (6.5) as

$$\partial f/\partial t + \dot{\mathbf{r}}.\nabla_{\mathbf{r}} f + \dot{\mathbf{v}}.\nabla_{\mathbf{v}} f = 0, \tag{6.7}$$

and simply write down the corresponding equation in the new coordinates (cf. Problem 1).

It is worth remarking that Eq. (6.6) is *equivalent* to Eq. (6.3), and not just deducible from it: if we seek a solution of Eq. (6.6) in the form $\delta(\mathbf{r} - \mathbf{r}(t), \mathbf{v} - \dot{\mathbf{r}}(t))$ then it turns out that $\mathbf{r}(t)$ must obey Eq. (6.3). The link between the two is also evident from a mathematical point of view. The standard method of solving a partial differential equation like Eq. (6.6) is by the method of characteristics (e.g. Garabedian 1986, Chapter 2), and it turns out that the characteristics of Eq. (6.6) are just the solutions of Eq. (6.3).

The analogy with ordinary fluids remains helpful in the interpretation of Eq. (6.5). If a fluid is incompressible, then its 'convective derivative' vanishes; the convective derivative is simply the derivative following the motion of the fluid, and so $\partial \rho/\partial t + \mathbf{v}.\nabla \rho = 0$. The resemblance to Eq. (6.5) shows that, in the phase-space description of a stellar system, f is constant if we follow the motion of a star in phase space; in other words, phase 'fluid' is incompressible. This property often allows us to visualise a solution of the collisionless Boltzmann equation that would be quite difficult to write down (Fig. 6.2).

The fact that f is invariant as we follow particles in phase space can sometimes be used to impose useful constraints on very complicated motions in stellar dynamics, provided that the collisionless Boltzmann equation is valid. For example, in studying collisions between galaxies, which usually lead to a single remnant, astrophysicists are interested in knowing conditions near the centre of the remnant. Conservation of phase-space density places an upper bound on the phase density there. In fact, other dynamical process such as 'phase mixing' (Chapter 10) finely intersperse regions of high and low phase density, and our incompressible phase fluid becomes more like a foam.

[1] Even here, however, the phrase 'Vlasov equation' commonly refers to the coupled system consisting of the collisionless Boltzmann equation and Maxwell's equations. Spelling also differs from one source to another; if you need to look it up in an index, try also *Wlasow*. See also Hénon (1982).

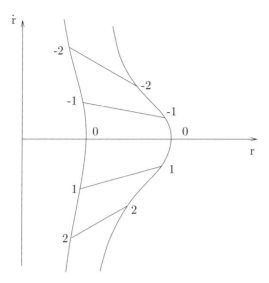

Fig. 6.2. Visualising a solution of the collisionless Boltzmann equation. The one-dimensional Kepler problem has conserved energy $E = \dot{r}^2/2 - GM/r$, and a two-dimensional phase space with coordinates $r, \dot{r}$. (The collision singularity at $r = 0$ is irrelevant to our purposes; besides, it could be removed by regularisation (Chapter 15).) Particles move downwards (i.e. $\dot{r}$ decreases, though r may either increase or decrease) on curves $E = $ constant. Particles at smaller E have shorter periods and move more quickly. The figure labels the positions of two particles at times $\ldots, -2, -1, 0, 1, 2, \ldots$. Suppose we have to solve the collisionless Boltzmann equation with initial conditions such that $f = $ constant inside the region $[-1, -1, 0, 0]$ and $f = 0$ outside. At time $t = 2$ the corners have moved to $[1, 1, 2, 2]$, and so $f = $ constant inside this region and $f = 0$ outside.

As we have just seen, f is constant along orbits in phase space. In some cases we know functions with this property. For example, the energy E is constant if ϕ is time-independent. If ϕ is spherically symmetric (see Chapter 7) then the angular momentum is constant. Since any function of these is also constant, it follows that any distribution of the form $f(E, J)$ solves the collisionless Boltzmann equation if ϕ is spherically symmetric and static. In the astrophysical literature this is a special case of what is often called Jeans' Theorem, after one of the founding fathers of stellar dynamics. But Jeans was a fine applied mathematician, and in his own writings (Jeans 1929) he attributed the result to Lagrange.

It is easy to be confused by the expression $f(E, J)$, because this is *not* the distribution function of E and J, but the distribution in phase space expressed as a function of these two variables. To relate the two concepts, let us define $N(E, J)dEdJ$ as the number of stars with E and J in small intervals dE, dJ, respectively. Then $N = \int f(E, J)\delta(v^2/2 + \phi - E)\delta(|\mathbf{r} \times \mathbf{v}| - J)d^3v d^3\mathbf{r}$. This is easily evaluated by first integrating with respect to $\mathbf{v}$, using $\mathbf{r}$ as polar axis. The result is that

$$N(E, J) = 8\pi^2 J P(E, J) f, \tag{6.8}$$

where $P(E, J)$ is the radial period of the orbit with parameters E and J (see Chapter 7). Similarly, if f depends on E alone, then the number of stars per unit energy is

$$N(E) = 16\pi^2 f(E) \int_0^{r_{\max}} r^2 \left(2(E - \phi)\right)^{1/2} dr, \tag{6.9}$$

where $\phi(r_{\max}) = E$.

What determines f? Beyond Jeans' Theorem, Eq. (6.6) is little guide, and information on this question comes from other considerations, most of which lie outside this chapter. First, the interpretation of f shows it must be non-negative. Another consideration, which is not as irrelevant as it might seem at first sight, is mathematical convenience. Much effort has gone into the construction of more-or-less simple distributions which can be used as models for the interpretation of observations or the construction of initial conditions for N-body models, and so on (see Chapter 8). Of deeper significance are distribution functions f with physical meaning. Since Eq. (6.6) shows that f evolves in time, there is at least a possibility that concepts of statistical mechanics might be applicable. If so, it might be possible to suggest particular forms (e.g. entropy maxima) towards which f would evolve after sufficient time (see Chapter 10). Finally, there is the dynamics which is not embodied in Eq. (6.6), i.e. the interactions between neighbouring particles. In a simple model of an ordinary gas this leads to relaxation to a Maxwellian distribution of velocities. In stellar dynamics the corresponding mechanism is called 'two-body relaxation' (Chapter 14). It is also known that three- and four-body interactions influence the distribution (Chapters 19, 26), but these involve 'internal degrees of freedom' which are not described by our single-particle distribution function f.

Jeans' instability

One of the standard applications of the formalism introduced in this chapter is the stellar-dynamic analogue of a problem in gas dynamics studied by Jeans. The calculation is accessible in Binney and Tremaine (1987), and we merely summarise the physical interpretation and touch on a historical curiosity. This is a real digression: we have little further use for the calculation in the remainder of the book.

The problem begins with the unlikely concept of a static, spatially infinite and uniform medium, described by a potential ϕ_0, density ρ_0 and distribution function f_0, all of which are independent of position and time. Next we suppose that these functions are subject to small (t-, $\mathbf{r}$- and $\mathbf{v}$-dependent) perturbations, denoted by subscript 1, and we proceed to linearise Eqs. (6.2) and (6.5). We find that the terms of zeroth and first order give the equations

$$\nabla^2 \phi_0 = 4\pi G \rho_0, \tag{6.10}$$

$$\nabla^2 \phi_1 = 4\pi G \rho_1, \tag{6.11}$$

$$-\nabla_{\mathbf{r}}\phi_0 . \nabla_{\mathbf{v}} f_0 = 0, \tag{6.12}$$

$$\frac{\partial f_1}{\partial t} + \mathbf{v}.\nabla_{\mathbf{r}} f_1 - \nabla_{\mathbf{r}}\phi_1 . \nabla_{\mathbf{v}} f_0 = \nabla_{\mathbf{r}}\phi_0 . \nabla_{\mathbf{v}} f_1. \tag{6.13}$$

Unfortunately there is no potential ϕ_0 satisfying both Eqs. (6.10) and (6.12). This should not surprise us: as shown in Problem 2.3, the gravitational force in an infinite uniform system can take any value we please.

We now have two choices. One choice is to revise the assumption of a uniform, infinite medium in dynamic equilibrium. The second is to invoke the 'Jeans swindle'. Let us consider these in turn.

The initial assumptions may be modified in various ways which allow progress. What we need is an unperturbed model whose potential does not accelerate (or, more fundamentally, change the momentum of) a perturbed particle. (This is the interpretation of the troublesome term on the right of Eq. (6.13).) An example is collapse of a 'cold' uniform sphere (Fig. 2.1). If we use coordinates which collapse along with the sphere, all particles are at rest in the absence of perturbation. There is a close analogy here with the theory of gravitational instability in the *expanding* universe, where the basic mechanism is well understood (see Peebles 1993). The stellar-dynamical problem is quite involved (Aarseth *et al.* 1988 and Problem 4),[2] but again the system is unstable, which accounts for the clumpiness visible in frames 6–8 in the cited figure. Another uniform model which does not accelerate a perturbed particle is described in Box 7.3, though there it is the angular momentum of the perturbed particle that is unaffected. Finally there is a very interesting (if highly artificial) system in which the gravitating matter is confined to a ring and the attraction is suitably doctored (Inagaki & Konishi 1993).

The other way of dealing with the troublesome term in Eq. (6.13) is simply to ignore it. Certainly this makes the problem tractable, and the resulting analysis leads to a stellar dynamical analogy to the Jeans instability of self-gravitating gases. We shall say a little more about this in Chapter 9.

We do not know who invented the term 'Jeans swindle', which invites the suspicion that Jeans cheated. To the best of our knowledge, Jeans tackled only the gas-dynamical analogue of the above problem, towards the end of a long paper (Jeans 1902) which also treated the stability of both spherical and slowly rotating equilibria. In the latter cases the equilibrium structure was not neglected, but when it came to the discussion

---

[2] Even its numerical simulation has some subtleties (Theis & Spurzem 1999).

of the infinite case Jeans argued that 'If space has no boundary there is presumably no need to satisfy a boundary condition at infinity, so that ρ may have any value', and simply asserted that $\phi_0 = 0$.

Problems

(1) Suppose the distribution function depends only on r, v_r and v_t, where r is the distance from the centre, v_r is the radial component of velocity, and v_t is the transverse component. Observe, by analogy with Eq. (6.7), that the collisionless Boltzmann equation takes the form

$$\partial f/\partial t + \dot{r}\frac{\partial f}{\partial r} + \dot{v}_r\frac{\partial f}{\partial v_r} + \dot{v}_t\frac{\partial f}{\partial v_t} = 0.$$

If the potential ϕ is spherically symmetric (Chapter 7), express $\dot{r}$, $\dot{v}_r$ and $\dot{v}_t$ in terms of r, v_r, v_t and ϕ, and deduce that

$$\partial f/\partial t + v_r\frac{\partial f}{\partial r} + \left(\frac{v_t^2}{r} - \frac{\partial \phi}{\partial r}\right)\frac{\partial f}{\partial v_r} - \frac{v_r v_t}{r}\frac{\partial f}{\partial v_t} = 0.$$

(2) If the potential is due to the attraction of a mass M, i.e. $\phi = -GM/r$, find the relation between $f(E)$ and $N(E)$ for $E < 0$, using Eq. (6.9).

(3) In the toy model of Chapter 5 (see Fig. 5.3), a particle of unit mass moves in the potential $\phi = -GM/r$ within a sphere of radius R. Show that the mean kinetic energy, averaged in the natural way over all particles of energy E, is

$$\langle T \rangle = \frac{\int_{r \leq R} \delta(v^2/2 + \phi - E)v^2 d^3r d^3\mathbf{v}}{2\int_{r \leq R} \delta(v^2/2 + \phi - E)d^3r d^3\mathbf{v}},$$

$$= \frac{\int_0^{r_{max}} \left(\frac{GM}{r} + E\right)^{3/2} r^2 dr}{\int_0^{r_{max}} \left(\frac{GM}{r} + E\right)^{1/2} r^2 dr},$$

where $r_{max} = R$ if $E > -GM/R$, and $-GM/E$ otherwise.
     If $E > 0$ show that

$$\frac{\langle T \rangle R}{GM} = x\frac{3x + 17x^2 + 22x^3 + 8x^4 - 3\sqrt{x(1+x)}\ln(\sqrt{1+x} + \sqrt{x})}{-3x - x^2 + 10x^3 + 8x^4 + 3\sqrt{x(1+x)}\ln(\sqrt{1+x} + \sqrt{x})},$$

where $x = RE/(GM)$. (We confess to using Maple for this.)
     Find similar results in the cases (a) $-GM/R < E < 0$ and (b) $E < -GM/R$. (They are plotted in Fig. 5.3).

(4) At time $t = 0$, N particles of mass m are distributed uniformly inside a sphere of radius R, and are at rest. Show that, because of Poisson fluctuations, two particles at the same initial radius R have accelerations which differ by a

relative amount of order $1/\sqrt{N}$. Deduce that the times at which they reach some smaller radius R' differ by a similar relative amount, and hence that their radii at the same time differ by an amount of order $V't'/\sqrt{N}$, where V' is the infall speed at time t'. By arguing that the minimum radius of the system is given by $R_{\min} \lesssim V't'/\sqrt{N}$, and estimating V' and t', deduce that $R_{\min} \sim RN^{-1/3}$.

7

Motion in a Central Potential

There are at least two reasons for studying this classical problem in a modern book on the gravitational N-body problem. First, it approximately describes the motion of individual stars in a nearly spherical system. Second, it describes the relative motion of two stars which happen to come close together. These applications are complementary. In the first we approximate the potential due to the other stars in the system by a smooth spherically symmetric field (Fig. 6.1). Occasionally, however, two stars come close together, and then the field of the rest of the cluster can be neglected temporarily. In this case it is the *relative* motion of the two stars that is described by a central force equation:

$$\ddot{\mathbf{r}} = -\frac{G(m_1 + m_2)}{r^3}\mathbf{r}, \tag{7.1}$$

where their masses are m_1 and m_2, and $\mathbf{r}$ is their relative position vector. The individual motion of each star can then be constructed if the motion of their barycentre is known.

In a central force problem motion is described by an equation of the form

$$\ddot{\mathbf{r}} = -\nabla\phi(r),$$

where $r = |\mathbf{r}|$. Thus, choosing $\phi = -G(m_1 + m_2)/r$ yields Eq. (7.1). We define the specific angular momentum, i.e. angular momentum per unit mass, by $\mathbf{J} = \mathbf{r} \times \dot{\mathbf{r}}$. It is immediately seen by differentiation that this is conserved during motion in a central field, even if the field is time-dependent. (Physically, a central force exerts no torque.) In turn this implies that motion is on a fixed plane passing through the centre of the force field, since $\mathbf{J} \cdot \mathbf{r} = 0$. The other important invariant (for a static potential)

Box 7.1. Noether's Theorem

Suppose a system is described by a Lagrangian $L(\mathbf{q}, \dot{\mathbf{q}})$, where $\mathbf{q}$ is a vector of generalised coordinates. Let $h^s(\mathbf{q})$ be a one-parameter (parameter s) group of transformations such that h^0 is the identity. Let the generalised velocities transform under the map $h^s_* : \dot{\mathbf{q}} \rightarrow \dfrac{\partial h^s}{\partial \mathbf{q}} \dot{\mathbf{q}}$. If L is invariant under these transformations, i.e. $L(\mathbf{q}, \dot{\mathbf{q}}) = L(h^s(\mathbf{q}), h^s_*(\dot{\mathbf{q}}))$, then the quantity $\dfrac{\partial L}{\partial \dot{\mathbf{q}}} \dfrac{d}{ds} h^s(\mathbf{q})|_{s=0}$ is invariant.

For motion in a spherical potential the Lagrangian is $L = \dfrac{1}{2}\dot{\mathbf{r}}^2 - \phi(r)$. Let h^s be rotation about the z-axis by angle s. Then $\dot{\mathbf{r}}$ is not invariant, but $\dot{\mathbf{r}}^2$ is, and so is r. For small s, $h^s(x, y, z) \simeq (x - ys, y + xs, z)$, and so $\dfrac{\partial L}{\partial \dot{\mathbf{q}}} \dfrac{d}{ds} h^s(\mathbf{q}) = (\dot{x}, \dot{y}, \dot{z}) \cdot (-y, x, 0)$, i.e. the z-component of angular momentum.

Bringing the energy integral within this framework requires some modest extensions (see Arnold 1978a).

---

is the specific energy, defined by

$$E = \frac{1}{2}\dot{\mathbf{r}}^2 + \phi(r).$$

These two invariants reflect symmetries of the dynamical problem, and are special cases of Noether's Theorem (Box 7.1). The invariance of $\mathbf{J}$ reflects only the spherical symmetry of ϕ, and is true even if the field is time-dependent. The constancy of E depends only on the assumed time-independence of ϕ, and not the spherical symmetry. Certain central forces have further symmetries, and further invariant quantities can be constructed. The Kepler problem (this chapter and Chapter 15) is one of these, and the three-dimensional simple harmonic oscillator is another, which is discussed further below.

The potentials corresponding to these two problems are approximately applicable to the dynamics of stellar systems. Far outside the bulk of the mass, $\phi \simeq -GM/r$, where M is the total mass. (The potential is arbitrary up to an additive constant, but in idealised models isolated from all external matter, it is convenient to choose the constant so that $\phi \rightarrow 0$ as $r \rightarrow \infty$.) Near the centre, on the other hand, if the potential is non-singular it can be expanded as

$$\phi \simeq \phi_0 + \frac{2\pi G\rho_0}{3}r^2, \tag{7.2}$$

where ρ is the density of matter and the subscript 0 refers to the centre. The coefficient here comes from the theory of the gravitational potential in spherical systems, for which Newton's Second Theorem (see Binney and Tremaine 1987) asserts that $-d\phi/dr = -GM(r)/r^2$, where $M(r)$ is the mass within an imaginary sphere of radius r. Near the centre, $M(r) \simeq 4\pi\rho_0 r^3/3$, whence Eq. (7.2) follows immediately.

From the properties of Keplerian and simple harmonic motions, several useful results follow. For example, comparison of Eq. (7.2) with the standard simple harmonic potential $\omega^2 r^2/2$ shows that the period of orbits near the centre is given approximately by

$$P \simeq \sqrt{3\pi/G\rho_0}. \tag{7.3}$$

Orbits which remain outside most of the mass will, on the other hand, resemble a conic section, by Kepler's First Law. If this is an ellipse of semi-major axis a, it follows from the theory of Keplerian motion (Box 7.2) that the period is approximately

$$P \simeq \sqrt{4\pi^2 a^3/(GM)}. \tag{7.4}$$

This can be written as

$$P \simeq \sqrt{3\pi/G\bar{\rho}}, \tag{7.5}$$

if $\bar{\rho}$ is thought of as the mean density in a sphere of radius a containing the entire mass of the cluster. The resemblance to Eq. (7.3) follows from dimensional arguments, but the equality of the coefficients may look like an undeserved bonus (Problem 1).

To investigate central orbits further we can use plane polar coordinates r, θ in the plane of motion. The angular momentum is then $J = r^2\dot\theta$, the energy integral becomes $E = \frac{1}{2}(\dot r^2 + r^2\dot\theta^2) + \phi$, and so

$$E = \frac{1}{2}(\dot r^2 + \frac{J^2}{r^2}) + \phi. \tag{7.6}$$

We assume that ϕ has the form of a finite potential well (Fig. 7.1), and that $\phi \to 0$ as $r \to \infty$. The function $\phi_{\text{eff}} = \frac{J^2}{2r^2} + \phi$, often called the 'effective potential', plays an important role in orbit theory. For a given non-zero value of J, ϕ_{eff} has a single minimum at some radius $a > 0$, where

$$-\frac{J^2}{a^3} + \phi'(a) = 0, \tag{7.7}$$

and a circular orbit of this radius and angular momentum is possible if $E = \phi_{\text{eff}}(a)$, since Eq. (7.6) then implies that $\dot r = 0$.

Box 7.2. Formulae for Keplerian motion

Proofs of the following can be found in numerous texts on dynamics, e.g. Goldstein (1980). We also like very much the vectorial treatment in Pollard (1976).

Consider the equation of planar motion $\ddot{\mathbf{r}} = -\dfrac{\mu\mathbf{r}}{r^3}$ (cf. Eq. (7.1)). Then the specific energy is $E = \dfrac{1}{2}\dot{\mathbf{r}}^2 - \dfrac{\mu}{r}$. If $E < 0$ then the motion is on an ellipse with the origin at one focus. The semi-major axis is given by $E = -\dfrac{\mu}{2a}$ and the period by $\dfrac{2\pi}{n}$, where the 'mean motion' is given by $n^2 a^3 = \mu$. The specific angular momentum has magnitude $J = \sqrt{\mu a(1 - e^2)}$, where e is the eccentricity of the ellipse. If the orbit is aligned with pericentre along the positive x-axis, then the position vector at time t is

$(a(\cos\psi - e), b\sin\psi)$, where $b = a\sqrt{1 - e^2}$ is the semi-minor axis, and the eccentric anomaly ψ is given by Kepler's equation $nt = \psi - e\sin\psi$, if the origin of t is taken at the time of pericentre. (We would prefer E for the eccentric anomaly, but that is already in use.) In plane polar coordinates the orbit is given by

$$r = \frac{a(1 - e^2)}{1 + e\cos\theta}. \tag{7.8}$$

Similar formulae are obtained for hyperbolic motion ($E > 0$), but are usually expressed in real form via hyperbolic functions instead of circular ones, and it is sometimes helpful to redefine a to make it positive, i.e. $E = \dfrac{G\mu}{2a}$. An important additional formula in the hyperbolic case is for the deflection angle α between the ingoing and outgoing asymptotes:

$$\tan(\alpha/2) = \sqrt{\mu a}/J. \tag{7.9}$$

Again the case of parabolic motion ($E = 0$) leads to a different suite of formulae. A unified treatment of all three types of motion is possible if specially designed variables are used (Stumpff 1962).

When these equations describe the relative motion of two bodies with masses m_1 and m_2 (cf. text), we have $\mu = G(m_1 + m_2)$, and the energy and angular momentum in their barycentric frame are obtained by multiplying by the *reduced mass* $m_1 m_2/(m_1 + m_2)$. Thus the energy becomes $E = \dfrac{1}{2}\dfrac{m_1 m_2}{m_1 + m_2}\dot{\mathbf{r}}^2 - \dfrac{Gm_1 m_2}{r}$.

For a given angular momentum J, if E slightly exceeds $\phi_{\text{eff}}(a)$ then motion in a small range of radii near radius a becomes possible. Differentiating Eq. (7.6), cancelling $\dot{r}$, expanding about $r = a$, and using Eq. (7.7), we find that $\ddot{r} = -\kappa^2(r - a)$ approximately, where $\kappa^2 = 3\phi'(a)/a + \phi''(a)$. This describes the radial motion. In addition, conservation of angular momentum shows that the star circulates at a rate which is nearly constant, except for an oscillation at frequency κ. The composition of these two motions can be represented as motion on a small ellipse whose centre circulates round the centre of attraction. Because of the resemblance of this description to the Ptolemaic model of the solar system, κ is referred to as the 'epicyclic' frequency.

The qualitative features of orbits which are not nearly circular can be easily deduced from Fig. 7.1. We continue to assume that $J \neq 0$. If E sufficiently exceeds $\phi_{\text{eff}}(a)$ that the epicyclic approximation is too inaccurate, but still $E < 0$, then Fig. 7.1 shows that motion is confined between two radii, which we denote by $r_{\min}$ and $r_{\max}$, and is periodic in radius. If $E \geq 0$ then the star escapes to infinity. In the case of the Keplerian potential the distinction between orbits of negative and positive energy is reflected in the transition from bound, elliptic motion to unbound motion on a hyperbola.

At the opposite extreme, as it were, from epicyclic orbits are nearly radial orbits. Such orbits, which pass close to the centre of the potential, are of particular interest for two reasons: first, stars which are ejected from the dense central parts of a stellar system travel on such orbits, and, second, they play a crucial role in one of the basic

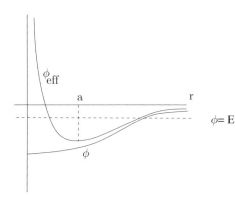

Fig. 7.1. Radial motion in a spherical potential. See the text for an explanation of the symbols. The case $\phi_{\text{eff}}(a) < E < 0$ is illustrated.

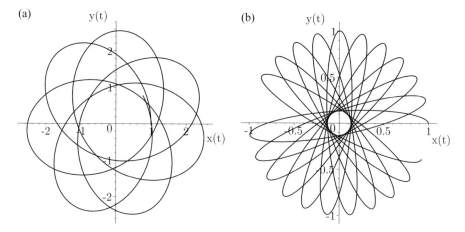

Fig. 7.2. Two orbits in the scaled Plummer potential $\phi = -1/\sqrt{1+r^2}$. They start at $r = 1$ on the x-axis, moving transversely upwards with speeds (a) 0.9 and (b) 0.1. Note that the angular separation of successive apocentres, when measured in the direction of motion, lies between the values for the simple harmonic oscillator (π) and Kepler motion (2π); see Problem 4.

instabilities to which stellar systems are liable (see, for example, Antonov 1973, Fridman & Polyachenko 1984, Palmer 1994).

Let us consider orbits of a given energy E and small angular momentum. For $J = 0$ the orbit is purely radial, and successive maxima in r are separated (in θ) by an angle exactly equal to π. If J is small the star passes close to the origin, and its speed then is nearly $\sqrt{2(E - \phi_0)}$. By angular momentum conservation it follows that the distance of closest approach to the origin is nearly proportional to J. It is also easy to see that the transverse deviation from a purely radial orbit is proportional to J, and therefore successive maxima in r are separated (in θ) by angles which differ from π by a small angle proportional to J (Fig. 7.2b and Problem 2).

It is this tiny angle by which nearly radial orbits rotate which allows them to cooperate with a non-radial perturbation in the potential to give rise to the 'radial orbit instability', especially in systems where such orbits are heavily populated. In such a system the pressure tensor is very anisotropic, and the transverse velocity dispersion is relatively low. In this sense such systems have a low 'transverse temperature'. We already remarked (Chapter 6) that cold stellar systems tend to clump, by a process analogous to the Jeans instability, and in much the same way systems with a low transverse temperature tend to clump into a bar-shaped configuration (see Barnes 1985). A proper understanding of the mechanism by which this happens involves a little more orbit theory, however, and is deferred to a later remark in this chapter about resonances, and Box 7.3.

Because an analysis of general orbits, i.e. those which are neither nearly radial nor nearly circular, can be reduced to motion in radius (cf. Fig. 7.1), the results can be expressed as quadratures. For example, the time between successive maxima in r is

$$2 \int_{r_{\min}}^{r_{\max}} dr/\dot{r} = 2 \int dr/\sqrt{2(E - \phi(r)) - J^2/r^2}, \tag{7.10}$$

by Eq. (7.6). Only a few cases are known where the appropriate integrals can be expressed in terms of elementary functions. Aside from the familiar Keplerian and simple harmonic cases, one potential with this property is the 'isochrone' potential $\phi = -\dfrac{GM}{b + \sqrt{b^2 + r^2}}$, which was so named by Hénon because it has the property that all orbits of the same energy have the same radial period (Problem 5).

Let us denote the frequency of radial motion by ω_r, and similarly use ω_θ to denote the mean angular velocity $\dot{\theta}$. Thus the angle through which the star rotates between two successive maxima in r is $2\pi\omega_\theta/\omega_r$. For nearly radial orbits this is nearly π, whence $\omega_r \simeq 2\omega_\theta$ for such orbits. Therefore such orbits exhibit a *resonance*, which has an important role in the 'radial orbit instability', as already mentioned. See Box 7.3.

We also mentioned that stars moving close to the centre of the cluster exhibit nearly simple harmonic motion with a certain period $\sqrt{3\pi/(G\rho_0)}$, and it might have been thought that such orbits would also resonate. However, the number of such stars is small. Consider circular orbits, for example. The period of a circular orbit of radius r is $2\pi\sqrt{r/\phi'}$. Now ϕ can be expressed as a power series in even powers of r (cf. Eq. (7.2)), and so, therefore, can the period. Therefore the period varies approximately linearly with the energy for low-energy orbits. Now in the most usual models of star clusters, the volume of phase space available to stars of energy less than E varies as $(E - \phi_0)^3$,[1] and so it follows that the number of stars of a given period increases rapidly as the period increases above the minimum value $\sqrt{3\pi/(G\rho_0)}$. Thus there is no significant population of resonant stars near this frequency.

---

[1] Consider, for example, a three-dimensional simple harmonic oscillator. The energy is $E = (\dot{\mathbf{r}}^2 + \omega^2\mathbf{r}^2)/2$, and the right side is the square of the radius of an ellipsoid in six-dimensional phase space.

Box 7.3. **The radial orbit instability**

Imagine a spherically symmetric system, with a slight aspherical potential perturbation. Does the response of the stars tend to reinforce this perturbation, or to suppress it?

To capture the essence of the problem we consider a perturbing potential with a plane of symmetry, so that a star initially orbiting in this plane remains there. With plane polar coordinates r, θ, suppose the perturbing potential has a form like $\phi_1 = -a(r)\cos(2\theta)$, where $a(r) > 0$ is some function and the subscript denotes a perturbation. (Note that we have assumed that ϕ_1 is invariant under rotation by π; the factor 2 in the argument of the cosine is often expressed by stating that we are examining an '$m = 2$' perturbation.)

As already stated, there is a near-resonance between radial and azimuthal motions for nearly radial orbits, i.e. a nearly vanishing linear combination of frequencies. If we think of frequency as rate of change of an angle, it turns out that the corresponding slowly varying angle can be interpreted as the orientation, ψ, of the long axis of the orbit (Fig. 7.3). The corresponding 'canonical momentum' (in the sense of Hamiltonian dynamics, cf. Goldstein 1980) is the angular momentum, J. As shown in Problem 2, in a spherical potential ψ advances in each radial oscillation by an amount of order J, and so the average rate of change is $\dot{\psi} \simeq AJ$, for some $A > 0$, plus a small contribution from the perturbing potential. It is also evident that J is altered only by the perturbation. If $0 < \psi < \pi/2$ the perturbation pulls the star towards the positive x-axis, and the angular momentum decreases; the orbit is 'pulled back' by the perturbation. Considering also the other quadrants we see that in general we may write $\dot{J} \simeq -B\sin 2\psi$, for some $B > 0$.

Now let $f(J, \psi)$ be the distribution of these two variables. Since ψ and J are conjugate, f approximately obeys a Vlasov equation

$$\frac{\partial f}{\partial t} + \dot{\psi}\frac{\partial f}{\partial \psi} + \dot{J}\frac{\partial f}{\partial J} = 0$$

(cf. Eq. (6.7)). We write f, $\dot{\psi}$ and $\dot{J}$ as a sum of a contribution from the underlying spherical cluster model and a contribution from the perturbation, e.g. $f = f_0(J) + f_1(J, \psi)$, and neglect all but linear terms involving the perturbation. Then

$$\frac{\partial f_1}{\partial t} + AJ\frac{\partial f_1}{\partial \psi} \simeq B\sin(2\psi)\frac{\partial f_0}{\partial J}. \tag{7.11}$$

(Notice, by the way, that the analogue of the term which gave us so much trouble in the study of the Jeans instability (Eq. (6.13)) would have the form $\dfrac{\partial \phi_0}{\partial \psi}\dfrac{\partial f_1}{\partial J}$, which vanishes because ϕ_0 is spherically symmetric.)

The forced solution of Eq. (7.11) is

$$f_1 = -\frac{B}{2AJ}\frac{\partial f_0}{\partial J}\cos 2\psi. \tag{7.12}$$

If f_1 is to be consistent with the assumed perturbation in the potential then we need an enhancement of orbits along the x-axis (near $\psi = 0$). By Eq. (7.12) we see that this will happen if $J\partial f_0/\partial J < 0$. It follows that, for the radial-orbit instability to proceed, it is necessary that the distribution function should increase as J decreases to zero from

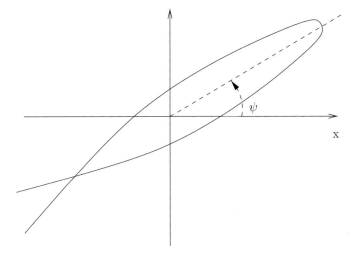

Fig. 7.3. A nearly radial orbit in a slightly non-spherical potential.

either side, i.e. $\partial f_0/\partial J$ is positive (negative) if J is negative (positive). This means that we expect to see the instability only in systems in which there is a strong radial anisotropy.

There are some general lessons here for several kinds of instabilities in stellar systems. In order for an instability to flourish, there must be a resonance, and there must be a sufficient gradient of f_0 across it (see, for example, Palmer 1994, and the classic Lynden-Bell and Kalnajs 1972).

These remarks are relevant to the radial orbit instability, but there are other instabilities which can afflict spherical stellar systems. For example, though it is easy to build a spherical system with an almost arbitrary mass profile $M(r)$, by setting particles on circular orbits with velocity $v(r) = \sqrt{GM(r)/r}$, there is no such model which is known to be stable (see Barnes *et al.* 1986).

Finally we turn to the behaviour of stars in central time-dependent potentials. This is important in the astrophysical applications of the million-body problem for two reasons. First, it is possible that stellar systems breed massive central black holes, and the resulting evolving potential is an example of the time-dependence we have in mind. A second example is the slow evolution of a stellar system caused by two-body relaxation (Chapter 14).

In time-dependent central forces the angular momentum J is still constant. All the interest is in the energy. Since $dE/dt = \partial\phi/\partial t$, it follows that the change during half of one radial oscillation is

$$\delta E = \int_{r_{\min}}^{r_{\max}} \frac{\partial \phi}{\partial t} \frac{dr}{\dot{r}}. \tag{7.13}$$

The integrand is evaluated at the time when the particle reaches each radius. If, however, ϕ changes only slowly (i.e. on a time scale much longer than the radial orbital period), to good approximation we can evaluate the right-hand side at some *fixed* time. Multiplying by half the radial period, we deduce that

$$\delta E \int_{r_{min}}^{r_{max}} \frac{dr}{\dot{r}} = \int_{r_{min}}^{r_{max}} \delta\phi(r) \frac{dr}{\dot{r}}, \tag{7.14}$$

where $\delta\phi(r)$ is the change at fixed r in half a radial period. Bringing the two sides together we now see that

$$\int_{r_{min}}^{r_{max}} \frac{\delta E - \delta\phi}{\sqrt{2(E - \phi) - \dfrac{J^2}{r^2}}} dr = 0,$$

and so

$$\delta \int_{r_{min}}^{r_{max}} \sqrt{2(E - \phi) - \frac{J^2}{r^2}} dr = 0. \tag{7.15}$$

Equation (7.15) shows that the quantity $I_r = \int \dot{r} dr$ is invariant (except for variations that cancel out over one period), provided that the potential varies on a long time scale. It is, in fact, an example of a standard result on the adiabatic invariance of action variables, since I_r is simply (except for a factor π) the radial action (see Palmer 1994). We shall make use of this property of I_r in a statistical description in Chapter 11.

Problems

(1)   Explain why Eqs. (7.3) and (7.5) have the same coefficient.

(2)   For motion close to the x-axis in the x,y plane of a spherically symmetric potential $\phi(x, y)$, show that we have $\ddot{y} = y(\ddot{x}/x)$ exactly, and
$\ddot{x} = -\dfrac{\partial}{\partial x}\phi(x, 0)$ approximately. By considering orbits with initial condition
$x = R, y = 0, \dot{x} = 0$ and $\dot{y} = J/R$, where R is a constant, show that the rate of rotation of the orbit is nearly proportional to J.

(3)   Consider motion of energy E and small angular momentum J in a smooth spherically symmetric potential $\phi(r)$ such that $\phi(0) = 0$. Show that successive pericentres are separated by an angle which is approximately

$$\pi + \frac{2J}{\sqrt{2E}} \left(-\frac{1}{r_{max}} + \int_0^{r_{max}} \frac{\sqrt{2E} - \sqrt{2(E - \phi(r))}}{r^2\sqrt{2(E - \phi(r))}} dr \right),$$

where $\phi(r_{max}) = E$.

(4)   (Suggested by a remark of H. Dejonghe.) Assuming that ρ is non-negative and is non-increasing with r, show that the angle between successive apocentres lies between π and 2π.

(5)   Compute the dependence of r and θ on t for an orbit of energy E and angular momentum J in the isochrone potential.

(6)   Compute the radial action (see Eq. (7.15)) for the bound Kepler problem, where $\phi = -GM/r$ and $E < 0$. Consider a circular Keplerian orbit. What happens if M is decreased by a factor f $(0 < f < 1)$ (a) slowly, and (b) instantaneously?

(7)   In the Kepler problem show that $\dot{\mathbf{r}}$ describes a circle in velocity space.

8

Some Equilibrium Models

In Chapter 6 we spent some time discussing the statistical description of an N-body system in terms of its one-particle distribution function, f. We also introduced an evolution equation for f, the 'collisionless' Boltzmann equation. We did not, however, dwell on any solutions of this equation, except to characterise them in terms of Jeans' Theorem. In fact *equilibrium* solutions have been known and studied for about 100 years, and they are of enduring importance.

The specific choice of f may be made for a variety of reasons. Plummer's model, for instance, is often used for starting a numerical calculation, because of its analytical convenience. The isothermal model is of importance in the study of thermodynamic stability (Chapter 17), while King's model has taken centre stage for many years in the interpretation of observations. Another approach to this particular topic, also based on the phase-space description, deserves a section on its own at the end of the chapter.

By Jeans' Theorem, f is a collisionless equilibrium solution if (but not only if) it depends on the energy, E. Most of the models we mention are of this kind. In what follows we shall usually characterise a model by its distribution function $f(E)$, but that is only part of the story, because f depends on the potential ϕ (via E), and we need to know how ϕ depends on the position in space $\mathbf{r}$ in order to determine f at any point in phase space. The space density is given by $\rho = m \int f d^3\mathbf{v}$, and also depends on ϕ. Therefore the potential must satisfy Poisson's equation in the form $\nabla^2\phi = 4\pi G\rho(\phi)$. If this equation can be solved, the end result is what is called a 'self-consistent' model of a stellar system.

The isothermal model

Let us suppose that

$$f = f_0 \exp(-2j^2 E), \tag{8.1}$$

where f_0 and j are constants. Then the distribution of velocities at any point is Maxwellian, with one-dimensional dispersion σ^2 given by

$$\sigma^2 = 1/(2j^2). \tag{8.2}$$

This is the distribution that is set up by collisions in simple gases, and two-body relaxation (Chapter 14) plays a similar role in the million-body problem.

Computing the density at a point where the potential is ϕ, we find that

$$\rho/\rho_0 = \exp(-m(\phi - \phi_0)/kT), \tag{8.3}$$

where $kT/m = \sigma^2$, and ρ_0 and ϕ_0 are the density and potential (respectively) at some point, often taken to be the centre. This equation already teaches us something important: that the density of more massive stars decreases with increasing ϕ more quickly than the density of less massive stars. This behaviour is a simple example of *mass segregation*, which we discuss more dynamically in Chapter 16. Note that it depends on the implicit assumption that all stars have the same temperature.

Now we turn to Poisson's equation $\nabla^2 \phi = 4\pi G\rho$, and introduce the scaled potential

$$u = -2j^2(\phi - \phi_0) \tag{8.4}$$

and the scaled radius

$$z = r\sqrt{8\pi G\rho_0 j^2}. \tag{8.5}$$

This leads to the *isothermal equation*

$$\frac{d^2u}{dz^2} + \frac{2}{z}\frac{du}{dz} + e^u = 0, \tag{8.6}$$

which is a spherically symmetric version of one of Liouville's equations,[1] by the way. Its solutions are thoroughly discussed in Chandrasekhar (1939). (This book is also well worth looking at for the marvellous isocline diagrams reproduced from Emden's *Gaskugeln* (Emden 1907).)

If we assume as above that $\phi(0)$ is the central potential, at the bottom of the potential well, the appropriate boundary conditions are

$$u(0) = 0, \quad u'(0) = 0. \tag{8.7}$$

---

[1] i.e. $\nabla^2 u + \exp u = 0$.

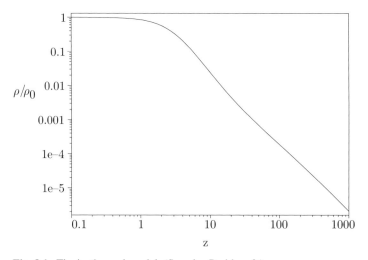

Fig. 8.1. The isothermal model. (See also Problem 2.)

Then the solution is uniquely specified (Fig. 8.1), and so we see that the family of isothermal models has two free dimensional parameters: the velocity dispersion σ^2 (which is related to j by Eq. (8.2)), and the central density. There is, however, also a limiting case, called the singular isothermal solution, which has infinite central density (Problem 1). We missed it by our implicit assumption that ρ_0 exists. Finite solutions approach the singular solution asymptotically at large radius (Problem 2 and Fig. 8.1).

Though it is important to understand the properties of this model because of its thermodynamic significance (Chapter 17), its practical applications are limited. If a model of a stellar system is in dynamic equilibrium there can be no escaping stars, and so f must vanish above the energy of escape. Now the isothermal distribution, Eq. (8.1), never vanishes, and this means that there is *no* escape energy. The reason for this is that the model turns out to have infinite mass (Problems 1 and 2).

King's model

The standard way of improving the behaviour of the isothermal model, at least in practical terms, is to 'lower' the Maxwellian distribution of Eq. (8.1) into

$$f = \begin{cases} f_0 \left(\exp(-2j^2 E) - \exp(-2j^2 E_0)\right) & \text{if } E < E_0, \\ 0 & \text{if } E > E_0, \end{cases} \tag{8.8}$$

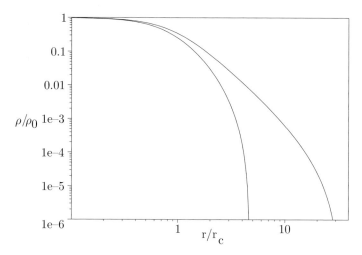

Fig. 8.2. The King models with scaled potential $W_0 = 3$ (below) and $W_0 = 7$.

where E_0 is the escape energy. This is King's model (King 1966),[2] and it has several attractive features, aside from finite mass and radius. Deep inside the system $E \ll E_0$, and so the distribution is nearly Maxwellian where the system may be expected to be nearly relaxed (since the time of relaxation is shortest where the density is highest, see Chapter 14). Also, the distribution function Eq. (8.8) turns out to be a good approximation to a solution of the Fokker–Planck equation (see Chapter 9). What is occasionally forgotten, however, is that this solution is not an equilibrium solution, but one in which stars escape at a rate governed essentially by the two-body relaxation time scale. Another pitfall is to imagine that the central velocity dispersion is still given by Eq. (8.2), but this is only good in the limit $E_0 \to \infty$.

King's models are characterised by a third parameter in addition to the two parameters needed for an isothermal model. This third parameter is often taken to be the quantity $W_0 = 2j^2(E_0 - E_c)$, where E_c is the energy of a star at rest at the centre. Fig. 8.2 illustrates the models with $W_0 = 3$ and 7. It is customary to characterise the size of the central region by what is called the 'core radius', r_c, though different definitions exist. We adopt the definition given by the physically appealing relation

$$\frac{4\pi G}{3}\rho_c r_c^2 = v_c^2, \tag{8.9}$$

where ρ_c and v_c^2 are the central density and mean square velocity, respectively; i.e. $v_c^2 = 3\sigma_c^2$. For an isothermal model, r_c is the radius at which the space density drops

---

[2] In another paper (King 1962) he proposed a simple mathematical function which fits the profile of the surface brightness Σ of globular clusters, and it is sometimes also referred to as King's model, unfortunately. The formula is $\Sigma = k\{[1 + (r/r_c)^2]^{-1/2} - [1 + (r_t/r_c)^2]^{-1/2}\}^2$. We propose to call this King's *Law*, by analogy with de Vaucouleurs' Law.

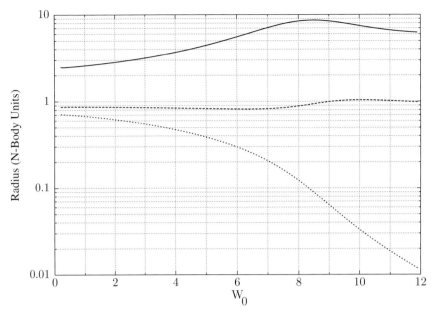

Fig. 8.3. The tidal radius (solid), half-mass radius (dashed) and the core radius (dotted) plotted against the scaled central potential W_0 for the sequence of King models, in N-body units (Box 1.1).

to about one third of its central value, and the projected (or surface) density falls to about one half of its central value. This and other characteristic radii are plotted in Fig. 8.3. A commonly used measure of how centrally concentrated a model is is the quantity c defined as $c = \log_{10}(r_t/r_c)$, although King himself defined c without the logarithm, and with a different definition of r_c.

King's model has remained a cornerstone of stellar dynamics for over 30 years. It has been extended to include anisotropy (Michie & Bodenheimer 1963), a mass spectrum (e.g. Gunn & Griffin 1979), and an external gravitational field (Heggie and Ramamani 1995), though these extensions do not have the same support from the Fokker–Planck equation. In addition, it quite unclear what is the appropriate theoretical definition of the core radius if stellar masses are unequal. Partly for this reason rather more empirical measures are in common use (Casertano and Hut 1985). On the theoretical side, substantial work has been done on the modes of oscillation of King models (e.g. Sobouti 1985). In particular, Weinberg (1994a) has pointed out how slowly some modes are damped.

Plummer's model

Our next exhibit is the world's favourite theoretical model, simply because its structure can be written down in terms of very simple functions; sufficiently simple, in fact,

Table 8.1. *Plummer's model*

Quantity	Symbol	Expression or value
Distribution function	f	$\dfrac{3.2^{7/2}}{7\pi^3}\dfrac{a^2}{G^5 M^4 m}(-E)^{7/2}$
Mass density	ρ	$\dfrac{3M}{4\pi a^3}\left(1+\dfrac{r^2}{a^2}\right)^{-5/2}$
Projected mass density	$\Sigma(d)$	$\dfrac{M}{\pi a^2}\left(1+\dfrac{d^2}{a^2}\right)^{-2}$
Mass within radius r	$M(r)$	$M\left(1+\dfrac{a^2}{r^2}\right)^{-3/2}$
Mass within projected radius d	$M(d)$	$M\left(1+\dfrac{a^2}{d^2}\right)^{-1}$
(Specific) potential	ϕ	$-\dfrac{GM}{a}\left(1+\dfrac{r^2}{a^2}\right)^{-1/2}$
One-dimensional velocity dispersion	σ^2	$-\dfrac{1}{6}\phi(r)$
Projected velocity dispersion	$\sigma_z^2(d)$	$\dfrac{3\pi}{64}\dfrac{GM}{a}\left(1+\dfrac{d^2}{a^2}\right)^{-1/2}$
Potential energy	W	$-\dfrac{3\pi}{32}\dfrac{GM^2}{a}$
Kinetic energy	T	$-\dfrac{W}{2}$
Total energy	E	$\dfrac{W}{2}$
Core radius	r_c	$\dfrac{a}{\sqrt{2}}$
Virial radius	R_v	$\dfrac{16}{3\pi}a$
Half-mass radius	R_h	$\dfrac{a}{\sqrt{2^{2/3}-1}}, \simeq 1.305a$
Half-mass relaxation time	t_{rh}	$\dfrac{0.206 N a^{3/2}}{\sqrt{GM}\ln\Lambda}$

that any theorist can memorise some of the basic data without difficulty (Table 8.1).
It is actually the stellar-dynamical analogue of an $n = 5$ polytrope (Plummer 1911;[3]
Box 8.1). Less well known is the fact that it is one of another series of analytical models which includes the isochrone model (Chapter 7; see Dejonghe 1984).

---

[3] He attributes the solution to Schuster, however.

Box 8.1. **Plummer's model**

A polytrope of index n is a gaseous model in which the pressure-density relation is of the form $p \propto \phi^n$. The $n = 5$ polytrope has some special properties. It is the last of the series (in increasing n) with finite mass, and the first with infinite radius. An examination of the polytropic equation shows why it is so special, as follows.

If all the physical constants are scaled out from Poisson's equation, with the above relation between ρ and ϕ, the resulting form is $y'' + (2/x)y' + y^n = 0$, where y is minus the scaled potential and x is the scaled radius. Now it is easy to see that this equation is invariant under the symmetry $y \to \lambda y$, $x \to \lambda^{(1-n)/2}x$, and this suggests use of the variable $Y = yx^{-2/(1-n)}$, which is symmetry-invariant. The resulting equation is

$$x^2 Y'' + \frac{6 - 2n}{1 - n}\left(xY' + \frac{Y}{1 - n}\right) + Y^n = 0,$$

and the form of the derivatives further suggests the transformation from x to $t = \ln x$. Then the equation takes the transparent form

$$\ddot{Y} + \frac{5 - n}{1 - n}\dot{Y} + \frac{6 - 2n}{(1 - n)^2}Y + Y^n = 0,$$

where a dot means a t-derivative. Now it is clear why $n = 5$ is so special: this is then the equation of a non-dissipative oscillator, and so

$$\frac{1}{2}\dot{Y}^2 - \frac{1}{8}Y^2 + \frac{1}{6}Y^6 = \text{constant}.$$

Even better, the boundary conditions $y(0)$ finite and $y'(0) = 0$ transform to the conditions $Y \to 0$ and $\dot{Y} \to 0$ as $t \to -\infty$, and so the constant is zero. In fact the required solutions emanate from and return to the equilibrium at $Y = 0$, and the solution which passes through $Y = (3/4)^{1/4}$ at $t = 0$ is easily found to be $Y^2 = \sqrt{3/4}\,\text{sech}\,t$. This transforms back to $y = 3^{1/4}/\sqrt{1 + x^2}$, which is the potential of a Plummer model, suitably scaled.

Evolutionary calculations starting with Plummer initial conditions have been carried out by many people (e.g. Spitzer & Shull 1975).

The free parameters of the model are the total mass M and the 'scale radius' a. Table 8.1 gives several quantities which we have not introduced before. These include the *projected mass density*, $\Sigma(d)$, which is the density per unit surface area of the sky at the distance of the object (so that, in astrophysical units, it would be measured in solar masses per square parsec). The line of sight passes at a distance d from the centre of the object. Analogous to σ^2, σ_z^2 is the mean square component of the velocity along the line of sight. The total mass is M, the mass within a sphere of radius r is $M(r)$, and $M(d)$ is the mass within a cylinder of radius d parallel to the line of sight.

Models of observational data

We have already mentioned that models such as these are used in the interpretation of observations, and now we consider just how much of a restriction the observations of a given stellar system actually place on a model. Throughout, we suppose that the system is spherically symmetric and in dynamical equilibrium, so that f can be regarded as a function of E and J.

Let us suppose first that the observations consist of data on the surface brightness of a stellar system. If it is supposed that 'mass follows light', this yields (within a constant of proportionality) the projected mass density, Σ, which can also be computed for any of the above theoretical models. The simplest approach which proves reasonably successful (e.g. Peterson & Reed 1987, Trager *et al.* 1995) is to choose the three free parameters of King's model (e.g. f_0, j^2 and E_0) to optimise the fit of the theoretical model to the observations. Enrico Fermi once said, 'Give me three parameters and I'll fit an elephant',[4] and indeed these models fit astonishingly well. One reason for this is that the process of turning a distribution function into a surface density is a sequence of integrations, and these tend to iron out differences between quite disparate models. (From f we integrate to get $\rho(\phi)$, integrate again to obtain $\phi(r)$ and hence $\rho(r)$, and integrate yet again to construct Σ.) Another reason may be that the profile of a star cluster is a lot simpler than that of an elephant.

An alternative approach is to attempt to invert this series of integrations more-or-less directly, and in principle this can be done. In other words, from Σ it is possible to invert the appropriate integral equation to obtain $\rho(r)$ (Merritt & Tremblay 1994), and then $\phi(r)$ comes from solving Poisson's equation. This gives $\rho(\phi)$, from which it is possible to construct a distribution function $f(E)$ depending on E alone, i.e. an isotropic model (Problem 5). In turn this suggests that the anisotropy is, in a sense, arbitrary, and that a whole sequence of models (differing in the degree of anisotropy) can be constructed to fit a given observed distribution Σ. Kinematic information is, therefore, essential to lift this degeneracy, and provide more unambiguous information on the distribution $f(E, J)$. Usually this is radial velocity data (of the sort catalogued in Pryor & Meylan 1993) but useful internal proper motion data exist for some clusters (e.g. van Leeuwen *et al.* 2000).

The problem of inverting the integral equations from observational data rather than from mathematical functions is another issue. Consider, for example, the problem of determining $\rho(r)$. If information on the distribution of the stars is given in the form of star counts, or even the positions of all the stars, the inversion usually magnifies the noisiness of the data. One way to cope with this is to determine a function $\rho(r)$ which best fits the data with the constraint that it is not too noisy. In practice, one represents ρ on a radial grid, writes down a corresponding expression for Σ at the

---

[4] Some people say it was four, but the point is the same.

location of each star, and then maximises the likelihood, with an additional penalty function which is large when the grid of values of ρ is too noisy, as judged by the differences in the tabular values (in a suitable sense.) This is not a purely academic discussion; Gebhardt & Fischer (1995) have given some interesting applications of this kind of method.

There are merits and disadvantages in both methods of fitting models to data. Parametric fitting can yield deceptively agreeable fits, while the results of non-parametric fitting take insufficient account of what is known on theoretical grounds about the distribution function, and the result depends on the choice of penalty function, leaving considerable freedom in the hands of the investigator. Nevertheless the velocity maps of the rotating cluster ω Centauri produced by Merritt *et al.* (1997) are a striking demonstration of the power of this approach.

Problems

(1)   Verify that $u = -\ln(z^2/2)$ is an exact solution of Eq. (8.6). Write down the corresponding density, and compute the mass inside radius r and the projected density.

(2)   Transform Eq. (8.6) to variables $v = u + \log(z^2/2)$ and $t = \log z$. Deduce that $v \to 0$ as $t \to \infty$, and hence show that all isothermal models have infinite mass. By linearising about the equilibrium $v = 0$, show that $u(z) \simeq -\ln(z^2/2) + az^{-1/2}\cos(\sqrt{7}(\ln z)/2 + b)$, where a and b are constants. The decaying oscillations are visible in Fig. 8.1.

(3)   Solve the one- and two-dimensional isothermal equations, with initial conditions (8.7). (The coefficient of $\dfrac{1}{z}\dfrac{du}{dz}$ in Eq. (8.6) is $n - 1$, where n is the number of dimensions.) See Fanelli *et al.* (2001) for an interesting *evolutionary* model of the one-dimensional case.

(4)   Find the first few terms of a series solution of Eq. (8.6), with boundary conditions (8.7).

(5)   If the mass density in phase space is $f = f(E)$, show that the space density at a point where the potential is ϕ is

$$\rho(\phi) = 4\pi \int_{\phi}^{\infty} \sqrt{2(E - \phi)}\, f\, dE.$$

Differentiate with respect to ϕ and solve the resulting Abel integral equation for f.

(6)   Show that, in the limit $W_0 \to 0$, a King model is a polytrope with index $n = 2.5$ (Chandrasekhar 1939). Deduce that the limiting radius in N-body units (Box 1.1) is $12/5$. (No Galactic globular clusters are known with very small W_0, and it might be thought that this is because the depth of the potential well is insufficient to retain stars. Remember, however, that W_0 is a

scaled potential. A King model with low W_0 enjoys as perfect a virial balance as any other.)

(7) Consider the distribution function $f = 2f_P$ if $J_z > 0$, 0 otherwise, where J_z is the z-component of angular momentum and f_P is the distribution of Plummer's model. Verify that this satisfies Jeans' Theorem (Chapter 6) if the potential is axisymmetric and stationary. Show that the density is the same as in Plummer's model. What is the angular momentum?

(8) Consider the distribution function

$$
f = \begin{cases} \dfrac{3}{4\pi^3 Gm}(-2E - J^2)^{3/2} & \text{if } 2E + J^2 < 0 \\ 0 & \text{otherwise,} \end{cases}
$$

where E and J are, respectively, total energy and angular momentum per unit mass. Deduce that the density at a point where the radius is r and the potential is $\phi(r)$ is $\rho = -\dfrac{3}{4\pi G}\dfrac{\phi^3}{1+r^2}$. Deduce that Poisson's equation is satisfied if ϕ is the scaled Plummer potential $\phi = -1/\sqrt{1+r^2}$. Show that the mean square radial and transverse velocities are, respectively, $\dfrac{1}{4}\dfrac{1}{\sqrt{1+r^2}}$ and $\dfrac{1}{2}\dfrac{1}{(1+r^2)^{3/2}}$. See Dejonghe (1987a).

9

Methods

In stellar dynamics we do not really study stellar systems like globular clusters and galaxies. That is the job of astronomers. What we do is study *models* of these systems. Just as in many branches of applied mathematics, a model is nothing other than a mathematical structure into which we try to incorporate our knowledge of the system at hand. Sometimes 'knowledge' is not the right word: it may be nothing better than a hunch about how things might work. Often, however, our knowledge will include physical laws, especially the ones we think are relevant. So far in this book, for example, we have implicitly thrown out almost everything we know about stellar systems except the gravitational dynamics.[1]

In the context of stellar dynamics, a model consists of two kinds of mathematical construct. One is the mathematical object used in the *description* of the system, and the other is the equation which determines its *evolution*. So far, for example, we have introduced the *N-body model*, where the system is *described* by N time-dependent vectors, which *evolve* according to Newton's Law of motion. We have also introduced a statistical model of collisionless stellar dynamics, where the system is *described* by the one-particle distribution function f, which *evolves* according to the collisionless Boltzmann equation.[2]

In this final chapter of Part II we give a foretaste of the full variety of models for the dynamics of *dense stellar systems*. It is a bridge between the essentially

---

[1] R.A. Lyttleton used to say that constructing a mathematical model was a bit like plucking a chicken, but at the end of the job the bird has still to be airworthy. Einstein said that everything should be made as simple as possible but not simpler.

[2] In the previous chapter we have also used the word 'model' in a different but traditional sense, to denote a specific solution.

simpler dynamics considered up to now and the more colourful variety of dynamical processes on which we focus in the rest of the book.

The four models we consider (Fig. 9.1) are taken in order of increasing detail and veracity, but also increasing complexity and difficulty. For any particular problem at hand, the choice of model is dictated by the usual balance of considerations: simplicity and veracity. The most important problems in the dynamics of the million-body problem have been studied using all four. Indeed it is a measure of the maturity of the subject that so much can be understood from such a wide variety of viewpoints.

The scaling description: evaporative models

The simplest model of a stellar system characterises it in terms of three fundamental parameters, which may be chosen to be its mass, its size and the mean square speed of the stars in it (cf. Box 1.1). Simple as it is, it can be used to discuss a variety of forms of dynamical evolution.

As an example, let us consider the virial theorem in the form

$$\ddot{I} = 4E - 2W, \tag{9.1}$$

where I is the 'moment of inertia', E is the total energy, and W is the potential energy (Box 9.1). This is very often applied to systems in dynamic equilibrium, when the average value of the left-hand side vanishes, and so it follows that $4E - 2W = 0$ on average. It is also possible, however, to use the virial theorem in a dynamical way, via the scaling description. If the system is specified by its total mass M and a characteristic radius R,[3] and if the assumption is made that the evolution of the distribution of mass is homologous, i.e. given simply by a change of scale, then there are 'form factors' α and β such that $I = \alpha M R^2$ and

$$W = -\beta G M^2 / R. \tag{9.2}$$

Substituting in Eq. (9.1) we see that there is an equilibrium radius $R_0 = -\dfrac{\beta G M^2}{2E}$, and that small oscillations around this radius have frequency ω given by $\omega^2 = \dfrac{\beta G M}{\alpha R_0^3}$. These are called 'virial oscillations' (Fig. 9.2). Note that the frequency involves the square root of G times a density, just as in Chapter 7.

It is clear that an N-body system has a tendency to collapse if the speeds of the stars are too slow (Fig. 2.1), and the above formulation of the virial theorem makes this quantitative. In fact $\ddot{I} < 0$ if $2T + W < 0$, i.e. if $\langle v^2 \rangle < \beta G M / R$, where $\langle v^2 \rangle$ is the

---

[3] For spherical systems it is common to take for R the *half-mass radius*, which contains half the mass. Then for many models (cf. Chapter 8) $\beta \simeq 0.4$. Another common choice is the *virial radius*, defined by taking $\beta = 1/2$.

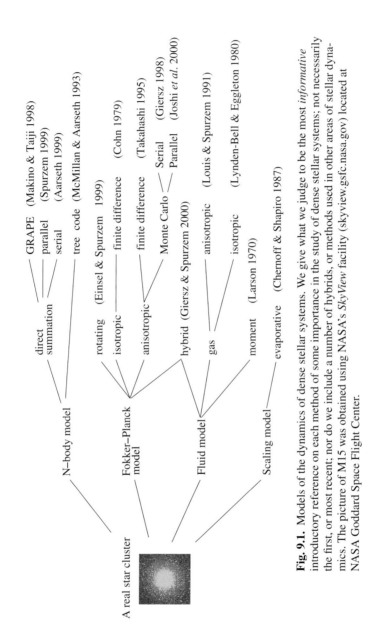

Fig. 9.1. Models of the dynamics of dense stellar systems. We give what we judge to be the most *informative* introductory reference on each method of some importance in the study of dense stellar systems; not necessarily the first, or most recent; nor do we include a number of hybrids, or methods used in other areas of stellar dynamics. The picture of M15 was obtained using NASA's *SkyView* facility (skyview.gsfc.nasa.gov) located at NASA Goddard Space Flight Center.

Box 9.1. The virial theorem

This has to be proved from one of the more fundamental models coming later in this chapter. As an illustration we start with the collisionless Boltzmann equation:

$$\frac{\partial f}{\partial t} + v_i \frac{\partial f}{\partial x_i} - \frac{\partial \phi}{\partial x_i} \frac{\partial f}{\partial v_i} = 0.$$

If we define

$$I = \int m f r^2 d^3 \mathbf{r} d^3 \mathbf{v},$$

the integral being taken over all of phase space, then

$$\dot{I} = \int m \left(-v_i \frac{\partial f}{\partial x_i} + \frac{\partial \phi}{\partial x_i} \frac{\partial f}{\partial v_i} \right) r^2 d^3 \mathbf{r} d^3 \mathbf{v}.$$

Integration of the second term with respect to v_i gives zero, if we assume that $f \to 0$ at infinity, while integration of the first term by parts with respect to x_i, with the same assumption, gives

$$\dot{I} = 2 \int m \mathbf{r} \cdot \mathbf{v} f d^3 \mathbf{r} d^3 \mathbf{v}.$$

Differentiating with respect to t again and applying similar tricks leads to the result

$$\ddot{I} = 2 \int m(v^2 - \mathbf{r} \cdot \nabla_{\mathbf{r}} \phi) f d^3 \mathbf{r} d^3 \mathbf{v}.$$

The first of these is evidently four times the kinetic energy, $4T$, while integration with respect to $\mathbf{v}$ shows that the second term is

$$-2 \int \rho(\mathbf{r}) \mathbf{r} \cdot \nabla \phi d^3 \mathbf{r} = -2G \int \int \rho(\mathbf{r}) \rho(\mathbf{r}') \frac{\mathbf{r} \cdot (\mathbf{r} - \mathbf{r}')}{|\mathbf{r} - \mathbf{r}'|^3} d^3 \mathbf{r} d^3 \mathbf{r}'.$$

Interchanging $\mathbf{r}$ and $\mathbf{r}'$ shows that either expression is

$$-G \int \int \frac{\rho(\mathbf{r}) \rho(\mathbf{r}')}{|\mathbf{r} - \mathbf{r}'|} d^3 \mathbf{r} d^3 \mathbf{r}' = 2W.$$

Finally, $E = T + W$, and so the virial theorem follows:

$$\ddot{I} = 4E - 2W.$$

For a system in dynamical equilibrium, $\ddot{I} = 0$ on average, and so $4E - 2W = 0$, or $2T + W = 0$. These are common forms in which the virial theorem is applied.

---

mean square speed. Aside from numerical factors, this is the same as the condition one obtains from naïve application of the Jeans stability criterion (Chapter 6) to a stellar medium whose density is estimated from M and R.

Crude though such models might seem, they capture much basic physics, and the above model can be extended to include rotation, flattening, external gravitational fields, loss of mass, escape of stars, and even several aspects of internal evolution

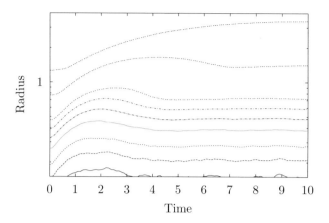

Fig. 9.2. Virial oscillations in an N-body model. Initial conditions: Plummer model (Chapter 8) with $N = 16\,384$, and velocities increased to make $2T/W = -1.4$. Units: N-body units (Box 1.1). Each curve is a *Lagrangian radius*, i.e. the radius of a sphere containing a fixed mass and centred at the densest part of the cluster. From the bottom the mass enclosed is 0.02, 0.05, 0.1, 0.2, 0.3, 0.4, 0.5, 0.75, 0.9. Except at large radii, where the time scale is larger, there is no evidence of more than one complete cycle: the oscillations are heavily damped, unless two-body relaxation (Chapter 14) is suppressed (cf. David & Theuns 1989). The form factor α (see text) is infinite (logarithmically divergent) for the Plummer model.

driven by relaxation. The phenomenon of stellar escape was one of the first processes to be discussed using this model (Ambartsumian 1938, Spitzer 1940). In this case the evolution equation is different. It is assumed that mass-loss takes place on a two-body relaxation time scale t_r (Chapter 14), and then

$$\frac{dM}{dt} = -\mu \frac{M}{t_r}, \tag{9.3}$$

where μ is a constant. It can be determined in various ways, e.g. by solving the Fokker–Planck equation (see below) in an idealised potential well (Spitzer & Härm 1958). Another equation is required for the evolution of E. For instance, if we suppose that stars escape following small-angle scattering, as in the theory of relaxation, they do so with negligible energy, and so we may suppose that E is constant (Problem 14.8). We shall not describe this problem further, but it gives its name to this whole class of models, which are often referred to loosely as 'the evaporative model'. It is possible to add further properties of the cluster, such as its *concentration*, which was defined for a King model in Chapter 8 (see, for example, Prata 1971, Chernoff & Shapiro 1987).

The fluid description: gas models

There is a close analogy between a cluster, regarded as a self-gravitating ball of stars, and a star, modelled as a self-gravitating ball of gas, and it leads to a more elaborate, but still highly simplified, description of the million-body problem (Hachisu *et al.* 1978, Lynden-Bell & Eggleton 1980, and Box 9.2). In the spherically symmetric case the structure is described by an r-dependent density and temperature, denoted by ρ and T respectively. Temperature here is used as a surrogate for the velocity dispersion σ^2, which endows the star-gas with a pressure $p = \rho\sigma^2$; this equation would be written as $p = nkT$ in the kinetic theory of a perfect gas. A more elaborate version of the description allows the pressure tensor to be anisotropic, corresponding to an anisotropic distribution of velocities.

Box 9.2. The equations of stellar structure and cluster evolution

In the simplest case of radiative transfer of energy in a spherical star consisting of a perfect gas in mechanical equilibrium, the equations are

$$\frac{\partial M}{\partial r} = 4\pi\rho r^2$$

$$\frac{\partial p}{\partial r} = -\frac{GM(r)}{r^2}\rho$$

$$\frac{\partial L}{\partial r} = -4\pi\rho r^2 \left(T\frac{DS}{Dt} - \epsilon \right) \tag{9.4}$$

$$\frac{\partial T}{\partial r} = -\frac{3}{4ac}\frac{\kappa\rho}{T^3}\frac{L(r)}{4\pi r^2},$$

(e.g. Clayton 1983, Chapter 6), where most of the variables have the same meaning as in the text, except that S is the specific entropy ($\propto \ln(T^{3/2}/\rho)$ for a perfect gas), ϵ is the specific rate of energy generation, a is the Stefan–Boltzmann constant, c the speed of light, and κ the opacity.

In the even simpler case of a star cluster, the equations are (Lynden-Bell & Eggleton 1980)

$$\frac{\partial M}{\partial r} = 4\pi\rho r^2$$

$$\frac{\partial p}{\partial r} = -\frac{GM(r)}{r^2}\rho$$

$$\frac{\partial L}{\partial r} = -4\pi\rho r^2 \left(\sigma^2\frac{DS}{Dt} - \epsilon \right)$$

$$\frac{\partial\sigma^2}{\partial r} = -\frac{1}{3GmC\ln\Lambda}\frac{\sigma}{\rho}\frac{L(r)}{4\pi r^2},$$

where σ is the root mean square one-dimensional speed, m is the individual stellar mass, C is a numerical constant determined by the theory of relaxation (Chapter 14, which also explains the *Coulomb logarithm* $\ln\Lambda$), $S = \ln(\sigma^3/\rho)$, and ϵ is the rate of generation of energy in three-body interactions (Eq. (23.5)).

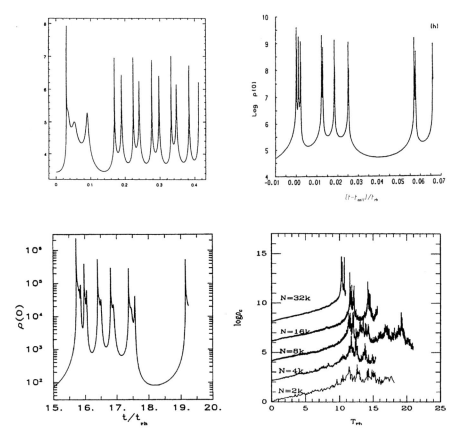

Fig. 9.3. Four models of core collapse and gravothermal oscillations: (top left) an extension of the evaporative model; (top right) the gaseous model; (bottom left) the Fokker–Planck model; (bottom right) the N-body model. After Allen & Heggie (1992),[4] Heggie & Ramamani (1989), Cohn *et al.* (1989) and Makino (1996). The density at the centre of a system is plotted against time. The initial rise to the first maximum corresponds to the phenomenon of 'core collapse', while the subsequent variations are referred to as 'gravothermal oscillations'.

Now we illustrate a little of the dynamics that can be associated with this description. For a stellar system in dynamic equilibrium the appropriate equation is

$$\frac{dp}{dr} = -\frac{GM(r)}{r^2}\rho, \tag{9.5}$$

where $M(r)$ is the mass within radius r; thus $dM/dr = 4\pi\rho r^2$. This equation is familiar from hydrostatics, as $-GM(r)/r^2$ is simply the gravitational acceleration. It can also be regarded as a moment equation (in the time-independent case), derived from the collisionless Boltzmann equation by taking the first moment with respect to the distribution of velocities (Problem 5). In the context of stellar dynamics it is usually referred to as (one of) 'Jeans' equations'.

[4] With reference to Eqs. (17) in that paper, the model parameters used were $T_{10}^* = 4.186$, $\xi_T = 0.00037$ and $\Gamma = 1.28$, and the initial values were 3.437 346, 3.442 962 and 1.258 366.

Beyond the condition of local hydrostatic equilibrium, the gas model can be extended to include a thermal conductivity (cf. Chapter 1) and the resulting slow evolution from one quasi-equilibrium to another. The intention is to model the effects of two-body relaxation (Chapter 14), and so the thermal conductivity must be chosen to ensure that transport of thermal energy takes place on the relaxation time scale. (In particular, the conductivity appropriate to a non-magnetised plasma yields the wrong time scale.)

The resulting model has been very successful in developing our understanding of the million-body problem. The phenomenon of *core collapse* (Chapter 18) was first understood in detail using models of this general kind. Even more impressively, it was with these models that the phenomenon of *gravothermal oscillations* (Chapter 28) was actually discovered. Fig. 9.3 shows this behaviour modelled with the four different methods discussed in this chapter.

Generally speaking, the gas model produces results which compare very well with those produced by other techniques (Aarseth & Lecar 1975), though the reason for this success has remained obscure. For example, it has not been demonstrated that *all* the equations may be obtained by taking moments of the Boltzmann equation. If one does take moments, however (Larson 1970, Louis 1990), it is possible to obtain a set of equations which are rather similar in complexity and effectiveness to the gas equations, provided that a suitable closure assumption is made, but they differ in detail (Box 9.3). One reason for this is that the mean free path in a stellar system is very long (compared to the typical size of the system; see Problem 14.5), whereas in an actual gas the mean free path is short compared to the length scale of the system. Perhaps the success of the gas equations may be explained if we suppose that so much of the behaviour of stellar systems depends more on the thermodynamics of self-gravitating spheres than on the details of the mechanism of energy transport.

Box 9.3. Gas and moment equations

The difference between the gas equations, which incorporate relaxation as thermal conduction, and the moment equations of the Fokker–Planck equation, can be illustrated by the following discussion, in the context of spherical symmetry. Both models involve a heat transport equation which can be written in the form

$$\frac{\partial L}{\partial r} = -4\pi \rho r^2 \sigma^2 \frac{D}{Dt} \ln \frac{\sigma^3}{\rho}, \tag{9.6}$$

where D/Dt denotes a 'convective' rate of change (i.e. following the radial motion of the material), and L is the conductive flux, relative to the material, of thermal energy across a sphere of radius r.

In the case of the gas model this is nothing other than the first law of thermodynamics in the form $dQ = TdS$, where dQ is an increment of thermal energy, and T and S are the temperature and entropy. For a spherical shell with radii r and $r + dr$, we have $dQ = (L(r) - L(r + dr))dt$, where dt is a short time interval. Similarly, using the expression for the specific entropy of a perfect gas, the entropy of the material in this shell is $4\pi \dfrac{\rho}{m} r^2 dr \times k \ln \dfrac{\sigma^3}{\rho}$, where k is Boltzmann's constant and m is the stellar mass. Then the gas model takes the flux per unit area, $L/(4\pi r^2)$, to be proportional to the radial gradient of the temperature, or of the mean square velocity dispersion in a single component, σ^2.

In the case of the moment method, Eq. (9.6) is obtained from the Fokker–Planck equation. For the present purpose this is best written in the form

$$\frac{\partial f}{\partial t} + v_i \frac{\partial f}{\partial x_i} - \frac{\partial \phi}{\partial x_i} \frac{\partial f}{\partial v_i} = \left(\frac{\partial f}{\partial t} \right)_{\text{enc}},$$

where the left-hand side is as in the collisionless Boltzmann equation, and the right-hand side is the 'encounter term', the form of which is discussed in Chapter 14. (Briefly, encounters alter the velocities of participating stars, and this changes f at a rate described by this term.) Next, take the moment with respect to $\mathbf{v}$ of $m(\mathbf{v} - \langle \mathbf{v} \rangle)^2$, where the mean velocity is defined by $\rho \langle \mathbf{v} \rangle = m \int f \mathbf{v} d^3 \mathbf{v}$. The moment of the right-hand side vanishes, as encounters do not change the total mass, momentum or kinetic energy of stars in a small spatial volume. Computing the moment of the left-hand side requires some care. It is also necessary to use the zeroth moment, or continuity equation, i.e. $\dfrac{\partial \rho}{\partial t} + \nabla \cdot (\rho \langle \mathbf{v} \rangle) = 0$, and the sort of tricks such as integration by parts that were adopted in the derivation of the virial theorem (Box 9.1). If it is assumed that the distribution function is isotropic in a frame moving with velocity $\langle \mathbf{v} \rangle$, the result is exactly Eq. (9.6), with an explicit form of L, i.e.

$$L = 4\pi r^2 \frac{1}{2} \rho \langle \hat{\mathbf{r}} \cdot (\mathbf{v} - \langle \mathbf{v} \rangle)(\mathbf{v} - \langle \mathbf{v} \rangle)^2 \rangle. \tag{9.7}$$

The term on the right, which involves a third moment of the velocity distribution, is clearly the flux of kinetic energy (in the rest frame of the material) crossing a sphere of radius r.

The equation looks the same, but the essential difference with the gas model is that, in the moment method, another equation must be written down for the evolution of the third moment. This in turn involves a fourth moment, whose evolution involves a fifth moment, and it is usual at this point to close the chain by assuming that this fifth moment can be expressed in terms of lower moments.

The foregoing discussion of the dynamics of the gas model has concentrated on the evolution due to relaxation processes. Stellar systems also evolve on the much shorter time scale of the orbital motions, but the gas model is not usually adopted for the study of such problems except in one sense. The dynamical stability of stellar systems is a difficult topic in general, but it is possible to show (under mild and reasonable hypotheses) that a stellar system is stable if the corresponding gaseous model is, and the latter question is often a good deal simpler (see Binney and Tremaine 1987 for an introduction to this area).

The phase space description: Boltzmann and Fokker–Planck equations

This approach has been introduced in Chapter 6, where we also described one of the appropriate dynamical equations – the collisionless Boltzmann equation. Like the gas equations, this can be extended to take account of the effects of two-body relaxation, and then the result is either the Boltzmann equation or the Fokker–Planck equation. The essential difference between the two models is whether or not one may neglect the occasional close two-body encounter in comparison with the cumulative effect of the more numerous distant encounters. Useful formulations including strong encounters have been developed (e.g. Kaliberda & Petrovskaya 1970), but it is the Fokker–Planck equation which is much the commoner in stellar dynamics, even though the relative errors in the Fokker–Planck description are no smaller than order $1/\log N$ (Chapter 14). Where it has been compared with evolution according to the Boltzmann equation, however (e.g. Goodman 1983), it has been found to yield very similar results. The reason may, again, be the same as that mentioned in the penultimate paragraph of the previous section.

The Fokker–Planck model is generally regarded as the most accurate model of a stellar system, short of the N-body model, and it is discussed in more detail in Chapter 14. Its general structure is illustrated by the form it takes for a spherical system in quasi-dynamic equilibrium, with an isotropic distribution of velocities. Here it is assumed that the distribution function f depends only on E and t, and then the Fokker–Planck equation is

$$\frac{\partial f}{\partial t}\frac{\partial s}{\partial E} - \frac{\partial f}{\partial E}\frac{\partial s}{\partial t} = -\frac{\partial}{\partial E}(D_E f) + \frac{1}{2}\frac{\partial^2}{\partial E^2}(D_{EE} f), \tag{9.8}$$

where D_E and D_{EE} are diffusion coefficients, representing aspects of the rate of change of energy of stars in consequence of numerous small-angle scatterings off other stars, and s is the volume in phase space inside the hypersurface with energy E, i.e. the surface $v^2/2 + \phi(r, t) = E$. The second term on the left side represents the change in f resulting from the fact that the energy of a star changes in consequence of the time-dependence of the potential.

There are several other forms of the Fokker–Planck equation, e.g. in the case where f is allowed to depend also on the angular momentum (Chapter 14). There are also different ways of solving these equations numerically, either using some flavour of a Monte Carlo approach, or finite differences (Fig. 9.1). The former has enough flexibility that it is somewhat easier to add more complicated dynamical processes such as physical collisions (Freitag & Benz 2001). In recent years the increasing attention paid to N-body models has been at the expense of Fokker–Planck models. And yet for the time being they remain the best source of evolutionary models of individual globular clusters (e.g. Drukier 1995).

The N-body description: simulation

This, the most detailed model, takes as its data the masses and time-dependent positions of the N stars. The dynamical equations are usually Newton's equations in the form

$$\ddot{\mathbf{r}}_i = -\sum_{j \neq i} \frac{Gm_j(\mathbf{r}_i - \mathbf{r}_j)}{|\mathbf{r}_i - \mathbf{r}_j|^3}. \tag{9.9}$$

In some problems the right-hand side is approximated by the gradient of a smooth potential (cf. Chapter 6). This is appropriate in the study of collisionless phenomena, e.g. dynamical stability. In this context a system may be simulated by an N-body model with a much smaller number of particles, provided that some precautions are taken. A very common trick is to replace $|\mathbf{r}_i - \mathbf{r}_j|$ in Eq. (9.9) by $\sqrt{|\mathbf{r}_i - \mathbf{r}_j|^2 + \epsilon^2}$, where ϵ is a cunningly chosen *smoothing parameter*. The effect of this is to circumvent the numerical errors which occur in close encounters of particles. The error made by using the wrong force is smaller than that due to using a small number of simulation particles to represent the behaviour of a much larger system, provided that ϵ is not too large (see Hernquist *et al.* 1993).

Another possibility is to use a tree code. Though one might be concerned whether the important process of two-body relaxation (Chapter 14) is correctly modelled in such codes, practical experience appears to confirm their reliability (McMillan & Aarseth 1993, Arabadjis & Richstone 1998). They may well be competitive for $N \gtrsim 10^4$ (McMillan & Aarseth 1993), though this depends on the formulation (cf. Jernigan & Porter 1989).

For an honest solution of an N-body problem, however, the problems are of a different order. As is evident from Problem 3.2, it is all too easy to code these equations inefficiently, and the present state of the art represents about 40 years of sustained improvement (Fig. 3.1). Different particles must be integrated with different time steps (or a hierarchy of time steps). Depending on the hardware, the force of distant particles can be summed less often than that from near neighbours (Ahmad & Cohen 1973). The long-range force itself is readily evaluated nowadays using special-purpose hardware (GRAPE). Special care is still needed to deal with the singularities of Eqs. (9.9) (see Chapter 15). Long-lived triple systems pose another batch of awkward problems. And all this before one has even figured out how to analyse the data (see Casertano & Hut 1985, Eisenstein & Hut 1998), or considered adding the effects of stellar evolution to the problem.

For some purposes there is no substitute for an honest N-body simulation. One shouldn't forget, however, that for a wide range of simpler and yet important problems, more approximate methods such as Fokker–Planck techniques may be quite adequate, and certainly much faster. For such problems all techniques give quite similar results (e.g. Aarseth *et al.* 1974, Giersz & Spurzem 1994, Takahashi & Portegies Zwart 1998).

Problems

(1) Use Eq. (9.5) to establish the virial theorem for a system enclosed in a sphere of radius r_e, in the form $2T + W = 4\pi r_e^3 p_e$, where p_e is the pressure at the boundary.

(2) Defining $I = \sum_{i=1}^{N} m_i r_i^2$, use Eq. (9.9) to establish the virial theorem. Repeat the exercise for two-dimensional gravity, in which the potential between two rods is $2Gm_i m_j \ln |\mathbf{r}_i - \mathbf{r}_j|$.

(3) Use an N-body code from Appendix A to look for virial oscillations. In the printed code the condition of virial balance is satisfied initially, and so one possibility is to multiply all initial velocities by a constant factor. The current virial radius, R_v, can be computed most easily from the current kinetic energy and the initial conditions, since $W = -GM^2/(2R_v) = W_0 + T_0 - T$. If only the initial velocities are altered in the code, $G = M = 1$ and $W_0 = -1/2$.

(4) Compute the half-mass and virial radii for Plummer's model, i.e. a model whose density is

$$\rho = \frac{\rho_0}{\left(1 + \dfrac{r^2}{a^2}\right)^{5/2}},$$

where ρ_0 is the central density and a is a constant.

(5) Starting from the collisionless Boltzmann equation, either (a) fill out the details of the calculation of Eqs. (9.6) and (9.7) in Box. 9.3, or (b) derive Eq. (9.5).

(6) Estimate the conductivity in gas models by equating the time scale of thermal energy transport with the relaxation time (Eq. (14.12)). Check your estimate against the model of Lynden-Bell & Eggleton (1980), who give $L = -12\pi GmC \ln \Lambda r^2 (\rho/\sigma)(\partial \sigma^2/\partial r)$, where C is a constant.

(7) Suppose that the potential is dominated by a massive central black hole, and that the density (ρ) and velocity dispersion are power laws in r. Estimate the energy of stars between r and $2r$, and the relaxation time. If the luminosity L is constant (independent of radius) show that $\rho \propto r^{-7/4}$. (For a thorough treatment of this problem see Bahcall & Wolf 1976.)

Part III

Mean Field Dynamics: $N = 10^6$

We continue the emphasis on *collective effects*, i.e. those in which individual interactions between stars are of no importance, but we increasingly focus on those effects that really matter in the million-body problem. Chapter 10 opens with a brief discussion of the notions of *equilibrium* and *stability* in this context, but is largely concerned with non-equilibrium phenomena: *phase mixing* and *'violent' relaxation*. Another mechanism for evolution of the distribution function, even in static potentials, is diffusion by chaotic motions.

Chapter 11 introduces a variant with a strong astrophysical motivation: the behaviour of N-body systems consisting of particles with time-dependent masses, and how this affects the energy and spatial scale of the system. Much depends on whether the variation is rapid or slow, and in the latter case we can easily study its effect on the distribution function itself.

Again motivated by the astrophysical setting, Chapter 12 introduces the effect of a steady external potential. The problem closely resembles an important idealised version of the motion of the Moon around the Earth under the external perturbing effect of the Sun (Hill's problem). We study the non-integrable motions in this potential, and the important problem of *escape*. The study is then extended to the case, even more important in applications, of an unsteady external potential.

10

Violent Relaxation

Equilibrium and stability

Mathematicians classify equilibria in various ways. There are, for example, *unstable* equilibria, which are rarely found in nature, but are important in the theoretical understanding of a complicated dynamical system. Of greater practical importance are *stable* equilibria. The definition of this concept amounts to saying that, if the system is disturbed slightly from the equilibrium, then it remains in the vicinity of the equilibrium. In nature, however, stable equilibria often exhibit a still stronger behaviour, which mathematicians classify as *asymptotic stability*. This means that the disturbed system *returns* to the equilibrium state from which it was disturbed. This happens commonly in nature because of dissipative forces. The process of returning to equilibrium is often referred to as *relaxation*, and it is one with which we are all familiar (late at night).

With this background it is astonishing that relaxation plays such a central role in stellar dynamics. Not only is there no dissipation in the gravitational many-body problem, there is no equilibrium either. It is true that one can think of some highly artificial solutions which can be regarded as equilibria. The Euler–Lagrange solutions of the three-body problem, in which the three stars appear to be at rest in a uniformly rotating reference frame, come into this class, and, from a more general point of view it may be fruitful to regard a periodic solution as a generalised equilibrium. But even where these solutions are stable, there is no question of *asymptotic stability*. It is true that, if an equilibrium is *unstable*, then there will be some special solutions which tend asymptotically to the equilibrium as $t \to \infty$; the mathematical definition, however, requires that the equilibrium be stable and that *all* solutions starting in its neighbourhood should tend towards it.

The concepts of equilibrium and relaxation begin to take on meaning when we replace an *N*-body system, described in terms of the position and velocity of the *N* stars, by a *distribution function*, *f* (Chapter 6). The dynamics of the *N*-body problem is then replaced by an appropriate evolution equation for *f*, such as the collisionless Boltzmann equation. Then the concept of equilibrium refers simply to a solution which is independent of time, *t*. A 'corresponding' *N*-body system (if we can refer to such a thing without explaining precisely what is meant) is not in equilibrium: the stars will be in constant motion, and we say that the system is in 'dynamic equilibrium'. Many such equilibria are known, even though, in the stellar-dynamic case, the problem of constructing *self-consistent* equilibria (Chapter 8) is non-trivial in general.

Observe that what we mean by 'equilibrium' cannot be dissociated from the way in which we model the system and its dynamics. The same will be true of the term 'relaxation', since this refers to the process by which a system evolves towards equilibrium. But, though we have now established a framework in which the concept of equilibrium is meaningful, we are no closer to seeing how a non-dissipative system can exhibit anything like relaxation.

Phase mixing

As a first concrete illustration, let us consider a system which is described by the collisionless Boltzmann equation (Chapter 6). In order to simplify the discussion to its utmost, let us consider instead the corresponding one-dimensional problem in a *time-independent* potential $\phi(x)$, i.e.

$$\frac{\partial f}{\partial t} + v\frac{\partial f}{\partial x} - \frac{\partial \phi}{\partial x}\frac{\partial f}{\partial v} = 0, \tag{10.1}$$

where x is the position, v is the velocity. Now the motion of a particle in a one-dimensional, time-independent potential is integrable, and instead of using coordinates x and v in phase space, we may be better to use *action-angle variables*, J and θ (e.g. Goldstein 1980). These are canonical variables such that J is constant and θ (the 'phase angle') increases linearly with time, its rate of change being some J-dependent frequency $\omega(J)$. Similarly, the distribution function may be written as $f(\theta, J, t)$, and then the collisionless Boltzmann equation, Eq. (10.1), takes the very simple form

$$\frac{\partial f}{\partial t} + \omega(J)\frac{\partial f}{\partial \theta} = 0. \tag{10.2}$$

(Note that Eqs. (10.1) and (10.2) are different forms of the general equation $\dot{f} + [H, f] = 0$, i.e. Eq. (6.6), where square brackets denote the usual Poisson bracket

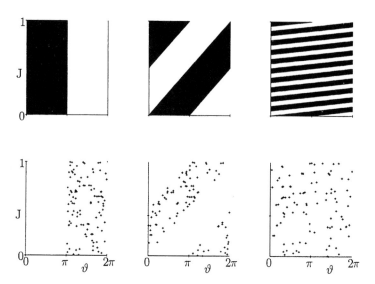

Fig. 10.1. Phase mixing. Here $\omega = J$. The top three frames give the initial conditions, as described in the text, and the solution at times $t = 2\pi$ and 20π. The lower three frames give the same information from 100 initial points.

and H is the Hamiltonian. In the variables of Eq. (10.1) this is $H = v^2/2 + \phi$, while $\omega = dH(J)/dJ$ if action-angle variables are used.)

The equilibrium solutions of Eq. (10.2) are easy to describe: since $\partial f/\partial t = 0$, Eq. (10.2) implies that $\partial f/\partial \phi = 0$ also, and so they are functions of J alone; but how do other solutions relax to a solution of this form? If the initial distribution (at $t = 0$) is $f_0(\theta, J)$, the solution of Eq. (10.2) is

$$f(\theta, J, t) = f_0(\theta - \omega t, J). \tag{10.3}$$

This never approaches an equilibrium solution (pointwise), unless f_0 is independent of θ, i.e. unless the system is already in equilibrium! At first sight, then, the very idea of relaxation seems irrelevant.

The top three frames of Fig. 10.1 illustrate the same points. They show several snapshots of the θ, J plane if the initial distribution f_0 is 1 (white) for $\pi < \theta < 2\pi$, and 0 (black) otherwise. The dynamics (in which J is fixed) causes a horizontal motion of the boundary of the region in which f is non-zero. As it moves off the right-hand side of the diagram at $\theta = 2\pi$ it reappears on the left, as θ is just a phase. The rate of motion depends on J (assuming that the potential is not simple harmonic), and so the region where $f \neq 0$ becomes increasingly stretched. It also becomes increasingly narrow, and the detail on this diagram reaches ever finer scales.

Let us recall that this diagram is meant to represent the evolution of a system with a large but finite system of particles (like a million). Suppose initially that all particles are in the region where $f_0 \neq 0$, and that we follow their motion in this diagram. At subsequent times the particles will still occupy the region where $f \neq 0$.

Eventually, however, the fine detail which is visible in the top row of Fig. 10.1 would not be discernible in the distribution of points; certainly this would be true by the time that the filaments are so fine that most do not contain any point. The lower three frames of Fig. 10.1 illustrate this. In the last frame the points are confined to 10 or 11 nearly horizontal strips, as in the frame above, but this is not evident unless one looks carefully for it (e.g. by looking at the figure along the page). By this stage we see that the *distribution* of points does not change perceptibly, even though the points themselves are in rapid motion. There is a sense in which equilibrium is approached, when we think in terms of particles rather than the distribution function; it is a *statistical* equilibrium.

The same idea can be re-expressed in terms of the distribution function. The distribution function is used to tell us how many particles there are in regions of phase space. Let R be such a region. When we compute $\int_R f \, d\theta \, dJ$ the fine filaments in f do not matter; we would obtain almost the same answer if f were smoothed over little regions of phase space which are small compared with R but large compared with the thickness of the filaments.

It is not difficult to see that, after some time, this smoothed distribution, which is often referred to as a 'coarse-grained distribution' in statistical mechanics, is given by averaging f_0 with respect to θ. In other words, in this smoothed or coarse-grained sense, f does indeed relax to $\langle f_0 \rangle_\theta$. In fact such an equilibrium solution is asymptotically stable, in the coarse-grained sense (Problem 1). And it is important to understand the evolution of the coarse-grained density. When one attempts to infer a phase density by looking at a galaxy with a telescope, or at an N-body system with a computer, the best one can hope to do is to estimate a coarse-grained density.

Curiously, the process we have discussed is not often referred to as relaxation, but as 'mixing'. In fact this word, like stability, is one which is overloaded with meanings, only one or two of which we touch on in this chapter. The particular type of mixing we have introduced is of a relatively mild kind, more like setting the tea in a cup in steady circulation (with a careful circular motion of the spoon) than churning it up with a vigorous stir. In the dynamical context it is referred to as 'phase mixing', for good reason. Consider two particles with neighbouring values of J and the same phase θ at some time t. The original phases of these particles, i.e. at $t = 0$, were $\theta - \omega(J)t$. Now ω depends on J in general, and so if t is large the original phases differed greatly, even though the particles now occupy neighbouring positions in phase space. The original phases have become mixed together, by the same mechanism which led to the stretched, narrow filaments in Fig. 10.1.

Notice that f in Eq. (10.3) depends on two integrals of motion (i.e. J and $\theta - \omega t$), which illustrates Jeans' Theorem (Chapter 6). The two integrals have rather different character, however. One (the first argument of f) is phase-dependent, and irrelevant for smooth solutions (at late times, in this case). It is an example of a *non-isolating* integral, because curves on which it is constant lie extremely close together (at late times); they do not 'isolate' gross regions of phase space from each

other. In all these respects it is quite different from J, which is an example of an
isolating integral.

The end point of violent relaxation

The gravitational million-body problem differs in two essentials from the simpler
problem we have considered so far: it is three-dimensional, and the potential is not
fixed. Fig. 2.1 gives an impression of what we have in mind.

Let us consider the second point first. One consequence of the fixed potential in
the one-dimensional case is that the energy of each particle, and hence the value of J,
is fixed. The distribution of the energies of the particles cannot vary. In an N-body
system, however, the potential depends, through Poisson's equation, on the evolving
spatial distribution of the particles, and therefore the distribution of energies may be
expected to evolve. The process by which a system reaches equilibrium when the
potential is time-dependent is referred to as 'violent relaxation'. The picture to have
in mind is that of a seething, pulsing distribution of stars, with swarms of stars (but
not the stars themselves) crashing into and passing through each other (Fig. 10.2).

There has been much discussion in the literature about what the end result of this
evolution may be. Certain rather general arguments can be given, however. As mixing
takes place, points where f is high become closely interwoven with points at which
f is low. The result is that the coarse-grained distribution, which we denote by F, is
a weighted average of high and low values of f; in this sense we write $F = \langle f \rangle$. It
follows that, for any convex function C, $C(F) \leq \langle C(f) \rangle$.[1] By summing this result
over the whole of phase space, it is not hard to arrive at a result which looks like
an 'H-theorem' (see, e.g., Reichl 1980): suppose that we equate f and F at some
time; then it may be shown that, at any later time, the value of $\int C(F) d\tau$ is less than
(or at least never exceeds) its initial value, where $d\tau$ is the element of phase volume
(Tolman 1938, Tremaine *et al.* 1986). Unfortunately, this is not quite the same thing
as proving an H-theorem, which states that there is a function H which decreases
with t (Dejonghe 1987b, Soker 1996).

Unfortunately there are other reasons why this does not get us very far: there is no
compelling reason to prefer one form for C over any other, and so we are no nearer
finding the function to which F should tend. But we can deploy a further argument.
The evolution is constrained by the fact that, as we follow the flow in phase space, the
value of f around us does not change. (Remember that this is what Eq. (10.1) means;
the phase fluid is incompressible (Chapter 6).) Therefore particles cannot crowd
together more closely than at time $t = 0$; they obey a kind of classical exclusion
principle. It is not too surprising, therefore, that the most probable distribution of

---

[1] As an example, let $C(f) = f^2$ (a convex function), and suppose that f takes two values
f_1 and f_2 with equal probability. Then $F = (f_1 + f_2)/2$, $\langle C(f) \rangle = (f_1^2 + f_2^2)/2$, and
$C(F) = \langle C(f) \rangle - (f_1 - f_2)^2/4$.

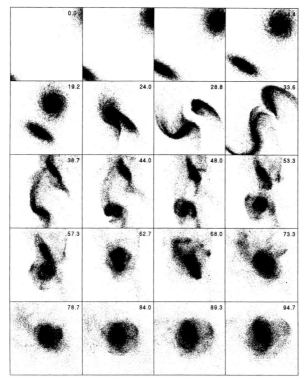

Fig. 10.2. Two colliding galaxies (from Mihos & Hernquist 1996). The time (in units of approximately 1.3 Myr) is given at top right of each frame.

particles subject to this constraint is one which resembles Fermi–Dirac statistics (Lynden-Bell 1967; see also Shu 1978).

Sadly, numerical experiments give at best ambiguous support to this idea (e.g. Cuperman & Harten 1972), and it is not hard to see one possible reason for this. We are relying on fluctuations in the potential to drive the evolution. But the potential depends on the density, which is an integral over the distribution function. We have seen, in the context of phase mixing, that such integrals settle down to the integral of a steady, coarse-grained distribution. Therefore one of the effects of the evolution is to quench the very potential fluctuations which drive the relaxation. It is therefore not surprising that the relaxation fails to complete its task of bringing F to the expected equilibrium form.

None of this theory really vividly brings to mind the picture that astrophysicists see when 'violent relaxation' is mentioned. They think of galaxies colliding in what is little short of an explosion (Fig. 10.2). The galaxies are shattered, and some fragments may even be ejected from the interaction region fast enough to escape. If this is so there will be almost as many fragments with small positive energies as there are with small negative energies. (The escape energy is a quantity set by conditions far away from the place where the energies of the particles are being violently relaxed, and the relaxation process does not know about it.) In other words, the number of stars per unit energy and angular momentum, $N(E, J)$, is almost independent of E near $E = 0$, at least for those stars whose angular momenta are small enough for them to have come from the ejection region. (If v_{esc} is the escape speed from the edge of the interaction region and R is its radius, the condition is $J < J_{max} = Rv_{esc}$.) Stars with $E \geq 0$ escape, whereas the remainder settle into dynamical equilibrium. We now set ourselves the task of determining their spatial distribution (von Hoerner 1957; this long-forgotten result was rediscovered decades later by Makino *et al.* 1990).

The orbits of these particles lie outside most of the mass, and are therefore nearly Keplerian, and so their period is $P \propto (-E)^{-3/2}$ when E is slightly negative. By Eq. (6.8) it follows that the phase space distribution is $f(E, J) \propto (-E)^{3/2}/J$ when $J \lesssim J_{max}$. By integrating over velocities we deduce that the density is $\rho \propto \int \frac{(-E)^{3/2}v}{r} \, dv \, d\theta$, where θ is the angle between $\mathbf{v}$ and $\mathbf{r}$, and integration is restricted to the domain where $J = rv \sin \theta < J_{max}$ and $v^2 < -2\phi$. When r is large the first condition reduces approximately to $\theta < J_{max}/(rv)$, whence $\rho \propto (-\phi)^2 J_{max}/r^2$, $\propto r^{-4}$. This result is borne out approximately in numerical experiments (see, e.g., van Albada 1982).

Motion in non-integrable potentials

Let us return once more to the problem of evolution in a *fixed* potential, but step up from the one-dimensional problem, where the only thing that happens is phase mixing, to a more realistic number of dimensions. In fact, two dimensions will do, as we shall then meet almost all the points we need consider. Nor is it such an unrealistic problem to tackle, a non-trivial example being motion in the meridional plane of an axisymmetric three-dimensional potential. Furthermore, it was precisely in this astronomical context that the seminal paper of Hénon & Heiles (1964) was set.

Let x, y be coordinates in this plane. Both coordinates oscillate, but in general the oscillations couple.[2] Poincaré gave a compact way of picturing the corresponding motion of a particle in phase space. Every time $y = 0$, plot the phase coordinates of the x-oscillation. We shall suppose that we use the action and phase for this purpose. Then, if there were no coupling between the two motions, our discussion of Fig. 10.1 shows us what to expect: the successive points representing one particle lie on horizontal lines (Fig. 10.3a).

Unless we are told the answer, it is much harder to guess what happens when the coupling is non-zero. In fact, if the coupling is small, the picture does not change much: successive points (representing the motion of one particle) still lie on curves in

---

[2] In cylindrical polar coordinates the equations are $\ddot{r} - J^2/r^3 = -\frac{\partial \phi}{\partial r}(r, z)$, $\ddot{z} = -\frac{\partial \phi}{\partial z}(r, z)$.
Unless $z = 0$ the oscillation in z influences the oscillation in r, and vice versa.

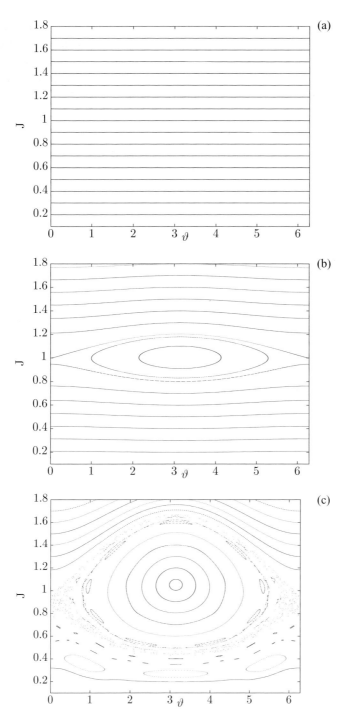

Fig. 10.3. Surface of section for two coupled oscillators in two dimensions. Details of the dynamics are given in Problem 3. For a problem better motivated by the million-body problem, see Fig. 12.2.

the plane (Fig. 10.3b). The curves are no longer quite straight, and also some of them distort into ovals instead of stretching all the way across the diagram (Problem 3). The latter are associated with *resonances* between the x- and y-motions.

When the coupling is larger still, more dramatic changes become apparent (Fig. 10.3c). For some particles, the points still lie on curves or ovals. In that case we talk of 'regular motion'. For other particles, however, the points are scattered on *regions* of the plane, and we speak of 'chaotic motion'. One orbit in Fig. 10.3c has this character. Incidentally, the difference between Figs. 10.3b and 10.3c is not as great as might be thought: it is just that, when the coupling is small (Fig. 10.3b), the chaotic regions are so slender as to be indistinguishable from curves.

Let us look (metaphorically) at one of these chaotic regions, where points appear to be scattered at random, and not on a curve. Instead of studying the motion of a single particle, let us consider what happens if, at time $t = 0$, we consider a set of particles occupying a little patch in phase space, much as in our consideration of Fig. 10.1. Again, this patch becomes stretched out as the evolution proceeds, but the rate at which it does so is very different. In Fig. 10.1, the stretching is caused by the linear increase of the phase with time. In the corresponding motion in Fig. 10.3c, however, the factor by which the patch of phase is stretched increases approximately *exponentially* with time.[3] There is mixing, and it is very different from phase mixing, but it leads to the same end result: in the region where this kind of behaviour occurs, the coarse-grained distribution function F will approach an equilibrium form.

Different chaotic zones may well be separated by curves. In this case, a particle which starts its evolution in one zone is forever separated from the other. The consequence of this is that the equilibrium value of F in one region may be quite different from that in another. We should picture a coarse-grained distribution function which varies smoothly across regions of regular motion, rather as in the one-dimensional case, but which is flat in each chaotic zone.

We have spoken of the curves between neighbouring chaotic zones as impermeable barriers. In fact some of them leak (and in three dimensions *all* of them do). A few leak like sieves, with large holes which let particles through readily; others are more like the paper in a coffee filter, and it takes a long time before the leakage becomes noticeable. It follows that the notion of an 'equilibrium coarse-grained distribution' depends on the interval of time, T, in which one is interested. If the boundary between two chaotic regions is easily breached on a time scale much shorter than T, then the appropriate distribution will be constant across both regions. It may be different in two regions if the barrier between them is unlikely to leak over a time of order T.

Unless the potential we are looking at is very simple, there will be many regions of chaotic motions, and the barriers between them may be breached on a wide variety

---

[3] This is different in character from the exponential divergence discussed in Chapter 13. A better picture to have in mind is Arnold's famous 'cat' map (see Lichtenberg & Liebermann 1992, p. 306).

of time scales. It would therefore follow that, no matter how long we waited, the coarse-grained distribution function would always be changing. There would always be barriers which could only be breached on a time scale comparable with the present age of the system, and it is leakage across those barriers that would be the most noticeable evolutionary process at that stage. The distribution function would already have reached equilibrium across barriers which could be breached on much shorter time scales. Those whose presence would only become noticeable on much longer time scales would be effectively impermeable.

Some time scales for diffusion are *exponentially* long (see Lichtenberg & Liebermann 1992, p. 398f). The so-called 'Arnold diffusion' appears to have this character, and it has even been said that 'if you can see it, then it is not Arnold diffusion'!

Relaxation, stability and evolution in dynamical astronomy

The remarks we have made are rather speculative, at least in the context of stellar dynamics, where their consequences are only just beginning to be worked out. Part of the reason for this is that the study of motion in a fixed potential, especially in the study of evolutionary phenomena, seems so far from reality, where the potential must evolve with the system. Nevertheless, the idea that a system will always be evolving on a time scale comparable with its age does seem to be a rather attractive and general one, when interpreted intelligently.

In recent years it has led to a reinterpretation of the age-old question of the stability of the solar system (Laskar 1996), a problem which has led to so much important dynamical theory. It is possible to argue that the solar system is always subject to instabilities acting on a time scale comparable to its lifetime. Instabilities acting on much shorter time scales will have either reached equilibrium or else have led to the ejection of objects on unstable orbits. Those acting on extremely long time scales will have had (as yet) little bearing on the evolution of the solar system.

In the case of stellar dynamics, certainly when applied to globular star clusters, it is usual to take a more simplistic point of view. It is usually supposed that the actions of processes such as phase mixing and violent relaxation play themselves out early on in the lifetime of the system. Thereafter the system is in dynamic equilibrium under the orbital motions of the stars. If the system is isolated and nearly spherical, then the sort of chaotic behaviour illustrated in Fig. 10.3c can be neglected. It does play a significant role in the escape of particles from clusters which are strongly perturbed by external fields (Chapter 12). That apart, however, one must look elsewhere for any mechanism which will cause the system to evolve slowly (or 'secularly') from one quasi-equilibrium to another. The mechanism usually invoked is two-body relaxation, and we turn to that topic in Chapter 14.

Problems

(1) Determine the t-dependence of f at a given point in Fig. 10.1. If f is aver-
     aged over a small interval δJ in J, show that $\langle f \rangle \to 1/2$, and determine a
     time after which $|\langle f \rangle - 1/2| < \epsilon$ for any given $\epsilon > 0$.

(2) Prove the quasi-H-theorem stated in the text for any convex function C, when
     evolution is governed by the collisionless Boltzmann equation.

(3) Figure 10.3 deals with the system $\dot{\theta} = J$, $\dot{J} = \epsilon \sin\theta \cos t$, where $\epsilon = 0, 0.02$
     and 0.2, respectively, in the three frames. Defining $\phi = \theta - t$, show that ϕ
     varies slowly near $J = 1$, and use the method of averaging to show that J and
     ϕ evolve approximately according to the equations

$$\dot{\phi} = J - 1$$

$$\dot{J} = \frac{\epsilon}{2} \sin\phi$$

if ϵ is small. Within this approximation deduce that $(J - 1)^2/2 + (\epsilon/2)\cos\phi$
is nearly constant, and hence determine the width (in J) of the resonant
region.

11

Internal Mass Loss

A stellar system in dynamic equilibrium loses neither mass nor energy. In fact the stellar systems in nature do both, and the reasons for this are both external and internal. In this chapter we consider the latter; that is, we consider a stellar system isolated from all external influences, including gravitational ones.

We have two processes in mind. One is caused by the internal evolution of the stars. Note that this is the first occasion on which we have abandoned the point mass model on which we have relied so far, at least to the extent that we now consider time-dependent masses. The other is caused by the gravitational interactions of pairs of stars, which is really the topic of Chapter 14, and will be discussed rather briefly in this chapter. We also deal with the effects in two ways. One is the scaling treatment (Chapter 9) and the other uses a phase space description.

Evolution of length scale

A single star evolves at a rate which is a rapidly increasing function of its initial mass. Therefore, if we examine the stars in an old stellar system, we find that only those with a sufficiently low mass are more-or-less unevolved, with masses close to those they were born with. Those which were born with higher masses will have evolved, and in the process will have lost mass, leaving a remnant which may take the form of a black hole, a neutron star, or a white dwarf. Simple prescriptions for these aspects of stellar evolution have been in use in stellar dynamics for a long time (see Terlevich 1987, Chernoff & Weinberg 1990 and Problem 1). The details, however, are constantly changing: the relation between initial and final mass is an

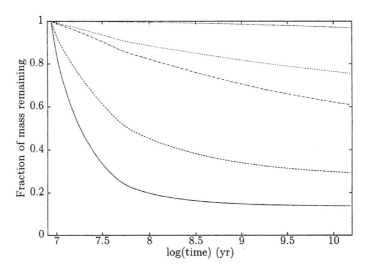

Fig. 11.1. Mass lost as a function of time for five possible model stellar populations. The middle curve corresponds to an analytical approximation of a Miller–Scalo mass function (Eggleton *et al.* 1989). The others give results for a power law mass function with probability density $f(m) \propto m^{-\alpha}$, with $\alpha = 3.35, 2.35, 1.35$ and 0.35. (The value 2.35 comes from Salpeter 1955.) In each case the maximum and minimum mass are 15 and 0.1 times the mass of the Sun, respectively. The prescription for stellar mass loss is taken from Chernoff & Weinberg (1990), though they applied it to different mass functions.

active area of research in stellar evolution (see, for example, Weidemann (2000) for white dwarfs and Timmes *et al.* (1996) for neutron stars and black holes).

In relation to the characteristic dynamical time scale in a stellar system, i.e. the crossing time (Box 1.1), these processes of mass loss may be either fast or slow. For example, the mass lost in a supernova event is expelled at speeds which exceed the orbital speeds in a stellar system by a factor of 100 or more, and the time scale is proportionately shorter. Mass expelled in a steady wind may be lost on a time scale much exceeding the crossing time. In either case the mass lost *by the system* is lost on a long time scale, since even the instantaneous loss of the entire mass of one star represents only a slight loss of mass by the entire system. Figure 11.1 shows the mass lost in several stellar populations representative of those often used in the study of old stellar systems.

The following argument shows why the time scale on which mass is lost is important (Hills 1980). Suppose a system is in dynamical equilibrium, so that its total, kinetic and potential energies are related by $T = -E$ and $W = 2E$ (see Chapter 9). If a fraction f of the mass is lost instantaneously then the new kinetic and potential energies are $T' = (1 - f)T$ and $W' = (1 - f)^2 W$, whence the new energy is $E' = E(1 - f)(1 - 2f)$. It follows that at most half of the mass can be lost if the system is to remain bound. In practice, if more mass is lost then the most bound stars

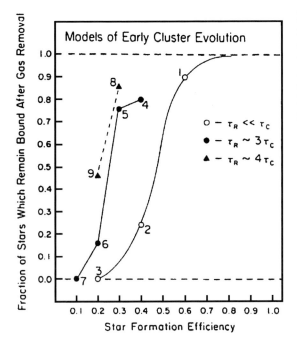

Fig. 11.2. Remaining bound mass following mass loss (after Lada *et al.* 1984). The time scale of mass loss is τ_R, and τ_C is a measure of the crossing time. Slow mass loss is less disruptive than fast mass loss.

remain as a small, bound system. Even if less than half the mass is lost, some stars escape (Fig. 11.2).

The role of dynamic equilibrium in this can be assessed by introducing the virial ratio $q = -T/W$. Thus $q = 1/2$ in equilibrium, and if the system is bound initially we require $0 < q < 1$. If the above calculation is repeated, the result is that $E' = E(1 - f)(1 - f - q)/(1 - q)$, and so the condition that the final system is bound is that $f < 1 - q$. Thus a 'cool' system can lose a relatively high fraction of its mass and still remain bound.

Now we return to systems in dynamic equilibrium, and consider what happens if only a tiny fraction of the mass is lost. Then the change in mass is $dM = -fM$, and so $dE = E' - E \simeq 3EdM/M$. If the loss of mass is sufficiently slow, the system is able to remain close to virial equilibrium. Therefore this relation is true for subsequent mass-loss events, and it follows that $E \propto M^3$. Thus the system can, in the end, lose an arbitrarily large fraction of its mass without becoming unbound. Also, the length scale of the system can be estimated by $R \sim -M^2/E$, and so $R \propto M^{-1}$.

Before passing on, it is worth stressing that all three scenarios lead to an increase in the energy of the system, and since the energy is initially negative, this means that the magnitude of the energy decreases. It might seem that this is obvious, as the mass of the system has also decreased. It is easy to see, however, that even the energy per unit mass increases, i.e. becomes less negative. In every sense, then, these mechanisms of slow or impulsive mass loss lead to a heating of the system.

We shall see in Part VIII that some heating mechanism is needed in order to lead a stellar system to a peaceful death, and nothing more complicated than loss of mass is required.

Now we turn our attention, more briefly, to another important mechanism of mass loss, whose effects on the energy of a system could not be more different. It occurs because of interactions of pairs of stars. In such interactions the two stars exchange energy, and in some cases the final energy of one of the stars may exceed the escape energy. Since the energy of this escaper is positive, the energy of the remaining (bound) members of the system is more negative than before. Let us assume, furthermore, that the average energy carried off by the escaper is a certain fraction of the energy of a single member of the stellar system. Therefore we may assume that the changes in mass and energy are related by $dE = -f dM(E/M)$, where f is a constant (unrelated to the constant f above). Note that dE, dM and E are all negative here, and so $f > 0$. It follows that $E \propto M^{-f}$, and so $R \propto M^{2+f}$. The results of this mass loss are therefore qualitatively very different from what we discussed previously: as the mass of the system decreases the system shrinks. Historically, this was the first suggestion of the evolution of stellar systems to a singular endpoint (Ambartsumian 1938, Spitzer 1940), an issue which will play a major role in later chapters.

Evolution of structure

So far the discussion has dealt with gross issues of mass, energy and radius, but has said nothing about how the detailed structure of a stellar system is altered. For this purpose we restrict attention to systems which remain spherical, and close to dynamical equilibrium. Then we may suppose the distribution function f depends only on E, J and t, where now E refers to the specific energy of a single star. How, then, does $f(E, J, t)$ change in response to the loss of mass, i.e. to the resulting changes in the potential?

One way of arriving at the correct answer is to substitute the expression $f(E, J, t)$ into the collisionless Boltzmann equation (Eq. 6.5), remembering that we must express E and J as functions of $\mathbf{r}$ and $\mathbf{v}$. For example, $E = v^2/2 + \phi(r)$. Note that E depends on t because ϕ is time-dependent. It follows that

$$\frac{\partial f}{\partial t} + \frac{dE}{dt}\frac{\partial f}{\partial E} = 0. \tag{11.1}$$

The argument we have just given is insecure. The rate of change of energy $dE/dt = \partial\phi(r, t)/\partial t$, and so depends on the location of the star. Thus Eq. (11.1) tells us that the rate of change of a function which depends on only E, J and t depends also on r. It is natural, perhaps, to interpret the term dE/dt in Eq. (11.1) as an average over the orbit of given energy and angular momentum, and Box 11.1 confirms this guess with a better argument.

Box 11.1. Evolution of the distribution function in a slowly evolving
        potential

To describe the motion of a star in a smooth, slowly varying, spherically symmetric
potential, we may introduce suitable action-angle variables. As actions we choose J,
J_z and I_r, where J_z is the component of angular momentum in the direction of any
suitable z-axis, and I_r is the radial action introduced at the end of Chapter 7 (see also
Palmer 1994). There are also three conjugate angles, but we do not need to concern
ourselves with them. The transformation from rectangular coordinates and momenta
to action-angle variables is canonical, and has unit Jacobian, and it follows that the
distribution of stars in action-angle space is equal to f. Also, it is clear that E may
be written (in principle) as a function of I_r and J, and so the distribution function may
be written as $f(E(I_r, J, t), J, t)$. Now two of the actions, viz. J and J_z, are constant,
while the third, i.e. I_r is an adiabatic invariant (Chapter 7). It follows that f, regarded as a
function of the actions and time, is independent of t. (In fact this is just the collisionless
Boltzmann equation in action-angle variables.) Hence

$$\frac{\partial f}{\partial t} + \frac{\partial E}{\partial t} \frac{\partial f}{\partial E} = 0,$$

which is just Eq. (11.1) again. The adiabatic invariance of I_r does indeed depend on
averaging over the fast radial motion of a star, and so it is now clear that $\partial E/\partial t$ here is
to be interpreted in an orbit-averaged sense. Incidentally, by Eq. (6.8), another way of
stating this result is to write

$$\left(\frac{\partial}{\partial t} \frac{N(E, J)}{P(E, J)} \right)_{I_r} = 0, \tag{11.2}$$

where $N\,dE\,dJ$ is the number of stars with energy and angular momentum in ranges
dE and dJ, respectively, and P is their radial period; the subscript indicates that the
time-derivative is to be taken at constant values of I_r.

We shall not stop to solve these equations, but draw attention to one or two
qualitative points which their solutions would exhibit. We have already seen that one
consequence of mass loss may be the expansion of the system. However, the angular
momentum of each star is constant, and determined by the size and velocity dispersion
of the system initially. By the time the system has expanded by a large factor, therefore,
most orbits will have low angular momentum for a system of such size. Even if the
initial distribution of stellar velocities had been isotropic, in the expanded system
nearly radial motions would predominate. It is also likely that this is one mechanism
by which the anisotropy of stellar systems grows in so-called post-collapse evolution
(Chapter 28), even though the mechanism of the expansion looks rather different.

Despite these remarks, frequent use is made in stellar dynamics of an analogous
equation for the evolution of f when the anisotropy is neglected, viz.

$$\frac{\partial f}{\partial t} \frac{\partial s}{\partial E} - \frac{\partial f}{\partial E} \frac{\partial s}{\partial t} = 0, \tag{11.3}$$

where s is the volume in phase space of the region in which the energy is less than E (see Chapter 9 and this chapter, Problem 4). Again this is based on Eq. (11.1), except that the expression for $\dot{E}$ is obtained by averaging over all points in phase space accessible to a star of energy E, instead of the points accessible to a star of energy E and angular momentum J.

Problems

(1)   Suppose the main sequence lifetime of a star of mass m is $t_{ms} = 10^{10} \times (m/M_\odot)^{-2.5}$ yr (see Hansen 1999), that the entire mass of the star is lost at this time, and that the mass function is as in the caption to Fig. 11.1. Compute the time scale of mass loss of the cluster, i.e. $-M/\dot{M}$. (For a more elaborate formula for the main sequence lifetime see Cassisi & Castellani 1993.)

(2)   Assume that the mass in Keplerian motion varies as $m(t) = m_r + (m_0 - m_r) \times \exp(-t/\tau)$, where m_0 and m_r are the initial and final (remanant) masses, respectively, and τ is constant. The radial equation of motion is

$$\ddot{r} = \frac{c^2}{r^3} - \frac{Gm(t)}{r^2},$$

where c is the specific angular momentum; we assume that this is constant. The orbit is initially circular with radius a, and we assume that τ is much smaller than the orbital period, but not infinitesimal. Show that, during the time when r does not vary significantly from its initial value,

$$\dot{r} \simeq \frac{G}{a^2}(m_0 - m_r)(t + \tau[e^{-t/\tau} - 1]).$$

We define the energy in terms of the potential of the remnant, i.e.

$$E = \frac{1}{2}\dot{r}^2 + \frac{c^2}{2r^2} - \frac{Gm_r}{r}.$$

Show that

$$\dot{E} = -\frac{G(m_0 - m_r)}{r^2}\dot{r}e^{-t/\tau}.$$

Substituting the approximate value of $\dot{r}$, deduce that the value of E immediately after completion of mass loss is

$$E_r \simeq \frac{c^2}{2a^2} - \frac{Gm_r}{a} - \frac{G^2\tau^2(m_0 - m_r)^2}{2a^4}.$$

Deduce that the condition for remaining bound is that

$$m_r \geq \frac{m_0}{2} - \frac{G\tau^2 m_0^2}{8a^3},$$

approximately.

For a more systematic solution of this problem see Hut & Verhulst (1981).

(3) Show that Eq. (11.3) may be written in the form

$$\left(\frac{\partial f}{\partial t}\right)_s = 0,$$

(11.4)

where the subscript indicates that the derivative is taken at fixed $s(E, t)$.

(4) The rate of change of potential, averaged over all phase space accessible to a star of energy E, is given by

$$\langle \dot{\phi} \rangle = \frac{\int \frac{\partial \phi}{\partial t} \delta \left(E - \frac{1}{2} v^2 - \phi \right) d^3 v d^3 \mathbf{r}}{\int \delta \left(E - \frac{1}{2} v^2 - \phi \right) d^3 v d^3 \mathbf{r}}.$$

(11.5)

In a spherical potential show that this may be written as

$$\langle \dot{\phi} \rangle = -\frac{\frac{\partial s}{\partial t}}{\frac{\partial s}{\partial E}},$$

where

$$s(E) = \frac{16\pi^2}{3} \int_{\phi(r,t) \leq E} r^2 \{2(E - \phi)\}^{3/2} dr.$$

(5) Suppose a stellar system is subject to a tidal cutoff, i.e. there is an absorbing spherical boundary with radius equal to the tidal radius $r_t \propto M^{1/3}$ (see Chapter 12 below), and that it loses almost all of its mass very slowly. Then the final range of values of s is tiny, and Eq. (11.4) implies that f must be the same as in the initial model near $s = 0$. What is the resulting model?

(6) (unsolved): repeat the previous problem for Eq. (11.2).

12

External Influences

In this chapter we add one ingredient to the topics discussed in the previous chapter. There we outlined what happens to a stellar system when it loses mass, by whatever mechanism. Implicitly, however, we assumed that the system was *isolated*. Now we add to the picture the fact that the stars in a stellar system are also affected by surrounding matter, and this is especially true of escaping stars. The picture we have in mind is of a system like a globular cluster, orbiting inside a galaxy, which is simply another stellar system, but much larger and more massive.

The way in which the galaxy affects the cluster depends on such factors as the orbit of the cluster, and the distribution of mass within the galaxy. We begin with the simplest non-trivial idealisation. We assume that the orbit of the barycentre of the cluster is circular of radius R. Clearly, this is possible only for certain types of galaxy, e.g. those with axisymmetric potentials ϕ_g. We use an accelerating and rotating frame of reference with origin at the barycentre of the cluster, such that the x-axis points radially outward, and the y-axis points in the direction of motion of the cluster. The acceleration of a star in the cluster has several terms, due to: (i) the field of the galaxy; (ii) the gravitational field, ϕ, of the cluster; (iii) inertial forces, i.e. Coriolis and centrifugal terms. As we show in Box 12.1 (see also Chandrasekhar 1942, Sec. 5.5, or Ogorodnikov 1965, Sec. 7.4) the resulting equations are

$$\ddot{x} - 2\omega\dot{y} + (\kappa^2 - 4\omega^2)x = -\frac{GMx}{r^3}$$

$$\ddot{y} + 2\omega\dot{x} \qquad\qquad = -\frac{GMy}{r^3} \qquad\qquad (12.1)$$

$$\ddot{z} \qquad\qquad + \omega_z^2 z = -\frac{GMz}{r^3}.$$

Box 12.1. Equations of motion of a star in a tidally perturbed cluster

We consider and model each of the forces in turn.

If we suppose the field of the galaxy is axisymmetric, with the cluster orbiting in a plane of symmetry, then we can write

$$\phi_g = \phi_g(\sqrt{(R + x)^2 + y^2}, z).$$

If the cluster is small compared with the radius of its orbit, we may approximate this by its Taylor expansion, i.e.

$$\phi_g \simeq \phi_g(R, 0) + \left(x + \frac{y^2}{2R}\right)\frac{\partial \phi_g(R, 0)}{\partial R}$$

$$+ \frac{x^2}{2}\frac{\partial^2}{\partial R^2}\phi_g(R, 0) + \frac{z^2}{2}\frac{\partial^2}{\partial z^2}\phi_g(R, 0). \tag{12.2}$$

(Terms with a single derivative with respect to z do not appear because of our assumption that the orbit is in a plane of symmetry.) By the elementary equations of circular motion,

$$R\omega^2 = \partial\phi_g/\partial R, \tag{12.3}$$

where ω is the angular velocity of the cluster about the galaxy.

The field of the cluster we approximate by that of a point mass M at the cluster centre. This is an inessential approximation, but it is adequate for escaping stars which do indeed lie outside the bulk of the mass, even if this will not be spherically symmetric because the tidal field, Eq. (12.2), is itself not symmetric. Thus $\phi = -GM/r$, where $r^2 = x^2 + y^2 + z^2$ and M is the mass of the cluster.

The centrifugal force is derived from the potential $-(1/2)\omega^2((R + x)^2 + y^2)$. The other inertial force is the Coriolis force, which has components $2\omega(\dot{y}, -\dot{x}, 0)$.

Now we can write down the equation of motion of an escaping star. By Eq. (12.3) there is no linear term in the sum of the galactic and centrifugal potentials, and the quadratic terms simplify to $(\kappa^2 - 4\omega^2)x^2/2 + \omega_z^2 z^2/2$, where κ is the epicyclic frequency (see Chapter 7) associated with circular motion of radius R in the galactic potential, and $\omega_z^2 = \frac{\partial^2 \phi_g}{\partial z^2}$.

Hence we obtain Eqs. (12.1).

These are the equations of a famous problem in dynamical astronomy, called Hill's problem (Hill 1878). G.W. Hill used them, towards the end of the 19th century, in pioneering studies of the motion of the Moon about the Earth, as perturbed by the Sun. Here, however, it describes the motion of a star about a cluster, perturbed by the Galaxy. It also arises in a beautiful modern problem: that of the mutual interaction of the coorbital satellites of Saturn when close to conjunction (see Petit & Hénon 1986). Finally, an identical force law occurs in the problem of the tides (with additional terms due to fluid forces), and this gives its name to the approximation used here: the 'tidal' approximation. Roughly speaking, it is applicable if the size of an object (in this case, the cluster) is small compared to the distance to the perturber. Thus a cluster affected by a large neighbouring mass is said to be 'tidally perturbed'. Just as the tides are raised on Earth in the directions towards and away from the Moon, in Eq. (12.1) we see that the tidal field accelerates stars away from the cluster in the $\pm x$ directions.

It is not surprising that κ appears here. If we neglect the right sides of Eqs. (12.1), the x, y motion is displaced simple harmonic motion with this frequency, i.e.

$$(x, y) = (a_0 + a_1 \cos \kappa t, b_0 + a_0(\kappa^2/\omega - 4\omega)t/2 - 2\omega a_1 \sin \kappa t/\kappa), \quad (12.4)$$

where the subscripted quantities are arbitrary constants. This is a linearised approximation of the epicyclic motion described in Chapter 7. The motion is retrograde (with respect to the motion of the cluster about the galaxy). Therefore it is easy for stars to remain in the vicinity of the cluster, if in retrograde motion.

Zero-velocity surfaces

The analysis of Eqs. (12.1) begins easily enough. There are two equilibrium points, on the x-axis at $x = \pm r_L$, where

$$r_L \equiv (GM/(4\omega^2 - \kappa^2))^{1/3}. \tag{12.5}$$

These are called Lagrangian points, as they correspond approximately to two exact solutions of the general three-body problem discovered by Lagrange (1772).

There is also an integral, called the Jacobi integral, given by $C = (\dot{x}^2 + \dot{y}^2 + \dot{z}^2)/2 + (\kappa^2 - 4\omega^2)x^2/2 + \omega_z^2 z^2/2 - GM/r$. Its existence stems from the fact that the potential is steady in this rotating, accelerating frame. For a star at rest at one of the Lagrange points, $C = C_L \equiv -3GM/2r_L$. The equipotential surfaces

$$(\kappa^2 - 4\omega^2)x^2/2 + \omega_z^2 z^2/2 - GM/r = \text{const.} \tag{12.6}$$

are called 'zero-velocity surfaces', and are generalisations of what are called 'Hill's curves' in the two-dimensional problem. These surfaces exclude the motion from certain regions of space (Fig. 12.1). For values of $C < C_L$ they consist of three pieces, the innermost of which confines a star to the vicinity of the cluster. Therefore, if a star is situated within the cluster and has $C < C_L$, it cannot escape. This is analogous to the existence of an escape energy for stars in an isolated cluster. The lowest value of C at which the surface opens out to infinity is $C = C_L$, and so it is usual to adopt the 'radius' of this last closed surface, i.e. r_L, as the limiting radius of a cluster in a galactic potential. It is usually called the *tidal radius*, denoted by r_t.

When $C \ll C_L$ the star is confined to the middle of the cluster, far from the reach of the tidal field. It behaves almost like a star moving in the potential of the cluster alone. As C increases the star can wander further and the effect of the tidal perturbation becomes stronger. Fig. 10.3 tells us what to expect, and helps to understand the surface

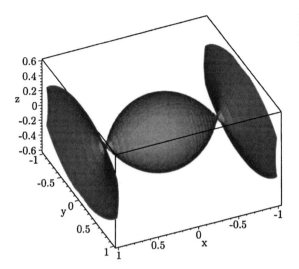

Fig. 12.1. The critical zero-velocity surface for Hill's problem.

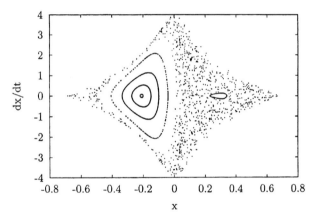

Fig. 12.2. Surface of section for a modified form of Hill's equations. Here the potential of the cluster is a Plummer potential with a suitably chosen scale radius. The equations of motion are obtained from Eqs. (12.1) by replacing r by $\sqrt{r^2 + a^2}$, where $a = 0.1$ is a suitably chosen scale radius, and setting $\kappa = \omega = GM = 1$. For further explanation see text.

of section actually at $C = C_L$ (Fig. 12.2). This shows several planar orbits ($z = 0$), and plots the phase space coordinates x and $\dot{x}$ whenever $y = 0$ and $\dot{y} > 0$. The points are confined by the Jacobi integral, and the orbits are a mixture of chaotic and regular ones. The large area of regular orbits on the left corresponds to retrograde motions, i.e. the stars move around the cluster in the opposite sense to the motion of the cluster around the galaxy. We have already seen why it is that retrograde motions are so special, and there is plenty of supporting numerical evidence (see, e.g., Keenan & Innanen 1975).

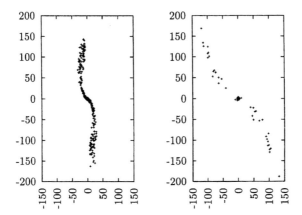

Fig. 12.3. Mass loss by a cluster in a tidal field. On the left a King model with concentration $W_0 = 3$ (Chapter 8) is placed on a circular galactic orbit, and is shown after one revolution. Stars are lost steadily. On the right the same cluster has been placed on an elliptical orbit of eccentricity 0.5; it starts at apogalacticon with a radius corresponding to the tidal radius at perigalacticon, and is shown after one revolution. In this case stars are lost mostly near perigalacticon, and there is a gap in the distribution of escapers near the cluster. Galactic centre is to the right.

Now we return to questions of escape. We have just stated that there is a limiting 'energy' below which escape is impossible, just as in the problem of escape from an isolated, spherical potential. There the resemblance ends. In an isolated cluster, a star with energy above the escape energy will almost certainly escape (unless, on its way out, it encounters another star which once again reduces its energy sufficiently). In a cluster in the tidal field of a galaxy, however, this is not guaranteed. We have already seen that stars in retrograde motion may remain near (but outside) the cluster indefinitely, and this may easily happen to cluster members in retrograde motion with $C > C_L$. (The link between epicyclic motion and retrograde escapers was drawn to our attention by Jeremy Goodman.) One might say that the tidal radius for retrograde orbits exceeds that for prograde orbits (Danilov 1997).

Even more fascinating is the fate of stars in *prograde* motion with $C > C_L$, and it is most easily discussed when $C - C_L$ is small. Then the openings in the equipotential surface through which a star must pass in order to escape are tiny. Stars which do escape must do so through these openings, and it is clear that the time they take to find them depends on the amount by which C exceeds C_L (Fukushige & Heggie 2000). This in turn depends on how much the energy can be changed by relaxation effects (Chapter 14) during the time taken to find the gaps. The result of this interplay is that the rate of escape from star clusters in a steady tidal field depends on N in a quite unexpected way (Baumgardt 2001).

When stars do escape, they do so along the directions of the Lagrange points, as expected (Fig. 2.2). But the Coriolis acceleration (i.e. the terms with coefficient $\pm 2\omega$ in Eq. (12.1)), bends the orbits along the direction of motion of the cluster round the galaxy (Fig. 12.3, left). 'Extratidal' extensions of this general nature are now known around a number of Galactic globular clusters (e.g. Grillmair *et al.* 1995) and can be modelled quite successfully (Combes *et al.* 1999).

When $C = C_L$ the width of the gaps through which stars escape reduces to zero, and they lie at the Lagrangian points. These equilibrium solutions exist only at this critical value of C. As C increases slightly, they give birth to a pair of periodic orbits, called Liapounov orbits. These orbits are unstable, but they control the way in which stars can escape. (Incidentally, the analogous orbits near the Earth, perturbed by the Sun, lead to a series of orbits which have a role in space dynamics. On December 2 1995 the international satellite SOHO, for observations of the Sun and heliosphere, was placed on an orbit of this general kind, with occasional adjustments to quench the instability ('station keeping'). An orbit in this location is particularly suited for continuous solar observation.)

The manner in which the Liapounov orbit controls escape is a little complicated, but is most easily understood for orbits in the x, y plane. For such orbits the Liapounov orbit behaves like a saddle point. An escaper, which necessarily has to come close to this orbit if $C - C_L$ is small and positive, must pass on one side or the other. On one side it is deflected by the Liapounov orbit out of the cluster, and it escapes in every sense. If it approaches on the other side it is deflected back into the cluster, and must try again later, when the same alternatives apply. Even a star which does escape may, if it approaches too close to the Liapounov orbit, spend a long time there before being deflected one way or the other. Furthermore, the number of times a star has to approach this orbit before escaping is an extremely irregular function of its position.

Incidentally, the Liapounov orbits (there is one near each Lagrangian point) also play a critical role in the behaviour of the coorbital satellites of Saturn, and there also give rise to sensitive dependence on initial conditions. It is an example of 'irregular scattering' (Hénon 1988).

Time-dependent tides

These details can be fascinating, but they live in a highly idealised problem, and the general lesson they should teach us is that escape is a complicated problem as soon as we step away from the spherical potential of an isolated cluster. And if we take a more realistic cluster orbit than the circular orbit discussed above, it is clear that almost nothing is understood about it.

For a cluster in an elongated galactic orbit, it is usual to assume that its limiting radius will be the minimum value of r_L around its orbit, which is at perigalacticon, where the tidal field is strongest. But we already know that this is an oversimplification for a circular orbit, and for an elongated orbit there is not even any analogue of the Jacobi constant, because the tidal field is now time-dependent, even in a rotating frame. Another way of expressing this is to say that the tide changes the energy of individual stars in a cluster on an elongated orbit, a mechanism referred to as 'tidal heating'. It should be clear from our explanation that this does not happen if the cluster is on a circular orbit, though it is a not uncommon misconception to suppose that tidal heating occurs in that situation also.

Two kinds of time-dependent tidal fields are of importance in the study of star clusters. One is associated with clusters on elongated orbits as they pass perigalacticon (Fig. 12.3). If this occurs close to the dense central parts of the galaxy the dynamical effects are referred to as *bulge shocking*. A local and short-lived effect is also caused by passage through the disk of the galaxy, and this is what we now study.

To see how different this is from the steady tides acting on a cluster in a circular orbit, consider the time scales involved. The scale height of the disk potential is of order 100pc, but a cluster passes through the plane with a normal speed of order 100km s^{-1}. On the other hand, the radius of a typical cluster is of order 10pc, while typical speeds for stars within the cluster are of order 10km s^{-1}. Therefore the time scales of passage through the disk and orbital motions within the cluster are comparable. This is why the process is often referred to as *disk shocking*.

While a cluster is traversing the disk, the relative effect of the disk and the cluster on the internal motions within the cluster are proportional to the differential accelerations, across the extent of the cluster, of the fields of the disk and cluster, respectively. Since the density in the disk ($\sim 0.1 M_\odot \mathrm{pc}^{-3}$) is much smaller than that in a typical cluster ($\sim 10^3 M_\odot \mathrm{pc}^{-3}$), and since the event is so short-lived, the effect of the disk is relatively minor. Nevertheless, clusters experience of order 100 such shocks in their lifetime, and so their cumulative effect is not so negligible.

Box 12.2 shows that each shock tends to *heat* the cluster (until it revirialises, once the shock is over), and finally gives an estimate of the time scale on which repeated shock heating becomes effective. Notice that none of the parameters of the cluster appear in that expression. Therefore, in a galaxy with a massive disk, all clusters will succumb to destruction by disk shocking on a comparable time scale, unless some other process acts even faster.[1] This is, admittedly, an oversimplification. The strength of shock heating does depend in detail on the structure, size and mass of the cluster, and the point in the plane where the crossing occurs: the density within the disk increases significantly towards the centre of a galaxy. Also, our estimate of the importance of this effect was a global one; in fact the effect of disk shocking is strongest on the outer stars of a cluster. Indeed for many clusters this is the dominant mechanism for changing the energies of the outermost stars, and has therefore been referred to as 'shock relaxation' (Kundić & Ostriker 1995). Detailed models suggest that it has a substantial effect on the escape rate (Spitzer & Chevalier 1973).

There is one circumstance in which this effect is suppressed, and that is when we are considering a cluster where the time scale of the disk crossing considerably

---

[1] J.P. Ostriker pointed this out to us.

Box 12.2. **Shock heating**

Imagine that the disk of the galaxy is modelled as an infinite sheet of matter of surface density Σ. Then the acceleration on opposite sides of the disk is $2\pi G\Sigma$, just as in the standard calculation of the electrostatic field due to a charge sheet.

Suppose now that a star in the cluster is on the opposite side of the plane from the centre of the cluster. It is accelerated relative to the centre by an acceleration of order $4\pi G\Sigma$, and this lasts for a time of order r/V, where r is the displacement of the star from the centre of the cluster and V is the speed of the cluster through the galaxy. (In fact we should measure both quantities in a direction perpendicular to the galactic plane, but the above estimate is good enough for order-of-magnitude purposes.) It follows that the internal speed of the star is altered by an amount δv given by $\delta v \sim 4\pi G\Sigma r/V$, and that the kinetic energy of the cluster is increased by an amount of order $M_c(\delta v)^2/2$, $\sim 8\pi^2 G^2 M_c \Sigma^2 r^2/V^2$, where M_c is the mass of the cluster.

Now tidal shocks occur at intervals of order $2\pi R/V$, where R is the radius of the orbit of the cluster around the galaxy. Therefore, if we write the kinetic energy of the cluster as $M_c v^2/2$, where v is the mass-weighted root mean square speed of its members, we find that the time on which tidal shocks may destroy a cluster is of order

$$t_{\rm sh} = \frac{v^2 R V}{8\pi G^2 \Sigma^2 r^2}. \tag{12.7}$$

Next, for a virialised cluster we may estimate that $v^2 \sim GM_c/r$ (Chapter 9). Also, if the cluster is tidally truncated, its size is given by the tidal radius, in order of magnitude. Therefore, by Eq. (12.5), where we may treat the galactic field as due to that of a point mass of mass M_g, we may write

$$r \sim R(M_c/M_g)^{1/3}. \tag{12.8}$$

It follows that

$$t_{\rm sh} \sim \frac{M_g V}{G\Sigma^2 R^2}. \tag{12.9}$$

exceeds that of orbital motions within the cluster. One might expect then that the energy of a star should behave as an *adiabatic invariant*, but in fact the situation is more complicated and interesting. In the first place the typical orbital time scales are longest in the outer parts of a cluster, and so the effects of a moderately slow disk crossing continue to be significant there. Secondly, in a general spherical potential the motion of a star is governed by two frequencies (of radial and azimuthal motion in the orbital plane). We have already seen (Chapter 7) that near-resonances can then occur, which means that there are combinations of frequencies corresponding to long time scales. Provided that the external potential is significantly coupled to these combinations of frequencies, even a slow crossing can have a substantial effect. This observation is due to Martin Weinberg (Weinberg 1994b; see also Gnedin & Ostriker 1999).

Problems

(1)   Show that the mean density within a tidally limited star cluster is constant, and comparable with that of the galaxy.

(2)   In a galaxy with a flat rotation curve, show that the tidal radius (for a cluster of given mass) varies with galactocentric distance as $r_t \propto R^{2/3}$. (See Surdin (1995) for a comparison with observational data.)

(3)   Explain qualitatively why seas on Earth exhibit two tides each day.

(4)   Consider a very large, planar epicycle, given by Eq. (12.4), which encircles the cluster. By using the method of variation of parameters on the full Eqs. (12.1), and averaging the resulting equations over the epicyclic motion, show that the centre of the epicycle oscillates stably.

(5)   By linearising Eqs. (12.1) in the vicinity of the Lagrangian point, find an approximation to the Liapounov orbits there, and examine their stability, all within the linear approximation.

Part IV

Microphysics: $N = 2$

The following three chapters are devoted to two-body interactions in the context of the million-body problem. Chapter 13 shows that these cause neighbouring orbits to diverge with an approximately exponential time dependence. Beginning with the three-body problem, we go on to investigate the N-dependence of the e-folding time scale for the divergence. We relate the phenomenon to the exponential divergence of geodesics in an alternative formulation of the problem.

Chapter 14 is quintessential collisional stellar dynamics. Here we consider the cumulative effect of many two-body encounters on the motion of a single star: the theory of *two-body relaxation*. We develop a number of standard formulae for the first and second moments of the cumulative change in its velocity. The first moment corresponds to the phenomenon of *dynamical friction*. We then go on to incorporate this theory into an evolution equation (the *Fokker–Planck equation*) for the distribution function. We approximate this equation in a form appropriate to the situation in stellar dynamics, when the time scale of relaxation is much longer than that of orbital motions. This also incorporates the evolution which may result from slow changes in the potential.

Chapter 15 takes a close look at the two-body problem itself. We show, in particular, that the two-body collision singularity is a removable singularity. This introduces a number of topics which might seem surprising in the context of the million-body problem: the Lenz vector, quaternions, the Hopf map, the simple harmonic oscillator, and even a transformation into four dimensions. The significance of all this theory is that it leads to an important computational tool: *KS regularisation*.

13

Exponential Orbit Instability

Up to this point in the book we have largely turned our back on the microscopic character of the million-body problem. Usually we have approximated the gravitational field by that of a smooth distribution of matter. Now we concentrate on the interactions between small numbers of stars in the system, often only two or three stars at a time. In later parts of the book we shall see how these microscopic processes influence the large-scale behaviour.

The purpose of this chapter is to look at the question of *sensitivity to initial conditions* in the million-body problem. Much current research in other dynamical problems is devoted to this question, because of its importance for prediction, for the foundations of statistical mechanics, and perhaps even for the survival of life on Earth. What lies behind this remark is the fact that the question of sensitivity is linked to stability, and the stability of the solar system is something we rely on implicitly. But the collision of comet Shoemaker–Levy with Jupiter in 1994 reminded us that the solar system is not the well regulated clock we often take it for. Less well known is the recent realisation that the rotation of the Earth (which itself influences climate strongly) appears to be stabilised by the presence of the Moon (see Laskar 1996).

Consider the one-body problem. The star proceeds with uniform rectilinear motion $\mathbf{r}_1(t)$. Another single star, not interacting with the first, and starting with a similar initial velocity and position, exhibits similar motion $\mathbf{r}_2(t)$. The distance between the two stars in space, i.e. $\sqrt{(\mathbf{r}_1 - \mathbf{r}_2)^2}$, or in phase space, i.e. $\sqrt{(\mathbf{r}_1 - \mathbf{r}_2)^2 + (\dot{\mathbf{r}}_1 - \dot{\mathbf{r}}_2)^2}$, eventually depends nearly linearly on t, after an initial transient, during which they may actually approach. Their separation in momentum space is constant, however.

To bring gravity into the picture, consider the relative motion of two stars, i.e. the Kepler problem, with equation of motion

$$\ddot{\mathbf{r}} = -\mu \mathbf{r}/r^3, \tag{13.1}$$

where μ is constant. Now we play the same game of considering two Keplerian motions started with similar initial conditions, but not influencing each other. Assuming they have negative energy, in general these two motions will have slightly different periods, and so their separation in space will grow nearly linearly while the separation remains small, except for an initial transient, and any periodic oscillation. While their separation $\delta\mathbf{r}$ is small, $\delta\mathbf{r}$ approximately obeys the *variational equation*

$$\ddot{\delta\mathbf{r}} = -\frac{\mu}{r^3}\left(\delta\mathbf{r} - \frac{3(\mathbf{r}.\delta\mathbf{r})\mathbf{r}}{r^2}\right), \tag{13.2}$$

where $\mathbf{r}$ is a solution of Eq. (13.1). Note that the coefficient on the right-hand side involves the *derivative* of the Keplerian acceleration, i.e. the 'tidal' acceleration of Chapter 12. (Observe the resemblance between the form of these equations in Problem 1 and Eqs. (12.1).)

For a while, Eq. (13.2) describes approximately the evolution of the displacement between the position on one orbit ($\mathbf{r}$) and that on the other ($\mathbf{r} + \delta\mathbf{r}$). For large times, however, both orbits remain bounded (being Keplerian ellipses), and so, therefore, does their separation. The approximation which leads to Eq. (13.2) breaks down, however, and the solution of this equation continues to increase nearly linearly (Problem 1).

The three-body problem is non-integrable (see Moser 1973), and it is harder to determine the rate at which solutions separate without recourse to numerical methods. One problem that has been studied analytically is Sitnikov's problem (see Chapter 20). It is a special case of the 'restricted' three-body problem, in which one star, denoted by m_3, is massless. The other two stars have equal masses and move as a binary on an elliptical relative orbit. As m_3 swings back and forth through the binary it is subject to the time-dependent acceleration of the binary components. Sometimes it gains energy and sometimes it loses energy, depending on the phase of the binary as m_3 passes through. It is easy to see that, if the amplitude of its motion is large, the time between successive passages is very large. In this case a small difference in the initial conditions leads to a large difference in the phase of the binary at the time of closest approach, and a large difference in the change of energy. The subsequent passage through the binary will then be very different for the two motions. The motion is very sensitive to initial conditions.

In fact for this example it is plausible to suppose that the difference between the two motions, while it is still small, is increased roughly by some factor during each passage through the binary, and therefore grows roughly exponentially with time. This growth does not necessarily persist, however, after the time at which the two

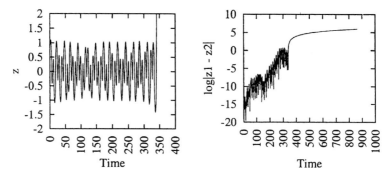

Fig. 13.1. Exponential divergence in Sitnikov's problem (see text and Chapter 20). The left-hand plot shows one solution of the equations of motion of this problem, and the right-hand plot shows (on a logarithmic scale) the absolute difference between two initially neighbouring solutions. Up to the time when the third body escapes ($t \simeq 350$) the growth of the difference is roughly exponential.

motions lose all resemblance. Moreover, in the long run it is likely that one and then the other motion results in escape. By that stage their separation grows no longer exponentially but nearly linearly (Fig. 13.1). The exponential growth of separation is also inapplicable at early times: before that first passage through the binary, the motions separate less rapidly.

While this example might seem ludicrously artificial, it encapsulates every essential aspect of these questions, even for the million-body problem. Here too the separation of solutions starting from neighbouring initial conditions is rather modest at first, and then there is a phase during which they separate exponentially, until all the stars in the two motions are separated by a distance of order the size of the system.[1] At still later times, when stars escape, it may be conjectured that the separation increases nearly linearly again. Overall, the factor by which the initial displacement is magnified up until this time can be enormous, even for a three-body system; factors of order 10^{200} have been measured in numerical experiments (Dejonghe & Hut 1986).

In applying this idea to larger N-body systems, all that is missing is an estimate of the time scale on which the exponential separation acts. After pioneering numerical work by Miller (1964) there was a long gap before the first theoretical attack on the problem. This was the work of V.G. Gurzadyan and G.K. Savvidi (1986), who used a formulation of the problem which goes back to work of N. Krylov. In this theory the trajectory of a set of N particles can be treated as a geodesic in their configuration space when it is endowed with a certain metric (Maupertuis' Principle). Then the divergence between two trajectories is analysed with the help of Jacobi's equation of geodesic deviation. The time scale on which trajectories diverge may be estimated by computing the scalar curvature of the manifold, which in turn involves the square

---

[1] More precisely, at a distance of order the size of the system divided by $\sqrt{N}$; see Box 13.1. Between that point and the time when the separation is of order the size of the system, the separation grows roughly as $\sqrt{t}$.

of the acceleration experienced by each body. Gurzadyan & Savvidi estimated this by using the mean square acceleration, though this is formally divergent, and it is necessary to impose a short-length cutoff, denoted by r_c. Unfortunately, the estimate of the time scale depends sensitively (as $r_c^{1/2}$) on this choice, and the choice made by Gurzadyan & Savvidi leads to a time scale which subsequent work has not confirmed.

The actual acceleration is never infinite, unless a collision occurs, and an alternative approach is to recognise that the acceleration on each particle varies with time, and to consider its cumulative effect. A rough argument is given in Box 13.1. It indicates that the encounters which are most effective in sustaining exponential divergence of neighbouring solutions are those which take place typically once per crossing time. Those which are closer are more effective, but too rare, while more distant encounters are just not powerful enough.

Box 13.1. Estimate of the Liapounov exponent for an N-body system

Suppose one star in an N-body system moves with relative speed V past another star, their distance of closest approach being p (Fig. 13.2). If m is a typical individual stellar mass, the change in velocity is of order $Gm/(pV)$, and the path of the first star is deflected through an angle of order $Gm/(pV^2)$ (cf. Box 7.2, Eq. (7.9), in the case $\alpha \ll 1$). Suppose this star now travels a distance D before its next encounter. The angular deflection caused by the first encounter changes its position during this encounter, and in particular its distance of closest approach, by an amount of order $GmD/(pV^2)$.

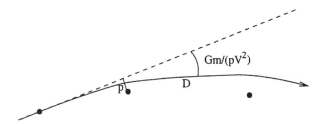

Fig. 13.2. Two-body deflection in an N-body system.

Now let us rerun these two encounters, but with a value of p differing by an amount δp. Then the distance of closest approach in the second encounter is changed by an amount of order $\delta p + GmD\delta p/(pV)^2$. (The first term here occurs because the closest encounter distance would have been affected even without the occurrence of the first encounter. The second term is the differential of the displacement caused by the encounter.) Thus the separation between the two orbits is magnified by a factor of order $1 + GmD/(pV)^2$ per encounter, or by a factor $1 + Gm\delta t/(p^2V)$, where $\delta t \sim D/V$ is the interval between encounters.

Now let's focus on the effect of repeated encounters at about this value of the impact parameter. To estimate how often these encounters occur, we use an '$n\sigma v$' argument: if we travel at speed v in a medium where the spatial number-density of stars is n, and we are trying to hit a target of cross sectional area σ around each star, the rate at which this happens is $n\sigma v$. Here $\sigma \sim p^2$ and $v = V$, and so the time between encounters is $\delta t \sim 1/(nVp^2)$. If we denote by s the separation between the two orbits, it follows that s increases by a factor of order $1 + Gmn\delta t^2$ between each encounter. After a time $t \gg \delta t$, then, we see that

$$\log s \sim \log s_0 + t \log(1 + Gmn\delta t^2)/\delta t, \qquad (13.3)$$

where s_0 is the initial value.

Now we can see which encounters are most effective in this process. The second term on the right of Eq. (13.3) is largest when $\delta t \sim 1/\sqrt{Gmn}$. This time scale is nothing other than the crossing time scale, which is of order the time scale of virial oscillations (Chapter 9) or typical orbital periods (Chapter 7).

Finally, we discuss what this result means. First, it is applicable while the separation between the orbits is small compared with the distance of closest approach, p. (Recall that the foregoing estimate was based on taking the differential of the expression $GmD/(pV^2)$.) The appropriate value of p here corresponds to the typical closest distance of approach in a time of order a crossing time. During this time a star traverses the entire system. If we project the positions of all stars onto a plane orthogonal to its trajectory we obtain N stars within a disk of radius R, i.e. the radius of the cluster. It is clear, therefore, that the typical closest encounter distance is of order $R/\sqrt{N}$. (This is another '$n\sigma v$' argument!) Because the growth of displacement is exponential while the displacement remains small enough, it takes only a few crossing times for any reasonable initial displacement to reach a size of this order; the dependence on the initial displacement is logarithmic. After the displacement becomes of this order, the exponential growth of s slows down (Problem 4).

The calculation in Box 13.1 is so rough that it is not even clear that a fully three-dimensional calculation would not change the result. Still, more careful studies do indicate that the result is correct except for a factor of order $\log(\log N)$. Indeed, for typical models such as Plummer's model (Chapter 8) the e-folding time for the separation of orbits is approximately $0.7/\ln(1.1\ln N)$ in standard units (Goodman, Heggie & Hut 1993). Incidentally, logs of logs are unusual in science, and it is not often that such a slow dependence on a parameter appears in a theory: for instance the time scale decreases by less than 35% between $N = 10^3$ and $N = 10^6$. Until recently it was impossible to verify such a slow dependence empirically (Hemsendorf & Merritt 2002).

It is important to notice that the time scale we have estimated is not a relaxation time scale in the sense which has been customary in stellar dynamics. Until recently (see the introduction to Chapter 14), relaxation has been a name given to any process which changes the energies of the stars (among other things). Close encounters do indeed change the energies of the stars, and this process is analysed in the next chapter. What we have considered in the present chapter, on the other hand, is how the evolution of two neighbouring systems differs, not how each evolves. Still, the exponential divergence continues only for as long as the linearised approximation (represented by the variational equations) remains valid. When the difference between two neighbouring orbits becomes large enough, the energies of the particles in the two systems evolve more independently, on a time scale which is indeed given by the relaxation time scale (Chapter 14). The distinction between relaxation and exponential divergence has been made particularly clear in the case of *one-dimensional* gravity, i.e. when we consider the perpendicular motion of infinite parallel sheets (Tsuchiya & Gouda 2000).

The fact that the motions in an N-body system exhibit such sensitivity to initial conditions does have important consequences, however. Though the following arguments have no rigorous foundation for the gravitational N-body problem, in each case there are analogous systems which also exhibit sensitive dependence and for which something rigorous can be said.

In the first place, sensitive dependence on initial conditions may imply that successive encounters between the same pair of stars (which may occur after the lapse of several crossing times) are effectively independent. The issue here is the *decay of correlations*, which are set up in a pair of stars each time they are involved in a close interaction. As we have seen, sensitive dependence is a symptom of the exponential divergence of neighbouring solutions. In the rigorous treatment of the foundations of statistical mechanics, this in turn is an essential element of what is termed 'hyperbolic' behaviour. In certain well studied models exhibiting this type of behaviour, it can be proved that correlations decay quickly (see, for example, the review by Liverani 2000). By analogy, then, such an argument may help to justify the *statistical* treatment of the cumulative effect of many encounters, which is the foundation for the use of the Fokker–Planck equation (Chapter 14).

Sensitive dependence also implies that the accurate computer simulation of N-body systems can never be guaranteed for more than a few crossing times. Indeed, attempts at the numerical integration of the same problem on different computers never yielded reproducible results on time scales of interest (e.g. Hayli 1970), except for the smallest N. Such integrations are often referred to as 'experiments', but reproducibility is of first importance in other areas of experimental science, and seems to be missing here. For a long time it has been assumed that, even though the positions and velocities of the stars in a simulation are certainly wrong, nevertheless the statistical results of a simulation are reliable. In a sense, each simulation is a running Monte Carlo simulation.

The demonstration of numerical shadowing theorems has done something to underpin this assumption (Quinlan & Tremaine 1992). What has been shown is that, even though numerical errors are made at each step of a simulation, nevertheless there exists an *exact* solution of the N-body problem which stays close to the computed solution, for a limited period of time at least. In practice, also, the statistical results of N-body calculations seem remarkably robust to errors (Smith 1979, Heggie 1991).

Problems

(1)   In Eq. (13.2), suppose that $\mathbf{r}$ corresponds to planar circular motion. In a frame
      rotating with the corresponding angular velocity, show that the coplanar
      components of Eq. (13.2) become

$$\ddot{x} - 2\omega\dot{y} - 3\omega^2 x = 0$$
$$\ddot{y} + 2\omega\dot{x} = 0,$$

where $\omega^2 = \mu/r^3$. Solve these equations.

(2)   Show that the variational equations for the N-body problem (analogous to
      Eq. (13.2)) are

$$\delta\ddot{\mathbf{r}}_i = -\sum_{j=1,j\neq i}^{N} \frac{Gm_j}{|\mathbf{r}_i - \mathbf{r}_j|^3}\left(\delta\mathbf{r}_i - \delta\mathbf{r}_j - \frac{3([\mathbf{r}_i - \mathbf{r}_j].[\delta\mathbf{r}_i - \delta\mathbf{r}_j])(\mathbf{r}_i - \mathbf{r}_j)}{|\mathbf{r}_i - \mathbf{r}_j|^2}\right).$$

By estimating the contribution from the typical nearest neighbour in a system
with density n, show that the corresponding e-folding time scale is of order
$1/\sqrt{Gmn}$, where m is a typical stellar mass.

(3)   (i) Let $V = -\sum_{a<b} \dfrac{Gm^2}{r_{ab}}$ be the potential function for the gravitational
      N-body problem with equal masses m. (This restriction is inessential;
      cf. Gurzadyan & Savvidi 1986.) Suppose the $3N$-dimensional
      configuration space, with coordinates x^i ($1 \leq i \leq 3N$), is given the
      metric $ds^2 = (E - V)\,dx^i\,dx^i$, where E is a constant, and summation
      convention is in use. Show that geodesics are orbits of the N-body
      problem with energy E.

      (ii) Show that the scalar curvature is

$$R = -\frac{3W_{,i}W_{,i}}{4W^3}(3N-1)(N-2) - \frac{W_{,ii}}{W^2}(3N-1),$$

      where $W \equiv E - V$ and a subscript denotes a partial derivative. Show
      that this vanishes if $N = 2$ (except at a singularity). Does this have
      anything to do with the lack of exponential divergence in the two-body
      problem?

      (iii) For the reduced planar Kepler problem in plane polar coordinates we
      have

$$ds^2 = (E + 1/r)(dr^2 + r^2 d\theta^2). \tag{13.4}$$

      Now consider an axisymmetric surface in R^3 of the form $(\rho(r), \theta, z(r))$,
      where (ρ, θ, z) are cylindrical polar coordinates. Try to find functions
      $\rho(r), z(r)$, so that the metric induced on the surface by the Euclidean
      metric in R^3 is as in Eq. (13.4).

(4) In the language of Box 13.1, suppose the separation s exceeds the impact
     parameter p of the most effective encounters ($p \sim RN^{-1/2}$, where R is the
     size of the N-body system). Ignoring encounters with $p \ll s$, show that
     Eq. (13.3) is replaced by $\ln s = \ln s_0 + Gmnt\delta t$, where $\delta t \sim 1/(nVs^2)$. Using
     this to obtain a differential equation for $s(t)$, show that $s \sim \sqrt{Gmt/V}$ asymp-
     totically. Assuming that the separation in energy, δE, may be estimated by
     $\delta E/E \sim s/R$, deduce that $\delta E/E \sim \sqrt{t/t_s}$, where $t_s \sim V^3/(G^2 m^2 n)$ for a
     system in virial equilibrium (Chapter 9).

14

Two-Body Relaxation

As first introduced by Maxwell, the term 'relaxation' meant the process by which a deformed elastic body returned to equilibrium. It was then extended to the dynamical theory of gases, where equilibrium is a statistical equilibrium characterised by the particular form of the distribution of energies, and then transferred by Jeans to stellar dynamics. In stellar dynamics equilibrium is never achieved, because particles escape, but one can still think of a 'quasi-equilibrium' on the time scale of many crossing times. Even so, in due course the term 'relaxation' gradually became applied to several mechanisms which alter such properties as the energies of the stars, whether or not they have anything to do with the approach to equilibrium. More recently it has been argued (Merritt 1999) that the term should be extended further to apply to any one of a variety of mechanisms which cause evolution of the distribution function, whether or not the quantities like energy or angular momentum are altered. The history of the word reflects the development of the subject, from its initial concern with equilibrium models to its modern concern with dynamical evolution.

This chapter deals with one mechanism of relaxation, in which the energy of one star is altered by its interaction with one other. It is often called 'collisional' relaxation, though the interaction is entirely gravitational; real collisions we do not discuss until Chapter 31. Chapter 10 describes 'violent' relaxation, in which the energy of a star is altered by the collective effect of many stars in the system, while Chapter 12 mentions 'shock' relaxation, which is caused by external fields.

Theory of two-body encounters

The link between two-body relaxation and orbital motions is quite subtle, and initially we consider a spatially uniform distribution of stars, in which somehow the stars all describe rectilinear orbits, except for occasional two-body encounters. We concentrate our attention on one star, of mass m_1 and velocity $\dot{\mathbf{r}}_1$, say, and consider how it is affected by encounters with other stars which it meets as it moves along. First, we consider just one encounter with another star, of mass m_2 and velocity $\dot{\mathbf{r}}_2$, say, assuming that the *impact parameter* (Box 14.1, Fig. 14.1) is sufficiently great that the deflection of the star is slight; this is referred to as the 'weak scattering

---

Box 14.1. Approximate theory of two-body relaxation

Let $\mathbf{r} = \mathbf{r}_1 - \mathbf{r}_2$ be the relative position vector of the two stars. If we neglect all other stars, it evolves according to

$$\ddot{\mathbf{r}} = -G(m_1 + m_2)\mathbf{r}/r^3. \tag{14.1}$$

Let V be their relative speed while still far apart, and p their distance of closest approach. (For small-angle scattering we need not distinguish between the impact parameter and the distance of closest approach.) In the weak scattering approximation the change in $\dot{\mathbf{r}}$ over the encounter may be evaluated by substituting rectilinear motion in the right side of Eq. (14.1) and integrating. The result is

$$\delta\dot{\mathbf{r}} = -2G(m_1 + m_2)\hat{\mathbf{p}}/(pV), \tag{14.2}$$

where $\hat{\mathbf{p}}$ is the unit vector in the direction of $\mathbf{r}$ at the instant of closest approach on the rectilinear motion (Fig. 14.1).

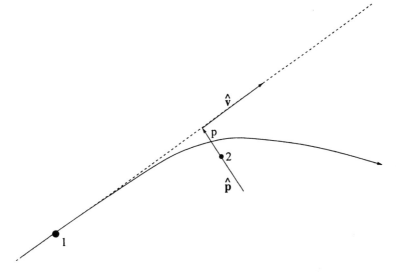

Fig. 14.1. Motion of an encounter in a frame in which particle 2 is fixed.

Eq. (14.2) is not quite accurate enough, even for small-angle scattering, but can be improved by the following trick. From Eq. (14.2) it follows that the angle of deflection is $\alpha \simeq 2G(m_1 + m_2)/(pV^2)$ (cf. Box 7.2, Eq. (7.9)), while energy conservation shows that the magnitude of $\dot{\mathbf{r}}$ equals V long after the encounter. Thus a vector of this magnitude has simply been rotated by this small angle, and so

$$\delta\dot{\mathbf{r}} = -\frac{2G(m_1 + m_2)}{pV}\hat{\mathbf{p}} - \frac{2G^2(m_1 + m_2)^2}{p^2V^3}\hat{\mathbf{V}}, \qquad (14.3)$$

where $\hat{\mathbf{V}}$ is a unit vector in the direction of the original relative velocity, and the second term is a small-angle approximation of $-(1 - \cos\alpha)V\hat{\mathbf{V}}$. The effect of the encounter on the velocity of m_1 is

$$\delta\dot{\mathbf{r}}_1 = m_2\delta\dot{\mathbf{r}}/(m_1 + m_2), \qquad (14.4)$$

since the barycentre of the two stars moves with constant velocity.

Now we sum the result over many encounters, assuming that different encounters are independent, as discussed in the text. It is evident that the average value of $\hat{\mathbf{p}}$ is zero, and so Eqs. (14.3) and (14.4) lead to

$$\sum \delta\dot{\mathbf{r}}_1 = -\sum \frac{2G^2m_2(m_1 + m_2)}{p^2|\dot{\mathbf{r}}_1 - \dot{\mathbf{r}}_2|^4}(\dot{\mathbf{r}}_1 - \dot{\mathbf{r}}_2).$$

In order to perform the sum, let $f(\dot{\mathbf{r}}_2)$ be the number density, in phase space, of stars of mass m_2 and velocity $\dot{\mathbf{r}}_2$. The rate at which encounters occur with stars moving with a range of velocities $d^3\dot{\mathbf{r}}_2$, with values of p in the range $(p, p + dp)$, is $2\pi p|\dot{\mathbf{r}}_1 - \dot{\mathbf{r}}_2|f(\dot{\mathbf{r}}_2)dpd^3\dot{\mathbf{r}}_2$, and so the mean value of the sum, over all encounters in time t, is given by Eq. (14.5).

approximation', or 'small-angle scattering'. (It turns out that it is the cumulative effect of many weak encounters that dominates, rather than the occasional stronger encounter.) Using the theory of Keplerian motion we can compute the change in velocity of the star, $\delta\dot{\mathbf{r}}_1$; see Box 14.1, Eqs. (14.3) and (14.4).

The next step in the calculation is to sum the result over all encounters which the star suffers in a time interval t, say, characterised by various values of the impact parameter p and the velocity $\dot{\mathbf{r}}_2$. In fact the star may experience several encounters simultaneously, but in the weak scattering approximation it is reasonable to suppose that successive (and even simultaneous) encounters can be treated as being independent. (After all, the result for a single encounter is obtained by using unperturbed motions of the two stars.)

The result (Box 14.1) is that

$$\sum \delta\dot{\mathbf{r}}_1 = -t \int \frac{4\pi G^2m_2(m_1 + m_2)}{p|\dot{\mathbf{r}}_1 - \dot{\mathbf{r}}_2|^3}(\dot{\mathbf{r}}_1 - \dot{\mathbf{r}}_2)f(\dot{\mathbf{r}}_2)dpd^3\dot{\mathbf{r}}_2, \qquad (14.5)$$

where p is the impact parameter and $f(\dot{\mathbf{r}}_2)$ is the number density, in phase space, of stars of mass m_2 and velocity $\dot{\mathbf{r}}_2$. (Because of the assumption of spatial homogeneity we need not mention any dependence on $\mathbf{r}_2$.) We shall see later that the above expression is not sufficient to characterise all important aspects of the encounters. Different encounters contribute different $\delta\dot{\mathbf{r}}_1$ with some probability density, and Eq. (14.5) gives an expression for the *first moment* of this distribution. The second moment will also be needed in due course.

We proceed with the further evaluation of the right side of Eq. (14.5). The integration with respect to p yields $\int dp/p = \log(p_{max}/p_{min})$, where p_{max} and p_{min} are hitherto unspecified limits on p. The choice of both limits exposes significant issues, which we now consider.

The minimum is easy; the small-angle approximation is valid only if $|\alpha| \ll 1$ (see Box 14.1), i.e. if $p \gg Gm/v^2$, where m and v are typical masses and speeds. It is not hard to include large-angle scattering, and in fact a more accurate theory shows that, when p is small, the value of $\delta\dot{\mathbf{r}}_1$ is *smaller* than the value given by the small-angle approximation. Thus this singularity in Eq. (14.5) as $p \to 0$ is just an artefact of the weak scattering approximation, and in a more exact theory the corresponding integral is no longer singular. We shall take care of this by simply cutting the integral off at $p \sim Gm/v^2$; it turns out that, when these results are applied to a star cluster with N stars in approximate dynamical equilibrium, the term neglected is smaller than the term we shall retain, by a factor of order $\log N$.

The choice of p_{max} has given greater difficulty. At one time (Jeans 1929, Chandrasekhar 1942) it was thought that encounters must take place one at a time, so that only the nearest neighbour of m_1 could contribute to its relaxation. Following more recent treatments, however (beginning with Cohen *et al.* 1950 and Hénon 1958), we have assumed implicitly that many relaxing encounters may occur simultaneously. This is supported by numerical experiments (Farouki & Salpeter 1994, Fukushige 1995). In a finite stellar system it is assumed now that encounters contribute up to a distance p at which the stellar density falls significantly below its local value.[1] Therefore for a star in the core, i.e. the densest central part of the system, p_{max} may be of order the radius of the core. For a typical star in a nearly spherical system, it may be of the order of a suitably defined radius, such as the *half-mass radius*, r_h, which is the radius of an imaginary sphere containing the innermost half of the mass. With this choice, $\int dp/p = \log(r_h v^2/Gm)$. Furthermore, for a system in virial equilibrium, we have typically $v^2 \sim GM/r_h$, where M is the total mass, and so it is usual to choose $\int dp/p = \log(\gamma N)$, where γ is a constant of order unity.

By a refinement of the above argument, it was customary for a long time to take $\gamma = 0.4$. More recently, however, comparison with N-body simulations has consistently suggested a smaller value, around $\gamma = 0.11$, for systems with equal masses (Giersz & Heggie 1994a). When there is a distribution of masses, considerably

---

[1] One could, however, *imagine* a system, however unrealistic, in which a vast low-density halo contributed most to relaxation.

smaller values are suggested by both theoretical arguments (Hénon 1975) and numerical experiments (Giersz & Heggie 1996). Whatever the value of γ, this factor is often referred to as the *Coulomb logarithm*, which betrays the analogy with the kinetic theory of plasmas, where the interaction is electrostatic.

By this stage we have arrived at the result

$$\sum \delta \dot{\mathbf{r}}_1 = -t \log \gamma N \int \frac{4\pi G^2 m_2 (m_1 + m_2)}{|\dot{\mathbf{r}}_1 - \dot{\mathbf{r}}_2|^3} (\dot{\mathbf{r}}_1 - \dot{\mathbf{r}}_2) f(\dot{\mathbf{r}}_2) d^3 \dot{\mathbf{r}}_2. \qquad (14.6)$$

The remaining integral has a familiar look. Apart from a factor, it is the same as the formula for the gravitational acceleration due to a mass distribution with spatial density f. Therefore, for an isotropic distribution of velocities (corresponding to a spherically symmetric distribution of matter), the result may be written

$$\frac{1}{t} \sum \delta \dot{\mathbf{r}}_1 = -4\pi G^2 m_2 (m_1 + m_2) \log \gamma N \frac{\dot{\mathbf{r}}_1}{|\dot{\mathbf{r}}_1|^3} \int_{|\dot{\mathbf{r}}_2| < |\dot{\mathbf{r}}_1|} f(\dot{\mathbf{r}}_2) d^3 \dot{\mathbf{r}}_2. \qquad (14.7)$$

The left side of this result is often written $d\dot{\mathbf{r}}_1/dt$, though it must be remembered that it has a statistical meaning. At any rate it is an important result. It implies that stars are subject to a deceleration, which is called 'dynamical friction'. Note that, in the case of isotropy, it is caused by stars moving more slowly than m_1. The strength of this deceleration increases with the mass m_1. This is unlike the gravitational acceleration due to a static external body, which would be independent of m_1, even though dynamical friction is caused by the gravitational attraction of other stars. To understand this, and to gain a physical feel for this phenomenon, imagine we move with a star whose velocity is high (Figs. 14.2, 14.3). Then the other stars approach predominantly from the direction in which we are moving. They are deflected by our mass (m_1) and this forms an enhancement in their density behind us. The larger our mass the larger the enhancement, and it is this density enhancement which decelerates us.

Since all stars are subject to dynamical friction, it would seem that all stars must slow down. Recall, however, that these results are statistical, and we shall see that they are not even sufficient to describe the main *statistical* effect of encounters (see Eq. (14.9), below). It should be added that the response of a finite, non-uniform stellar system can be a lot more complicated than Eq. (14.7), which is a local approximation. Even if an object orbits a stellar system outside all the members of the system, so that the local value of f is zero, it will still experience an effective deceleration caused by the response of the system. A theory dealing with this and other issues was developed by Tremaine & Weinberg (1984).

Just as Eq. (14.6) sums the changes in $\dot{\mathbf{r}}_1$, we may compute the sum of their squares, or, more generally, the tensor $\sum \delta \dot{\mathbf{r}}_{1i} \delta \dot{\mathbf{r}}_{1j}$. Looking at Eq. (14.3) in Box 14.1, we see that this involves terms which vary with various powers of p. Only terms in p^{-2} yield the Coulomb logarithm; terms which decrease even more rapidly do not do so, and in fact are smaller by a factor of this order. It is, however, necessary to treat

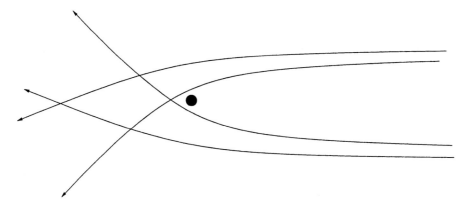

Fig. 14.2. Creation of a wake behind a fast-moving star.

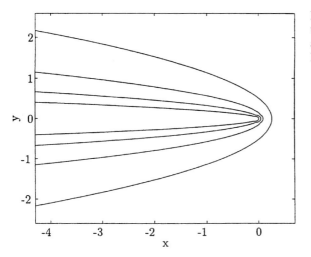

Fig. 14.3. Density contours in the wake behind a fast-moving star. (See Problem 1.)

large-angle scatterings more exactly to establish this. Such terms are called 'non-dominant' terms, and are usually neglected in the standard theory.

The one result about the tensor which we shall quote is its trace. For an isotropic distribution $f(\dot{\mathbf{r}}_2)$ this is given by

$$\frac{1}{t}\sum(\delta\dot{\mathbf{r}}_1)^2 = 8\pi G^2 m_2^2 \log \gamma N$$

$$\times \left(\frac{1}{v_1} \int_{|\dot{\mathbf{r}}_2|<|\dot{\mathbf{r}}_1|} f(\dot{\mathbf{r}}_2)d^3\dot{\mathbf{r}}_2 + \int_{|\dot{\mathbf{r}}_2|>|\dot{\mathbf{r}}_1|} \frac{1}{v_2} f(\dot{\mathbf{r}}_2)d^3\dot{\mathbf{r}}_2 \right). \quad (14.8)$$

Unlike dynamical friction, this result does not involve the mass m_1. This will be important in Chapter 16. (By the way, just as the integral in Eq. (14.7) stems from one analogous to gravitational acceleration, that in Eq. (14.8) resembles the expression for the gravitational potential in a spherical distribution. Indeed, these integrals are sometimes referred to as 'Rosenbluth Potentials' (see Rosenbluth *et al.* 1957).)

One may continue to characterise the statistical effect of small-angle scattering by computing the sums of products of degree three (and even higher) in $\delta \dot{\mathbf{r}}_1$. The corresponding integrands decrease sufficiently rapidly with increasing p, however, that all such terms are non-dominant, in the sense introduced above.

With Eq. (14.7) and similar results, other important effects of relaxation can be analysed. For example, the sum of the changes in the specific kinetic energy of m_1 caused by encounters is

$$\sum \delta E_1 = \dot{\mathbf{r}}_1 . \sum \delta \dot{\mathbf{r}}_1 + \sum (\delta \dot{\mathbf{r}}_1)^2 / 2, \tag{14.9}$$

and so the sum may be computed from expressions which have already been given. Notice that the two terms have opposite sign. The dynamical friction term leads to a decrease in energy, which can be compensated statistically by the second. It is the balance between these two terms which attempts to bring m_1 into thermal equilibrium with the distribution of stars m_2.

Kinetic equations

The information we have obtained describes the statistical effect of relaxation due to encounters on a single star. Now we must consider how this information can be extended to describe its effect on a distribution of stars. In practice this may be done by starting with the Kolmogorov–Feller equation

$$\frac{\partial}{\partial t} f(\dot{\mathbf{r}}_1, t) = \int \left(f(\dot{\mathbf{r}}_1 - \delta \dot{\mathbf{r}}_1) R(\dot{\mathbf{r}}_1 - \delta \dot{\mathbf{r}}_1, \delta \dot{\mathbf{r}}_1) - f(\dot{\mathbf{r}}_1) R(\dot{\mathbf{r}}_1, \delta \dot{\mathbf{r}}_1) \right) d^3(\delta \dot{\mathbf{r}}_1),$$

$$\tag{14.10}$$

where $R(\dot{\mathbf{r}}_1, \delta \dot{\mathbf{r}}_1)$ denotes the rate at which a star with velocity $\dot{\mathbf{r}}_1$ suffers encounters which change its velocity by $\delta \dot{\mathbf{r}}_1$. The first term on the right shows how the distribution function at a particular velocity is increased by encounters which cause the velocity of a star to change from some other value, while the second gives the rate at which it decreases as stars with velocity $\dot{\mathbf{r}}_1$ are scattered to other values of the velocity.

Intuitively obvious though the meaning of this equation is, some fundamental assumptions are needed to justify it. For example, it describes relaxation as a (continuous) *Markov* process, in which the evolution over an infinitesimal time interval depends only on the state of f at the beginning of that interval. The considerations of the previous chapter provide some intuitive foundation for this assumption, though without rigour. On the other hand, the Kolmogorov–Feller equation clearly does not apply on time scales shorter than the duration of a typical encounter.

It is not yet clear how our statistical information on the cumulative effect of many encounters may be exploited in Eq. (14.10), but in fact this is rather easy. Our small-angle approximation is satisfactory if $\delta \dot{\mathbf{r}}_1$ is small, and a truncated Taylor expansion of Eq. (14.10) leads to

$$\frac{\partial}{\partial t} f(\dot{\mathbf{r}}_1, t) = -\frac{\partial}{\partial \dot{r}_{1i}} \int \delta \dot{r}_{1i} f(\dot{\mathbf{r}}_1) R(\dot{\mathbf{r}}_1, \delta\dot{\mathbf{r}}_1) d^3(\delta\dot{\mathbf{r}}_1)$$

$$+ \frac{1}{2} \frac{\partial^2}{\partial \dot{r}_{1i} \partial \dot{r}_{1j}} \int \delta\dot{r}_{1i} \delta\dot{r}_{1j} f(\dot{\mathbf{r}}_1) R(\dot{\mathbf{r}}_1, \delta\dot{\mathbf{r}}_1) d^3(\delta\dot{\mathbf{r}}_1).$$

Now, except for the factor $f(\dot{\mathbf{r}}_1)$, the integrals here are the statistical sums calculated earlier, i.e. this equation may be written

$$\frac{\partial}{\partial t} f(\dot{\mathbf{r}}_1, t) = -\frac{\partial}{\partial \dot{r}_{1i}} \left(\frac{1}{t} \sum \delta\dot{r}_{1i} f(\dot{\mathbf{r}}_1) \right)$$

$$+ \frac{1}{2} \frac{\partial^2}{\partial \dot{r}_{1i} \partial \dot{r}_{1j}} \left(\frac{1}{t} \sum (\delta\dot{r}_{1i} \delta\dot{r}_{1j}) f(\dot{\mathbf{r}}_1) \right). \qquad (14.11)$$

Note that the terms we have neglected in the Taylor expansion (of degree three and higher) are smaller than these terms by a factor of order the Coulomb logarithm. This is not a very big factor, even for the million-body problem: for $\gamma = 0.11$, $\ln(\gamma N)$ varies from about 7 to about 12 for N in the range from 10^4 to 10^6. Nevertheless, formulations exist where some of the neglected effects can be accounted for (e.g. Ipser & Semenzato 1983). Where it has been possible to check, the omission of these terms seems to make little difference to the resulting behaviour of a stellar system (Goodman 1983).

Eq. (14.11) is a form of Fokker–Planck equation. It is a diffusion equation with a drift term, the first term on the right side resulting from the action of dynamical friction. An important property of this equation is that a Maxwellian distribution is a solution. This solution does not, however, provide a thermal equilibrium solution to the structure of a stellar system, for two reasons. One is that, as the system relaxes, the evolution of the distribution of velocities leads to changes in the spatial distribution of matter; in turn these cause the potential to vary, and this changes the energies of the stars in addition to the changes due to two-body relaxation. The problem of incorporating the effect of this process on the evolution of the distribution function will be discussed in a moment. The second reason why the Maxwellian distribution fails to provide an equilibrium model is the boundary conditions: in a stellar system stars may escape, especially those with high energy. The way in which a system reconciles these two inconsistent requirements, thermal equilibrium and boundary conditions, is the subject of Chapter 17.

The most important lesson to learn from Eq. (14.11) is the time scale on which two-body relaxation acts. In fact, from Eqs. (14.7), (14.8) and (14.11) it can be seen that this is of order $v^3/(G^2 m^2 n \log \gamma N)$, where v is a typical speed, m is a typical mass, and n is the number density of stars.[2] This is an estimate of what is called the

---

[2] Recall that f is the number density in phase space, so that $n = \int f d^3 \dot{\mathbf{r}}$.

'time of relaxation', though the precise definition is a matter of convention; most authors follow Lyman Spitzer, who introduced the definition

$$t_r = \frac{0.065v^3}{G^2m^2n\ln\Lambda} \tag{14.12}$$

(see Spitzer & Hart 1971), where v is the root mean square speed, m is the mean stellar mass, and Λ is written for the argument of the Coulomb logarithm. The curious numerical coefficient here comes from an assumption that the distribution of stellar velocities is Maxwellian, and from other conventions (see Problem 4), but the expression itself is often used even when the distribution is different from Maxwellian.

The numerical factor in Eq. (14.12) gives a spurious impression of precision: in fact this formula is only a guide. In particular, in systems in which stars of widely differing mass occur, the time scale on which two-body encounters operate may vary widely. For example, though dynamical friction on a star of mass m_1 acts on a time scale comparable with t_r if $m_1 \simeq m$, where m is the average mass, we have already seen (in the discussion of Eq. (14.7)) that its strength increases with m_1. Therefore the time scale on which it acts varies roughly as mt_r/m_1 for $m_1 \gg m$. This helps to explain why mass segregation occurs so rapidly in N-body systems (Chapter 16; see, for example, Fig. 16.1).

Eq. (14.12) makes it clear that t_r varies a lot from place to place, and it is often useful to have some more global measure of the relaxation time. Following Spitzer (1987), it is usual to try to evaluate Eq. (14.12) for average quantities inside the half-mass radius, r_h, which yields what is called the *half-mass relaxation time*

$$\boxed{t_{rh} \simeq \frac{0.138N^{1/2}r_h^{3/2}}{(Gm)^{1/2}\ln\Lambda}.} \tag{14.13}$$

This result is boxed. Anyone who uses this book for reference will consult this formula more than any other.

The relaxation time is very long compared with the other fundamental time scale in a stellar system, i.e. the crossing time, t_{cr} (Box 1.1). In fact, for a system in dynamic equilibrium it is easily shown from the virial theorem that $t_r/t_{cr} \sim N/\log\Lambda$ (Problem 5). This implies that the mean free path of the stars greatly exceeds the size of the system. Equivalently, the stars make many orbits almost unaffected by two-body encounters. If the distribution of density is nearly spherical, for instance, the theory of Chapter 7 is applicable: stars describe rosette-shaped planar orbits at nearly constant angular momentum and energy. Gradually, however, the effects of two-body encounters make themselves felt, and the orbit of a star will be seen to evolve slowly and stochastically. (The N-dependence of this process is illustrated in Kandrup & Sideris 2001.) Such behaviour contrasts strikingly with the effect of collisions in an ordinary gas, where the mean free path is tiny.

Relaxation and orbital motion

These considerations reveal a complication. Our theory of relaxation was based on a spatially uniform distribution of stars. In fact, a typical star moves between regions of high and low density, where relaxation is (respectively) rapid and slow, before the net effect of relaxation becomes significant. We shall assume that the theory we have described applies locally, but must take into account the fact that the Fokker–Planck coefficients vary widely as the star moves. Furthermore, as the distribution function evolves by relaxation, so does the density in space. In turn this alters the potential, and there is a further resulting evolution of stellar orbits and the distribution function in phase space. We have to consider how to incorporate these factors into the Fokker–Planck equation.

We have seen that the relaxation time much exceeds the crossing time. If, therefore, the distribution function is evolving on the time scale t_r, it is nearly constant on the crossing time scale. Therefore it is nearly a solution of the collisionless Boltzmann equation (Chapter 6). If, moreover, the potential is nearly stationary, we may exploit Jeans' Theorem. Thus, in the case of a spherically symmetric system we suppose that $f = f(E, J, t)$, where the explicit time-dependent evolution takes place on the time scale t_r. This deals with the orbital motions.

We have already seen in Chapter 11 how the distribution function evolves in response to slow changes in the potential. It turns out to be most convenient to express the result in terms of $N(E_1, J_1, t)$, which was defined to be the number density of stars with energy E_1 and angular momentum J_1. It is related to f by Eq. (6.8), and it evolves according to Eq. (11.2), which may be written in the form

$$\left(\frac{\partial N}{\partial t}\right)_{I_r} - \frac{N}{P}\left(\frac{\partial P}{\partial t}\right)_{I_r} = 0. \tag{14.14}$$

Here P is the period of radial motion of a star with these parameters, and the subscript indicates that the time derivative is taken with the radial action I_r fixed.

The effect of two-body relaxation is simply to contribute an additional *encounter* term to the right-hand side of Eq. (14.14). We cannot simply express Eq. (14.14) in terms of f and then add the right side of Eq. (14.11), however, as Eq. (14.11) deals with functions of velocity and position, whereas the left side of Eq. (14.14) involves functions of E_1 and J_1. One way of proceeding is to average the collision term around an orbit of a star with energy E_1 and angular momentum J_1. It is, however, easier to approach the problem less directly (Box 14.2), though the result is the same. It is

---

Box 14.2. The Fokker–Planck equation in a slowly varying potential

If we include encounters only, $N(E, J)$ evolves in an analogous way to f in Eq. (14.11), i.e.

$$\frac{\partial N}{\partial t} = -\frac{\partial}{\partial E_1}\left(\frac{1}{t}\sum\langle\delta E_1\rangle N\right) - \frac{\partial}{\partial J_1}\left(\frac{1}{t}\sum\langle\delta J_1\rangle N\right)$$
$$+\frac{1}{2}\frac{\partial^2}{\partial E_1^2}\left(\frac{1}{t}\sum\langle\delta E_1^2\rangle N\right) + \frac{\partial^2}{\partial E_1\partial J_1}\left(\frac{1}{t}\sum\langle\delta E_1\delta J_1\rangle N\right)$$
$$+\frac{1}{2}\frac{\partial^2}{\partial J_1^2}\left(\frac{1}{t}\sum\langle\delta J_1^2\rangle N\right). \tag{14.15}$$

The angle brackets here signify that the moments of the changes in energy and angular momentum are obtained by orbit-averaging expressions like Eq. (14.9). Then all we have to do is replace the left side of this equation by the left side of Eq. (14.14), and we are done.

The case of an isotropic distribution of velocities, where we suppose that $f = f(E, t)$, is handled similarly. The analogue of Eq. (14.14), when expressed in terms of f, is Eq. (11.3), and the analogue of Eq. (14.15) above is Eq. (9.8), which is important enough to repeat:

$$\frac{\partial f}{\partial t}\frac{\partial s}{\partial E} - \frac{\partial f}{\partial E}\frac{\partial s}{\partial t} = -\frac{\partial}{\partial E}(D_E f) + \frac{1}{2}\frac{\partial^2}{\partial E^2}(D_{EE} f). \tag{14.16}$$

Here $s(E, t)$ is the volume in phase space in which the energy is less than E, and D_E and D_{EE} are diffusion coefficients analogous to those of Eq. (14.15), but incorporating the conversion between f and the number of stars per unit energy; this is $N(E) = f(E)\frac{\partial s}{\partial E}$, and

$$D_E = \frac{1}{t}\sum\langle\delta E_1\rangle\frac{\partial s}{\partial E}. \tag{14.17}$$

This coefficient is an orbit-averaged version, for equal masses, of Eq. (16.13).

---

referred to as the *orbit-averaged Fokker–Planck equation*, though the essential idea goes back to Kuzmin (1957).

There are two formulations of the resulting equation which look relatively elegant and memorable. One, which we learned about from Donald Lynden-Bell, is the form appropriate to a spherically symmetric system with an isotropic distribution of velocities and equal masses, for which the Fokker–Planck equation reduces to

$$\left(\frac{\partial f_1}{\partial t}\right)_s = 16\pi^2 G^2 m^2 \ln\Lambda \frac{\partial}{\partial s_1}\int \min(s_1, s_2)f_1 f_2\left(\frac{f_1'}{f_1} - \frac{f_2'}{f_2}\right)dE_2, \tag{14.18}$$

where $s_i \equiv s(E_i)$, defined in Box 14.2, and $f_i \equiv f(E_i)$, $i = 1, 2$. Box 14.3 presents an outline of a derivation of this equation. The beauty of the end result

---

Box 14.3. **Outline derivation of the orbit-averaged isotropic Fokker–Planck equation for spherical systems with equal masses**

Our task is to compute the diffusion coefficients D_E and D_{EE} in Box 14.2, Eq. (14.16). We shall give a fairly complete derivation for the first of these and leave the second as an exercise. We assume that all masses are equal ($m_1 = m_2 = m$), that the distribution $f(\dot{\mathbf{r}}_2)$ is isotropic in velocity, and that the system is spherically symmetric with potential $\phi(r)$.

Substituting Eqs. (14.7) and (14.8) into Eq. (14.9) we readily find that

$$\frac{1}{t}\sum \delta E_1 = 16\pi^2 G^2 m^2 \ln \Lambda \left(-\frac{1}{v_1}\int_0^{v_1} v_2^2 f(v_2)dv_2 + \int_{v_1}^{\infty} v_2 f(v_2)dv_2\right),$$

where $\Lambda \equiv \gamma N$ and $v_i \equiv |\dot{\mathbf{r}}_i|$. This expression depends on v_1 and also on the position in the cluster, since $f(v_2)$ depends on position. As we intend to average over all parts of phase space accessible to a star with a given energy, we re-express the result in terms of $E_i \equiv v_i^2/2 + \phi$, $i = 1, 2$. The result is

$$\frac{1}{t}\sum \delta E_1 = 16\pi^2 G^2 m^2 \ln \Lambda$$

$$\times \left(-\frac{1}{\sqrt{E_1 - \phi}}\int_{\phi}^{E_1}\sqrt{E_2 - \phi}\, f(E_2)dE_2 + \int_{E_1}^{\infty} f(E_2)dE_2\right).$$

(14.19)

The effective upper limit of the second integral will usually be finite, because f will vanish above a certain energy.

Averaging a function $F(r)$ at energy E_1 (i.e. over an energy hypersurface in phase space) means performing the integral

$$\langle F \rangle = \frac{\int F\delta\left(E_1 - v_1^2/2 - \phi\right) d^3\mathbf{r}_1 d^3\mathbf{v}_1}{\int \delta\left(E_1 - v_1^2/2 - \phi\right) d^3\mathbf{r}_1 d^3\mathbf{v}_1}.$$

The denominator is just $\partial s/\partial E_1$, and cancels in the expression for D_E (Box 14.2, Eq. (14.17)). Integrating with respect to $\mathbf{v}_1$ and then leaving only the integration with respect to $r \equiv |\mathbf{r}_1|$ results in

$$\frac{\partial s}{\partial E_1}\langle F \rangle = 16\pi^2 \int_0^{\phi^{-1}(E_1)} \sqrt{2(E_1 - \phi)}Fr^2 dr.$$

Setting $F = 1$ gives a convenient expression for $\dfrac{\partial s}{\partial E_1}$.

Applying this averaging procedure to Eq. (14.19), we readily find that the factor $\sqrt{E_1 - \phi}$ cancels out, and the first term in Eq. (14.19) yields the expression for $\dfrac{\partial s}{\partial E_2}$, i.e. the derivative of the phase-space volume at energy E_2. The result is the surprisingly compact and symmetric expression

$$D_E = 16\pi^2 G^2 m^2 \ln \Lambda \left(-\int_{\phi(0)}^{E_1} \frac{\partial s}{\partial E_2} f(E_2)dE_2 + \frac{\partial s}{\partial E_1}\int_{E_1}^{\infty} f(E_2)dE_2\right).$$

A similar sequence of calculations needs to be performed for the coefficient D_{EE}. There are no surprises, but it helps to have the formula

$$s(E) = \frac{16\pi^2}{3} \int_0^{\phi^{-1}(E)} \{2(E - \phi)\}^{3/2} r^2 dr \tag{14.20}$$

at hand. The result is that

$$D_{EE} = 32\pi^2 G^2 m^2 \ln \Lambda \left\{ \int_{\phi(0)}^{E_1} s(E_2) f(E_2) dE_2 + s(E_1) \int_{E_1}^{\infty} f(E_2) dE_2 \right\}. \tag{14.21}$$

These results are now to be substituted into Eq. (14.16). It is again surprising how much the right-hand side simplifies, if one performs integrations by parts to remove derivatives of s, and a little further creative manipulation. Also, the left-hand side may be written as $\left(\dfrac{\partial f}{\partial t} \right)_s \dfrac{\partial s}{\partial E}$, where the subscript denotes a derivative at fixed s, but then we divide by the second factor to reach Eq. (14.18).

makes up for the ugliness of the derivation, but also hints at the presence of some as yet undiscovered insight. Incidentally, the result is quite reminiscent of Landau's form of the Fokker–Planck equation in plasma kinetic theory (see Thompson 1962). As well as being elegant and memorable it is also useful, as it is the basis of much of the Fokker–Planck industry which, starting with Hénon's classic study (Hénon 1961), has been the foundation of so much that we know about the million-body problem.

The other setting in which the orbit-averaged Fokker–Planck equation looks relatively elegant invokes action-angle variables. If we assume that f is a slowly varying function of the actions I_i, $i = 1, 2, 3$, the result is

$$\frac{\partial f}{\partial t} = -\frac{\partial}{\partial I_i} \left(\frac{1}{t} \sum \langle \delta I_i \rangle f \right) + \frac{1}{2} \frac{\partial^2}{\partial I_i I_j} \left(\frac{1}{t} \sum \langle \delta I_i \delta I_j \rangle f \right).$$

There is no term on the left caused by the slowly varying potential, as the actions are adiabatic invariants, and the phase-space volume factor analogous to the partial derivative in Eq. (14.17) can be ignored; action-angle variables are canonical, and so the factor is a constant.

Problems

(1)   In Fig. 14.2 let r, θ, ϕ be spherical polar coordinates centred on the massive body, which has mass m_1. Suppose stars are treated as test particles and approach at speed V on paths parallel to the axis $\theta = 0$ when at a large distance. Show that the path of a star with impact parameter p is given by

$$r = \frac{p^2}{p \sin \theta + a(1 - \cos \theta)},$$

where $a = Gm_1/V^2$. (Note that two values of p, of opposite sign, correspond to each value of the pair (r, θ).) By using conservation of flux between paths with impact parameter p and $p + dp$, and conservation of energy, show that the density at r, θ is given by

$$\frac{\rho}{\rho_0} = \frac{|p|\sqrt{1 + \left(\frac{1}{r}\frac{\partial r}{\partial \theta}\right)^2}}{\sqrt{1 + \frac{2a}{r} r \sin\theta} \left|\frac{\partial r}{\partial p}\right|},$$

where ρ_0 is the density far upstream. Adding the results for the two values of p, show that

$$\frac{\rho}{\rho_0} = \frac{r^2 \sin^2\theta + 2ra(1 - \cos\theta)}{r \sin\theta \sqrt{r^2 \sin^2\theta + 4ra(1 - \cos\theta)}}.$$

(2) Write down an expression for $\sum(\delta E_1)^2$ which is analogous to Eq. (14.9). By means of Eqs. (14.3) and (14.4), remove all non-dominant terms. For the case in which the distribution $f(\dot{\mathbf{r}}_2)$ is isotropic and both masses are equal, deduce that

$$\frac{1}{t}\sum(\delta E_1)^2 = \frac{32\pi^2 G^2 m^2 \ln \Lambda}{3}$$

$$\times \left(\frac{1}{v_1}\int_0^{v_1} v_2^4 f(v_2)dv_2 + v_1^2 \int_{v_1}^{\infty} v_2 f(v_2)dv_2\right).$$

(3) Show that the length scale in Fig. 14.2 is proportional to the mass, m_1, of the fast-moving star (when this is large). Deduce that the mass of the wake is proportional to m_1^3 and that the deceleration varies as m_1. What would the corresponding result be for particles moving in a plane, or for a system of rods extended orthogonally to their common plane of motion? By a simple extension of the argument, obtain the asymptotic form for the dependence on v_1, when this is large. Compare with Eq. (14.7).

(4) Compute the result of Problem 2 for a Maxwellian distribution

$$f(v_2) = \frac{n_2}{(2\pi\sigma^2)^{3/2}} \exp\left(-\frac{v_2^2}{2\sigma^2}\right).$$

Note also that, when non-dominant terms are neglected, the result can be used to establish the value of $\frac{1}{t}\sum(\delta v_{1\parallel})^2$, where $\delta v_{1\parallel}$ is the component of $\delta\dot{\mathbf{r}}_1$ parallel to $\dot{\mathbf{r}}_1$. Hence find the time when $\sum(\delta v_{1\parallel})^2 = \sigma^2$, assuming that $v_1^2 = 3\sigma^2$. Compare your result with Eq. (14.12).

(5) For a system in virial equilibrium, estimate the mean square speed by using the fact that, for a wide range of models, the potential energy is approximately $-0.4GM^2/r_h$, where r_h is the half-mass radius (Spitzer 1987, p. 13f; see also

Fig. 8.3 and Table 8.1). By adopting the mean density inside r_h as a typical density, use Eq. (14.12) to derive Eq. (14.13). Deduce that $t_{rh} \sim \dfrac{N}{\ln \Lambda} t_{cr}$, where t_{cr} is the crossing time. (This implies that a star moves a distance much greater than the size of the system before its energy is significantly altered.)

(6) Starting from the result of Problem 2, obtain the formula for D_{EE} (see Eq. (14.21)).

(7) Verify that a Maxwell–Boltzmann distribution $f_1 = f_2 \propto \exp(-\beta E)$ is an equilibrium solution of Eq. (14.18).

(8) Suppose that a system evolves *at constant total energy* according to the evaporative model (Eq. (9.3)). Using results of Problem 5, show that the current relaxation time may be written as

$$t_r = t_{r0} \left(\frac{N}{N_0} \right)^{7/2} \frac{\ln \Lambda_0}{\ln \Lambda},$$

where a subscript 0 denotes an initial value. If the variation of the Coulomb logarithm is neglected, show that $N \to 0$ as $t \to \dfrac{2}{7\mu} t_{r0}$.

15

From Kepler to Kustaanheimo

The previous chapter was concerned with the consequences of two-body interactions, but made use of nothing more than an approximate solution of the two-body problem. Here we consider the classical two-body problem without approximations. It is one of the oldest solved problems of dynamics, and so, as we mentioned in the preface, it is no longer really a problem. Yet its structure is of enduring interest, and offers new surprises each time we view it from a fresh angle.

Along with the simple harmonic oscillator, the Kepler problem is to dynamics what the Platonic solids are to geometry. And, just as there is a duality among the latter (for example the cube, with six faces and eight vertices, is dual to the octahedron, with six vertices and eight faces), we shall see that there is an intimate link between these two dynamical problems. This chapter may *look* self-indulgent compared with the serious issues of stellar dynamics in the surrounding chapters, and should perhaps be in a box of its own, but in fact some of the results we shall survey have important applications to the million-body problem. The reason is that we shall be taking a close look at the singularity of the two-body equations, where numerical methods cause a lot of trouble.

Removing the collision singularity

Consider first the one-dimensional Kepler problem. With a suitable scaling, the equation of motion is

$$\ddot{x} = -1/x^2. \tag{15.1}$$

This equation is singular, corresponding to a collision in the Kepler problem. Even so, its solution is routine. The energy is $h = \dot{x}^2/2 - 1/x$ if $x > 0$, whence $\dot{x} = \pm\sqrt{2(h + 1/x)}$. This is still singular when $x = 0$, but we can sidestep the singularity in the following way. The solution requires the integral $\int dx/\sqrt{h + 1/x}$, and if $h < 0$ this suggests the change of variable

$$x = -\sin^2\theta/h, \tag{15.2}$$

and then we find that

$$\dot{\theta} = \frac{(-2h)^{3/2}}{4\sin^2\theta} \tag{15.3}$$

if we make an appropriate choice of sign. Hence

$$2\theta - \sin 2\theta = (-2h)^{3/2}(t - t_0), \tag{15.4}$$

where t_0 is a constant. (This is a form of Kepler's equation.)

Eqs. (15.2) and (15.4) are a parametric form of solution. They may not look very elegant, but this solution has some remarkable features. First, the singularity at $x = 0$ is no longer evident: the solution can be computed for any time. Second, if we think of θ as the independent variable, we may also write the solution (Eq. (15.2)) as $x = -(1 - \cos 2\theta)/(2h)$, which is simple harmonic motion with the centre of attraction displaced from the origin. Therefore, if we transform from t to θ in the equation of motion we may expect to arrive at the equation of a displaced simple harmonic oscillator. In fact, Eqs. (15.2) and (15.3) show that this transformation is given in differential form by $dt/d\theta = x\sqrt{-2/h}$. Though the square root is unpleasant, we can easily scale θ to remove it.

All this suggests the introduction of a new independent variable τ defined by

$$dt/d\tau = r, \tag{15.5}$$

where $r = |x|$. Though τ looks like a scaled version of θ its status is actually quite different: τ will play the role of a transformed time-variable, whereas θ was a transformed spatial coordinate. Anyway, the transformation of Eq. (15.1) is very simple, and indeed leads to an equation of the expected form:

$$x'' = 2hx + 1,$$

where a prime means $d/d\tau$. This also holds when $h \geq 0$, and again the remarkable feature of this equation is the removal of the singularity at $x = 0$. Though the expression for h is singular at $x = 0$, it need not be evaluated anywhere except for the initial conditions, since it is constant.

The procedure we have outlined is easily extended to the perturbed case, and to three dimensions. Indeed it yields a practical equation for the solution of the perturbed two-body problem (Burdet 1967). If we simply transform from t to τ, defined by Eq. (15.5), it is a straightforward exercise to show that the three-dimensional unperturbed Kepler problem transforms to

$$\mathbf{r}'' = 2h\mathbf{r} - \mathbf{e},$$ (15.6)

where

$$\mathbf{e} = \dot{\mathbf{r}} \times (\mathbf{r} \times \dot{\mathbf{r}}) - \mathbf{r}/r.$$ (15.7)

This vector, like h, is a constant of the motion. It is usually called the Lenz or Runge–Lenz vector. Again we notice that Eq. (15.6) is displaced simple harmonic motion.

Most of this chapter is concerned with a different link between the Kepler problem and the simple harmonic oscillator. Again it is suggested by the solution of the one-dimensional problem, Eq. (15.2), where we notice that $\sqrt{x(\theta)}$ also yields simple harmonic oscillation (this time without displacement) if θ is used as independent variable. This suggests the introduction of a new *dependent* variable X defined by

$$x = X^2,$$ (15.8)

in addition to the transformation from t to τ defined by Eq. (15.5). The transformations are most easily carried out in the energy equation, and then differentiation leads to the expected equation

$$\frac{d^2X}{d\tau^2} = \frac{1}{2}hX.$$ (15.9)

Note that the frequency is half that obtained in the previous transformation. In a way this is obvious from Eq. (15.8): one oscillation of X gives two oscillations of x. There is also, however, a geometric significance to this observation, as will become clear as we proceed. It also accounts for the fact that Eqs. (15.6) and (15.7) are not the preferred route for dealing with two-body encounters in N-body simulations: Eq. (15.9) and its descendents are, in this sense, twice as efficient.

The tricky step with the transformation leading to Eq. (15.9) is its extension to three dimensions, and we proceed by way of the planar case. In two dimensions Eq. (15.8) is easily generalised if we think of a two-vector as a complex number, z. Then the equation of motion is $\ddot{z} = -z/|z|^3$, the energy is $h = |\dot{z}|^2/2 - 1/|z|$, and we define new variables by $z = Z^2$ and $dt/d\tau = |z|$. Direct substitution yields the rather unpromising equation

$$\frac{d^2Z}{d\tau^2} = -\frac{Z}{2|Z|^2} + \frac{1}{\bar{Z}}\left|\frac{dZ}{d\tau}\right|^2,$$

but we see that the right side simplifies if we transform also the expression for h. In fact, if a prime denotes differentiation with respect to τ, we find that $Z'' = hZ/2$, in precise analogy with Eq. (15.9).

The transformation we introduced (a step due to Levi Civita 1906) has a strong geometrical flavour. Without this transformation, Kepler orbits passing close to the

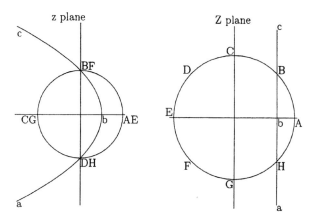

Fig. 15.1. Keplerian orbits in two dimensions (left) and in the transformed plane (right). One circular and one parabolic orbit are shown, and the letters in the two diagrams show corresponding points.

origin are nearly parabolas. The singularity in the transformation $z = Z^2$ at $z = 0$ approximately straightens these curves out.[1] Indeed, when $h = 0$ the curves are exactly parabolas, and the motion in the Z-plane is uniform and rectilinear (Fig. 15.1). Furthermore, if the orbit has $h < 0$, one revolution in the Z-plane corresponds to two in the z-plane: after a single revolution in the z-plane the Z-orbit moves from one Riemann sheet of the transformation to the other, and after two revolutions it moves back again.

KS regularisation

It took about 50 years before Levi Civita's transformation was extended to the practical case of the three-dimensional Kepler problem. This delay seems remarkable when one learns that the requisite tool (quaternions) had been invented as long ago as 1843. Furthermore, they were invented by Hamilton (1853) in order to provide a means of performing algebra with three-vectors; our requirements here are not dissimilar, i.e. to be able to take the square root of a three-vector. Another point of historical interest is that, when the three-dimensional extension eventually was carried out, it was not done in the language of quaternions, but in terms of spinors. So strongly did this approach hold sway that it was even said that the requisite transformations could not be expressed in quaternions. Only in 1991[2] was the full story told, almost 150 years after Hamilton's discovery. In the following account of the theory we shall try to convince the reader that he could have found his way there with only the most rudimentary knowledge of quaternions (as in the outline solution to Problem 1, see Appendix B).

---

[1] We might say that it 'unfolds' the orbit, but this word is loaded mathematically. It would be more accurate to say that it 'blows up' the singularity, as McGehee (1974) did so famously for the collinear three-body problem. See also Waldvogel (1982) for a similar approach to the non-collinear problem.

[2] Neutsch (1991). The account in Vivarelli (1985) omits only the role of our Eq. (15.15).

The usual convention is that a three-vector (x, y, z) corresponds to the quaternion $ix + jy + kz$, where $\{1, i, j, k\}$ are four basis vectors of the space of quaternions. Now the real form of Levi Civita's transform, in obvious notation, is $x = X^2 - Y^2$, $y = 2XY$. Let us try to treat the space of quaternions spanned by i and j in the same way, i.e. we write $ix + jy = i(X^2 - Y^2) + 2jXY$. The right side is easily factored as $i(X - kY)^2$, which can be written in the form

$$ix + jy = (X + kY)i(X - kY). \tag{15.10}$$

The latter form is very convenient for handling rotations, and, now that we know how to handle one plane with quaternions, we can handle any plane, simply by rotating it into position. For example, to rotate a quaternion $ix + jy$ into the quaternion $ix + ky$ (in the plane spanned by i and k) we pre- and post-multiply as follows (cf. Problem 1):

$$\frac{1 + i}{\sqrt{2}}(ix + jy)\frac{1 - i}{\sqrt{2}} = ix + ky. \tag{15.11}$$

This fits beautifully with Eq. (15.10), as we deduce that

$$ix + ky = \frac{1}{\sqrt{2}}(X + iX - jY + kY)i\frac{1}{\sqrt{2}}(X - iX + jY - kY).$$

In general, this suggests that any quaternion representing a vector should be written as

$$ix + jy + kz = qi\bar{q}, \tag{15.12}$$

where q is a quaternion and $\bar{q}$ its conjugate (obtained by changing the sign of the i-, j- and k-components).

There is a difficulty here, because this map from quaternions to physical space is many-to-one. This implies that the initial conditions of a Kepler orbit do not specify unique initial conditions for motion in the space of quaternions. For example, each point in real space corresponds to a 'circle' of quaternions, given by

$$q = q_0 \exp(i\phi), \tag{15.13}$$

where ϕ is any angle and q_0 any quaternion satisfying Eq. (15.12).[3] Furthermore, the relation between the velocities is

$$i\dot{x} + j\dot{y} + k\dot{z} = \dot{q}i\bar{q} + qi\dot{\bar{q}}, \tag{15.14}$$

which has to be solved for the four components of $\dot{q}$ in terms of the three quantities $\dot{x}$, $\dot{y}$ and $\dot{z}$. Another condition would be needed in order to specify $\dot{q}$ uniquely. Again, we can use the special case of motion in the x, y plane as a guide. In this case we may take $q = X + kY$, and it is easily seen that the two terms on the right of Eq. (15.14) are equal. This suggests that the extra condition to be added is

$$\dot{q}i\bar{q} - qi\dot{\bar{q}} = 0. \tag{15.15}$$

---

[3] Note the analogy with the local Abelian U(1)-gauge invariance of quantum electro-dynamics.

There are three other reasons why this choice is appealing. First, it has a lovely geometrical interpretation, as it implies that $\dot{q}$ is orthogonal to the tangent to the curve given parametrically by Eq. (15.13). Second, we are hoping that the equation of motion in q-space will be

$$q'' = hq/2, \tag{15.16}$$

and it turns out that the expression on the left of Eq. (15.15) is a first integral of this equation; thus if Eq. (15.15) holds initially it holds forever. Third, Eq. (15.15) simplifies Eq. (15.14).

Now it is straightforward, if slightly tedious, to demonstrate the link between the simple harmonic oscillator equation in quaternions and the Kepler problem in real space. We assume that (15.15) and (15.16) hold, and transform from q to $\mathbf{r} = (x, y, z)$ using Eq. (15.12), and from τ to t using the obvious definition $dt/d\tau = |\mathbf{r}|$. The result is the familiar equation of motion of the Kepler problem.

This transformation has many interesting properties (Stiefel & Scheifele 1971), of which we mention two. In two dimensions we saw that the transform variable Z moved from one Riemann sheet to another (and back again) as z traversed one orbit. In the three-dimensional case the corresponding relation has some added complexity. After one complete revolution in space, the moving ring of quaternions returns to its original position, corresponding to the quaternion $q(0)$. However, at this time the quaternion that is the solution of Eq. (15.16) is $-q(0)$, and it is only after a second revolution in space that it returns to its original position. Fig. 15.2 (Problem 3) shows two of the loops for motion in space on a circle of unit radius. In this case the quaternions sit on the unit sphere, which is three-dimensional, and there is a familiar map which takes this sphere into ordinary three-dimensional real space.

The second interesting point we shall mention stems from an obvious integral of Eq. (15.16), viz. $e = hqi\bar{q} - 2q'i\bar{q}'$. This resembles the energy integral of the harmonic oscillator, but the latter is obtained (up to a numerical factor) by omitting the factors i. At any rate, it is only another tedious but mechanical exercise to show that this is equivalent to the famous Runge–Lenz vector defined in Eq. (15.7). Its appearance in the Kepler problem is a good deal less obvious.

While we have Eq. (15.12) in front of us let us write it out in coordinates. If u_i, $i = 1, \ldots, 4$ are the four coordinates of q, we find immediately that

$$x = u_1^2 + u_2^2 - u_3^2 - u_4^2,$$
$$y = 2(u_1 u_4 + u_2 u_3),$$
$$z = 2(-u_1 u_3 + u_2 u_4). \tag{15.17}$$

These differ in only one or two inconsequential signs from the so-called KS transformation, named after its discoverers Kustaanheimo and Stiefel (1965). They found their transformation via spinors, which are quaternions in modern dress. Spinors entered physics in the 1930s, to describe elementary particles with spin one half.

One of the remarkable properties of this theory is that a $360°$ rotation in space corresponds to a sign-reversal of the spin amplitudes, i.e. a rotation by $180°$ (see Feynman 1965). This is reminiscent of a property of Levi Civita's regularisation to which we already drew attention. In fact, even the prolific Euler would have recognised these formulae and their characteristics; they occur in the kinematics of a rigid body when expressed in terms of the four Euler parameters (Euler 1776; cf. Goldstein 1980, who spells out the relation with spinors). Truly, each generation must reinvent for itself the discoveries of the past.

If the next generation wants to try something new, it might look beyond the quaternions. Complex numbers and quaternions are examples of *Clifford algebras* (see Peacock 1999 for a snappy introduction). After quaternions the next in line is the eight-dimensional algebra of *Cayley numbers*, where multiplication is not even associative (i.e. it matters where you place the brackets; see, for example, Lambert & Kibler 1988). No-one has yet put them to good use in stellar dynamics, though there is a known regularisation of the *five*-dimensional Kepler problem using eight regularising variables (see Mladenov 1991)!

Numerical regularisation

Though collisions of point masses should not occur in computer simulations of the million-body problem, it is a common experience that numerical methods behave badly even in the vicinity of a singularity. In N-body simulations a moderately eccentric close binary is close enough to cause trouble. The most common approach is to replace the position vectors of the two components, $\mathbf{r}_1$ and $\mathbf{r}_2$, by their relative position vector $\mathbf{r} = \mathbf{r}_1 - \mathbf{r}_2$ and the position vector of their barycentre, $\mathbf{R}$, and then to apply the KS transformation to $\mathbf{r}$. Binaries form and disintegrate, however, and so codes include appropriate procedures for deciding when regularisation is desirable and when it should terminate.

This technique is very successful, but it has never been clear just which aspects of the technique are the vital ones. Even the transformation to $\mathbf{r}$ and $\mathbf{R}$ alone does much to combat rounding error. It has been found that simply ensuring that the integration of a close binary is carried out by a time-reversible algorithm (without the KS transformation) produces roughly comparable performance (Hut *et al.* 1995).

Another property of an integrator which seems relevant is symplecticness (Chapter 22). This is a property of the N-body equations which not all numerical integrators preserve. But there do exist very simple symplectic integrators which can integrate even a collision orbit (Mikkola & Tanikawa 1999), albeit with phase errors. In their method it is not a coordinate transformation that is pulled out of the hat but a whole new Hamiltonian. Let $\mathbf{r}$ be the position vector for Kepler motion, $\mathbf{p}$ the conjugate momentum and h the energy. Then, for suitable scaling, the usual Hamiltonian is $H = \mathbf{p}^2/2 - 1/r$ and h is the value of H. Now it is easily seen (but nevertheless astonishing) that the Hamiltonian $\Lambda = \ln(\mathbf{p}^2/2 - h) + \ln r$ gives canonical equations which correspond to the same Kepler motion, with an independent variable τ defined by $d\tau/dt = \mathbf{p}^2/2 - h$. A leapfrog integrator of these equations (Chapter 22) preserves everything except phase, even for a collision orbit.

Problems

(1) If $\alpha = ix + jy$, show that $x = -(\alpha i + i\alpha)/2$ and $ky = (i\alpha - \alpha i)/2$. Hence obtain Eq. (15.11).

(2) Let $q_a = ia_1 + ja_2 + ka_3$, where a_1, a_2 and a_3 are real, and similarly define $q_b = ib_1 + jb_2 + kb_3$. Show that $q_a q_b = -\mathbf{a}.\mathbf{b} + [\mathbf{a}, \mathbf{b}]$, where $\mathbf{a}$ is the three-vector (a_1, a_2, a_3), $\mathbf{b}$ is defined similarly, $\mathbf{a}.\mathbf{b}$ is the usual dot (scalar) product, and $[\mathbf{a}, \mathbf{b}]$ is the quaternion corresponding to the usual vector product.

(3) *A quaternionic view of circular motion*
(i) By squaring Eq. (15.12) show that $|\mathbf{r}| = \|q\|^2$, with obvious notation. (This shows that motion on the unit sphere $|\mathbf{r}| = 1$ in space maps to the sphere $\|q\| = 1$ of unit quaternions. Now we construct a way of visualising the latter.)
(ii) Show that stereographic projection from the quaternion 1 to the space of imaginary quaternions (spanned by i, j, k) maps $q_0 + iq_1 + jq_2 + kq_3$ to $(iq_1 + jq_2 + kq_3)/(1 - q_0)$. If the unit sphere $(q_0^2 + q_1^2) + (q_2^2 + q_3^2) = 1$ is parameterised in an obvious way as

$$
\begin{aligned}
q_0 &= \cos\theta\cos\phi \\
q_1 &= \cos\theta\sin\phi \\
q_2 &= \sin\theta\cos\psi \\
q_3 &= \sin\theta\sin\psi,
\end{aligned}
\tag{15.18}
$$

show that this point maps to the quaternion equivalent to the three-vector

$$\mathbf{R} = (\cos\theta\sin\phi, \sin\theta\cos\psi, \sin\theta\sin\psi)/(1 - \cos\theta\cos\phi). \tag{15.19}$$

This gives us a way of visualising a unit quaternion, but note that the coordinates are singular when θ is a multiple of $\pi/2$.
(iii) Show that the quaternion given by Eq. (15.18) maps, via Eq. (15.12), to the quaternion equivalent to the three-vector

$$\mathbf{r} = (\cos 2\theta, \sin 2\theta\sin(\phi + \psi), -\sin 2\theta\cos(\phi + \psi)).$$

(Thus for unit vectors and quaternions, the KS map can be thought of as the transformation from $\mathbf{R}$ to $\mathbf{r}$, given by these formulae.)

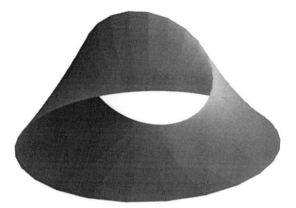

Fig. 15.2. Circular Kepler motion in the space of unit quaternions, represented in three-dimensional space as in Eq. (15.19). Each point on a Kepler motion corresponds to a circle in the space of unit quaternions, and the figure shows how this circle sweeps out a surface as the Kepler motion describes one quarter of an orbit. The edges of the surface shown here correspond to Eq. (15.20) for $\theta = \pi/4$ and $3\pi/4$. Note that these two curves loop through one another.

(iv) Show that the point

$$\mathbf{r} = (\cos 2\theta, \sin 2\theta, 0) \tag{15.20}$$

maps (under the inverse KS map) to the curve C defined by

$$\mathbf{R} = (\cos\theta \sin\phi, \sin\theta \sin\phi, \sin\theta \cos\phi)/(1 - \cos\theta \cos\phi), \quad 0 < \phi < 2\pi$$

(cf. Fig. 15.2. Incidentally, the final result is correct even where θ is a multiple of $\pi/2$, though a separate argument is needed).

(4) *Any* quaternion may be represented as in Eq. (15.18) if the factors $\cos\theta$ and $\sin\theta$ are replaced by variables r and ρ, respectively. Hence establish a link between the KS-like transformation Eq. (15.17) and the parabolic coordinates used in the quantum mechanical treatment of the Stark effect:

$$x = \sqrt{\xi\eta}\cos\phi,$$
$$y = \sqrt{\xi\eta}\sin\phi,$$
$$z = \frac{1}{2}(\xi - \eta)$$

(cf. Landau & Lifshitz 1977). Taken in reverse, this is a pleasing route to the KS transformation (Yoshida 1982).

(5) For the three-body problem in the plane with equal masses, let the complex numbers x_i, $i = 1, 2, 3$, be the position vectors of the three bodies. Interpret geometrically the numbers

$$z_1 = x_3 - x_2$$
$$z_2 = x_1 - \frac{1}{2}(x_2 + x_3).$$

Finally, define three real numbers u_i by

$$u_1 = |z_1|^2 - |z_2|^2$$
$$u_2 + iu_3 = 2\bar{z}_1 z_2.$$

Show that the u_i are independent of rotation (and are therefore an appropriate parameterisation of the shape of the triangle formed by the three bodies), and that the map $(z_1, z_2) \rightarrow (u_1, u_2, u_3)$ is essentially the KS map.

Part V

Gravothermodynamics: $N = 10^6$

The following three chapters begin the application of earlier results to the million-body problem itself. Chapter 16 discusses two effects of two-body gravitational encounters: *escape* and *mass segregation*. The first of these actually develops the theory of two-body relaxation further, as we cannot, in this context, approximate encounters by any small-angle scattering approximation. This approach is, however, applicable to mass segregation, which is an effect of the tendency to equipartition of energies in two-body encounters. It also has an important influence on the stability of the million-body problem (the 'mass stratification instability').

Chapter 17 is also concerned with instability, but an instability which even exhibits itself in systems with equal masses. It was first discovered through a remarkable thermodynamic result obtained by Antonov, which helps to explain the relevance of the term 'gravothermodynamics'. This chapter deals with extrema of the entropy, and the stability of *linear series* of equilibria.

Chapter 18 follows up the previous two chapters by tracing the consequences of the mass stratification and gravothermal instabilities. This is the process referred to as *core collapse*. In other contexts this would be referred to as an example of 'finite-time blow-up' and, in common with other examples of this behaviour, it can be described asymptotically by approximate self-similar solutions of the governing equations.

16

Escape and Mass Segregation

This chapter deals with two effects of two-body encounters. In a general way this process was discussed in Chapter 14, but now we begin to study the effects on the system itself. Furthermore, the theory described there is applicable only to one of the two topics of this chapter. That theory describes the cumulative effects of many weak scatterings, and is perfectly adequate for an understanding of mass segregation. The escape of stars from an isolated stellar system, however, is controlled by single, more energetic encounters, and a better theory is necessary. The theory we shall describe is illustrative of a whole body of theory which improves on that of Chapter 14, though for most purposes (e.g. mass segregation) the improvements are unimportant.

Escape

We consider the case of an isolated stellar system. For this case, a star with speed v will almost certainly escape if $v^2/2 + \phi > 0$, where ϕ is the smoothed potential of the system at the location of the star, with the convention that $\phi \to 0$ at infinity. The exceptions are binary components (for which the true potential differs significantly from the smoothed potential), and an escaper which, on its way out, interacts with another star in such a way that its energy once again becomes negative. The latter possibility is rare in large systems (King 1959), precisely because two-body relaxation takes place on a much longer time scale than orbital motions (Chapter 14).

Several mechanisms may be responsible for the generation of an escaper. It may result from three- or four-body encounters involving binaries (Chapter 23). It may happen that loss of mass by stars through their internal evolution makes the potential

154

well of the cluster sufficiently shallow that weakly bound stars become escapers. Here, however, we consider the most classical escape mechanism: two-body interactions, in which the slingshot effect of an encounter leads to escape of one participant.

The condition for escape is that

$$\mathbf{v}_1 \cdot \delta\mathbf{v}_1 + (\delta\mathbf{v}_1)^2/2 > -E_1, \tag{16.1}$$

where $\mathbf{v}_1$ is the initial velocity of the star, $\delta\mathbf{v}_1$ is the change in this velocity over the encounter, and E_1 is its initial energy. The change $\delta\mathbf{v}_1$ is given by Eq. (14.4), where the quantity $\delta\dot{\mathbf{r}}$ on the right side, in the small-angle scattering approximation, is given by Eq. (14.3). In the derivation of that equation it was stated that the relative velocity $\dot{\mathbf{r}}$ rotates through an angle α, for which an approximate formula was given, but a more accurate formula (which is exact for two-body Keplerian motion) is

$$\tan(\alpha/2) = G(m_1 + m_2)/(pV^2). \tag{16.2}$$

It follows by simple geometry that

$$\delta\mathbf{v}_1 = \frac{2m_2 V}{m_1 + m_2} \sin\frac{\alpha}{2} \hat{\delta v}_1, \tag{16.3}$$

where

$$\hat{\delta v}_1 = -\hat{\mathbf{p}} \cos\frac{\alpha}{2} - \hat{\mathbf{V}} \sin\frac{\alpha}{2}. \tag{16.4}$$

(Recall that a hat denotes a unit vector, $\mathbf{V}$ is the initial relative velocity of the participants, and $\hat{\mathbf{p}}$ is a unit vector orthogonal to $\mathbf{V}$.)

An important result can be deduced right away, though it is also obvious intuitively. If m_2 is negligible, then so is $\delta\mathbf{v}_1$, and so escape of the star with mass m_1 is unlikely. More generally, the likelihood of escape is greater, the greater is the mass m_2. This already shows that two-body encounters are likely to lead to the preferential escape of the stars of lower mass, which is an extreme form of mass segregation.

Now we consider the case of equal masses. In order to derive the escape rate a slightly intricate sequence of transformations, due to Michel Hénon (1960), must be carried out. As usual, we relegate most of the details to a box (Box 16.1), but we would like to draw the attention of even the most hurried reader to Eq. (16.9) there, which is an amazingly simple formula for the rate at which changes of velocity of any given magnitude occur. The next step of the calculation is to add the condition that the change must be big enough, and eventually we obtain the result

$$Q = \frac{8\pi^2 G^2 m^2}{3v_1 E_1^2} \int_{\sqrt{-2E_1}} \left(2E_1 + v_2^2\right)^{3/2} v_2 f(v_2) dv_2 \tag{16.5}$$

for the rate of escape of a star of energy E_1 and speed v_1.

Box 16.1. Computing the escape rate from a star cluster

We first compute the rate at which a star of given mass, velocity and position experiences encounters which would cause it to escape. This is

$$Q = \int V f(v_2) |d\hat{\mathbf{p}}| \, p \, dp \, d^3\mathbf{v}_2, \tag{16.6}$$

the integral being taken over the domain defined in a very implicit way by Eq. (16.1); here f is the distribution of stellar velocities, normalised to the space density of stars. To make progress, we notice first (from Eqs. (16.2) and (16.4)) that, for given initial velocities of the two stars, $\hat{\delta}\mathbf{v}_1$ is determined by $\hat{\mathbf{p}}$ and p. In fact in Eq. (16.6) we can transform the integration with respect to these two variables to give

$$Q = 4G^2 m^2 \int \frac{f(v_2)}{(\delta v_1)^3} d^2\hat{\delta}\mathbf{v}_1 d^3\mathbf{v}_2, \tag{16.7}$$

where we have also used Eq. (16.3).

Escape obviously depends on the magnitude of $\delta\mathbf{v}_1$, which we denote by δv_1, though it is not the change in v_1. Now from Eqs. (16.3) and (16.4) it follows that

$$(\mathbf{v}_1 - \mathbf{v}_2) \cdot \delta\mathbf{v}_1 = -V^2 \sin^2 \frac{\alpha}{2},$$

whence

$$v_{2\parallel} = \mathbf{v}_1 \cdot \hat{\delta}\mathbf{v}_1 + \delta v_1, \tag{16.8}$$

where the subscript $\parallel$ denotes the component parallel to $\hat{\delta}\mathbf{v}_1$. This suggests the use of cylindrical polar coordinates for $\mathbf{v}_2$, with $\hat{\delta}\mathbf{v}_1$ as polar axis. But since we assume that the distribution of $\mathbf{v}_2$ is isotropic, it is also advisable to adopt v_2 as one of the variables. In fact an unorthodox but useful choice is $v_{2\parallel}$ (or δv_1), v_2 and ϕ, where ϕ defines the orientation of $\mathbf{v}_2$ around $\hat{\delta}\mathbf{v}_1$ (i.e. the cylindrical polar angle), and so from Eqs. (16.7) and (16.8) we find that

$$Q = 4G^2 m^2 \int \frac{f(v_2)}{(\delta v_1)^3} d^2\hat{\delta}\mathbf{v}_1 d\delta v_1 d\phi v_2 dv_2, \tag{16.9}$$

though we have to add the restriction $v_2 > |v_{2\parallel}|$, i.e.

$$|\mathbf{v}_1 \cdot \hat{\delta}\mathbf{v}_1 + \delta v_1| < v_2 \tag{16.10}$$

by Eq. (16.8). The integration with respect to ϕ is trivial.

So far we have not used the fact that we wish to compute the escape rate. In fact this expression gives a general result for the rate of encounters leading to a given change in the velocity of the star. Now we add the condition, Eq. (16.1), and the integral with respect to $\delta\mathbf{v}_1$ is most easily carried out using spherical polar coordinates with $\hat{\mathbf{v}}_1$ as polar axis. If the polar angle is denoted by θ, the conditions (16.10) and Eq. (16.1) may be expressed as

$$-v_1 \cos\theta + \sqrt{v_1^2 \cos^2\theta - 2E_1} < \delta v_1 < -v_1 \cos\theta + v_2. \tag{16.11}$$

Integration with respect to δv_1 now yields

$$Q = 4\pi G^2 m^2 \int f(v_2) v_2 dv_2 2\pi \sin\theta d\theta \left(\frac{1}{\delta v_{1\min}^2} - \frac{1}{\delta v_{1\max}^2} \right),$$

where $\delta v_{1\min/\max}$ are the limits in Eq. (16.11), and the integration with respect to θ is restricted to the range $\cos^2\theta < \left(2E_1 + v_2^2\right)/v_1^2$. In this way we arrive at text Eq. (16.5).

In order to evaluate the total escape rate over the entire cluster, we must integrate over the initial velocity and position of the escaper. These integrations hold no terrors, and the result may be expressed as

$$\frac{dN}{dt} = -\frac{256\pi^4 G^2 m^2}{3}\sqrt{2}$$
$$\times \int r^2 dr \int_{E_1+E_2>\phi} \frac{f_1 f_2 (E_1 + E_2 - \phi)^{3/2}}{E_1^2} dE_1 dE_2, \qquad (16.12)$$

where the distribution function f has been written as a function of the energy E for both stars, and $\phi(r)$ is now the potential at radius r.

This result is more elegant than its proof, which quickly becomes intractable if one sets a foot wrong anywhere along the route. It also reveals the competing effects which control the net escape rate. The denominator favours stars which are nearly unbound, as does the large spatial volume which such stars occupy. On the other hand, f is very small near $E = 0$. Nevertheless, in models for which the calculations have been carried out, the bulk of escapers are already weakly bound just prior to the encounter which ejects them.

Hénon's formula appears to work well when the assumptions behind it are satisfied, i.e. the distribution of velocities is isotropic. It differs qualitatively, however, from many other formulae for the escape rate (those summarised in Giersz & Heggie 1994a, for example), which are based on the time of relaxation. For example, for a Plummer model Hénon's result is $\dot N = -0.00088N/(t_{rh} \ln \Lambda)$, while the classic result of Spitzer & Härm (1958) for a square-well potential is $\dot N = -0.0085N/t_r$ (Spitzer 1987). The difference in the coefficient (even though this cannot reasonably be explained by different definitions of the relaxation time) is not the issue. The issue of principle is indicated by the fact that the Coulomb logarithm only appears in one of the formulae. Hénon argued that escape by diffusion does not work, because an increasingly weakly bound star spends longer and longer on an orbit well outside the core, and relaxation is ineffective; therefore escape is dominated by single-encounter events. In fact a synthesis of the two basic ideas seems to work best: relaxation gives rise to a halo of loosely bound particles (on the relaxation time scale), which can then become unbound in a single passage through the core (Spitzer & Shapiro 1972).

What is certain is that, in evolving systems, the spatial structure and the velocity distribution of the stars both change with time. Even if we continue to assume that the distribution of velocities is isotropic, it is found that the rate of escape increases as the core contracts and becomes denser (Chapter 18; see Giersz & Heggie 1994a). In fact, the distribution also becomes increasingly anisotropic (if it was isotropic initially), and this enhances the escape rate, in the sense that, in a series of models which have the same spatial structure but different anisotropy, the escape rate increases with the amount of radial anisotropy. For these reasons, if one computes Hénon's escape rate for a Plummer model (Problem 1), it is a poor guide to the escape rate from a system which starts life as a Plummer model.

This discussion summarises the case of isolated systems of equal masses. As we have seen, escape favours stars of low mass. For a system consisting of equal numbers of stars of some fixed mass and some very small mass, the low-mass stars escape about 30 times as fast as the high-mass stars (Hénon 1969). The *total* escape rate is enhanced by a similar factor (compared to the case of equal masses) for more realistic mass functions. When we place the cluster in a tidal field the rate of escape is further enhanced, but it also becomes a more complicated dynamical problem (Chapter 12).

Mass segregation

The preferential escape of stars of low mass is an extreme manifestation of a tendency in all two-body encounters between stars of different mass, even when neither escapes. It is most easily understood as the tendency towards equipartition of kinetic energies, which ideally leads to the condition $m_1 \langle v_1^2 \rangle = m_2 \langle v_2^2 \rangle$: heavy stars move more slowly, on average. In stellar systems, the massive stars then drop lower in the potential well, while the stars of smaller mass move out and may even escape. In other words, the primary mechanism is the tendency to equipartition of kinetic energies, which leads to segregation of masses by energy, and this in turn leads to spatial segregation (Fig. 16.1).

In more detail, the tendency to equipartition of kinetic energies may be seen as the result of the competing tendencies of relaxation and dynamical friction, which give rise to the two terms on the right of Eq. (14.9). The first term, which decreases the energy on average, is proportional to $m_2(m_1 + m_2)$, by Eq. (14.7), whereas by Eq. (14.8) the second varies as m_2^2; it is this term which causes the kinetic energy to increase on average. A balance between these two effects can only be struck if stars of high mass move more slowly on average.

This idea may be quantified, in terms of the kinetic energy per unit mass, E_1,[1] using these three equations from Chapter 14. They lead to the result

$$\frac{1}{t} \sum \delta E_1 = 4\pi G^2 m_2 \log \gamma N \left(-\frac{m_1}{v_1} \int_{v_2 < v_1} f(v_2) d^3 \mathbf{v}_2 \right.$$
$$\left. + m_2 \int_{v_2 > v_1} \frac{1}{v_2} f(v_2) d^3 \mathbf{v}_2 \right). \qquad (16.13)$$

---

[1] Earlier in the chapter E denoted the total energy (per unit mass).

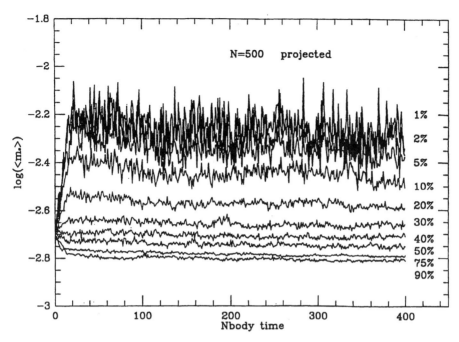

Fig. 16.1. Mass segregation in an isolated N-body simulation. Each graph gives the mean stellar mass within a 'Lagrangian shell', i.e. a shell containing a fixed proportion of the total mass. This plot shows 'projected' data, i.e. the shell is a cylinder along the line of sight; its axis passes through the densest part of the system. Results are averaged over a large number of simulations, each having 500 particles. The initial model is a Plummer model, the distribution of masses is $dN(m) \propto m^{-2.5}dm$, and the ratio of the extreme masses is 37.5. Initially all masses have the same spatial distribution, and the mean mass is the same (within statistical fluctuations) in all shells. Very quickly, however, the mean mass becomes much higher within the central part of the cluster, and somewhat smaller further out. After this phase of rapid mass segregation, however, the subsequent evolution is very slow, for reasons which are not understood. Even at late times there are significant departures from equipartition well within the half-mass radius, and these do not diminish. From Giersz & Heggie (1996).

Multiplying by m_1 and the distribution of velocities of stars of mass m_1, and integrating with respect to $\mathbf{v}_1$, we obtain an expression for the mean rate of change of the mean kinetic energy of stars of mass m_1, which is

$$\left\langle \frac{d}{dt}(m_1 E_1) \right\rangle = \frac{4\sqrt{3\pi}\, G^2 m_1 m_2 n_2 \log(\gamma N)}{(\langle E_1 \rangle + \langle E_2 \rangle)^{3/2}} (m_2 \langle E_2 \rangle - m_1 \langle E_1 \rangle)$$

in the case of Maxwellian distributions; n_2 is the number density of stars of the second species. This strikingly simple result shows that encounters drive the species towards equipartition – if $m_1 \langle E_1 \rangle$ is smaller than $m_2 \langle E_2 \rangle$ then the former increases and the latter decreases. We can also easily estimate the time scale for this to happen; by computing the rate of change of $m_2 \langle E_2 \rangle - m_1 \langle E_1 \rangle$ we easily find that the e-folding time for the tendency to equipartition is

$$t_e = \frac{(\langle E_1 \rangle + \langle E_2 \rangle)^{3/2}}{4\sqrt{3\pi}\, G^2 m_1 m_2 \log(\gamma N)(n_1 + n_2)}. \tag{16.14}$$

This result bears an interesting resemblance to the relaxation time (Chapter 14). For many Galactic globular clusters the latter is considerably shorter than the age of the cluster, and so we might expect to see evidence for mass segregation in them. For a long time the observational difficulty of counting low-mass (and therefore faint) stars in globular clusters was the main obstacle, but nowadays the evidence is robust (e.g. King et $al.$ 1995).

These results show that encounters tend to drive the system towards equipartition, which means that the 'temperatures' of all species are the same. It does not follow, however, that this is what happens, because the distribution of velocities is affected by another process: the change of potential caused by evolution of the spatial distribution of the stars (cf. Eq. (11.3)). Calculations with various models (e.g. Inagaki & Wiyanto 1984) show that different species can either move towards or away from equipartition in different circumstances. Indeed, there are results which show that equipartition is impossible, under reasonable assumptions (Saslaw & De Young 1971).

To see physically what the difficulty is, let us turn first to an analogous problem for a system of equal masses. Here encounters tend to set up a Maxwellian distribution of velocities, but the system cannot reach the necessary isothermal equilibrium. This can be understood from several points of view. On the one hand, we know that isothermal systems must have infinite mass (Chapter 8). On the other, a Maxwellian distribution must allow arbitrarily high speeds for a small fraction of stars, but stars with speeds above the escape speed cannot remain in the system. Finally, it is known that isothermal systems are unstable (Chapter 17). How a system of equal gravitating point masses reconciles these features with the demands placed on it by two-body encounters is the subject of Chapter 18, but it already illustrates that the behaviour is likely to be less straightforward than that of a perfect gas in a box.

Let us return to a system with stars of two different masses, and ask how it responds to the encounters which are driving it towards equipartition. For simplicity we imagine that, initially, the distribution of stellar velocities at a given location in the system is the same for both species. Let v^2 be the mean square speed of both species at this time, and R the 'radius' of the system (e.g. the half-mass radius). We also suppose that the heavier species (of individual mass m_1) is a minor constituent of the system, i.e. $M_1 \ll M_2$, where M_i denotes the total mass of the ith species.

As encounters take place, the stars of the heavier species tend to lose energy, and sink in the potential well of the lighter stars. They will have reached equipartition by the time their mean square speed has fallen to a value of order $m_2 v^2 / m_1$. If the

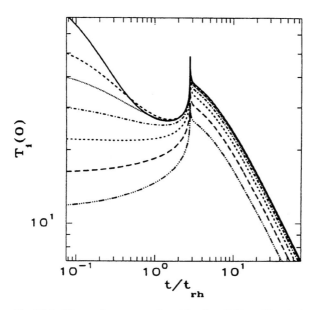

Fig. 16.2. The tendency to equipartition in a Fokker–Planck simulation. The initial model is a Plummer model, with an initial mass spectrum $dN(m) \propto m^{-3.5}dm$ within the range $0.1M_{\odot} < m < 0.75M_{\odot}$ approximately, though it is represented by a number of discrete components; in addition there is a population of heavier 'remnants' of mass $1.2M_{\odot}$. The central 'temperature' of each component, which is proportional to the mean kinetic energy, is plotted against time. Initially all components have the same mean square speed, and so $T \propto m$. Just as in Fig. 16.1, there is an initial phase of rapid evolution (note the logarithmic scale of t), after which the evolution is nearly self-similar, i.e. the ratio of temperature between two species is nearly constant. Note that the heavier components nearly reach equipartition in the initial phase, but the lighter particles remain cooler. From Murphy *et al.* (1990).

potential well of the lighter stars is modelled by a parabolic profile, as in a simple harmonic oscillator, it is easily seen that the spatial distribution of the heavier stars is confined within a region whose size is of order $R\sqrt{m_2/m_1}$. The total mass of the two species within this region is thus M_1 for the heavier species and $M_2(m_2/m_1)^{3/2}$ for the lighter species. Now it is no longer clear that the lighter stars are the dominant mass in this region. If they are, then our estimates are valid in order of magnitude, and approximate equipartition may well be possible. If they are not, then the stars of the heavier species will not only be moving in the potential well of the lighter species, but will also be increasingly affected by their own self-gravity. If this happens, it is less clear that they can contract sufficiently to bring themselves into equipartition with the lighter stars. Indeed, a self-gravitating system heats as it contracts (Chapter 5), and it may not be possible to reach the low mean square speed required (see Fig. 16.2).

These arguments suggest that it may not be possible to achieve equipartition if

$$M_1 \geq M_2(m_2/m_1)^{3/2}. \tag{16.15}$$

More refined calculations confirm that a condition of this form (with a suitable numerical coefficient) closely represents a real limit on the composition of a self-gravitating system which can achieve equipartition. The really interesting issue remaining is to find out what happens if this criterion is violated. In that case the heavier species already becomes self-gravitating before it has reached the low speeds required by equipartition. It therefore heats up: interactions with the light species cause the heavy stars to lose energy, but the resulting contraction in their own potential well causes an increase in kinetic energy which exceeds the loss to the lighter stars. This simply exacerbates the absence of equipartition, which is thus self-perpetuating. Eventually, the heavy stars have contracted so much, ex-pelling lighter stars as they do so, that the latter no longer have a significant role. From this point on the heavier species behave like a self-gravitating system with stars of the same large mass. How such a system behaves is the topic of the next chapter.

Before going on, however, it is worth remarking that we have (as usual) considered an idealised problem: a stellar system with no initial (or 'primordial') mass segregation. There is evidence that the dynamical processes we have discussed do not act fast enough to explain the mass segregation observed in young clusters in Nature (e.g. Kontizas *et al.* 1998, Fischer *et al.* 1998). Presumably they were born that way.

Problems

(1) Use Eq. (16.12) to compute the escape rate from a Plummer model of mass M (Table 8.1). Deduce that the fraction lost per half-mass relaxation time (assuming that the escape rate does not change) is about $0.00088/\ln \Lambda$.

(2) Explain why the effective relaxation time at a distance r in the halo of an isolated system varies as $r^{-1/2}$. Hence estimate the mass flux there if the density is n (cf. Problem 9.7). If the mass flux is independent of r, obtain the density profile.

(3) Use the virial theorem to show that the mass-weighted mean square escape speed is related to the mean square speed by $\langle v_e^2 \rangle = 4 \langle v^2 \rangle$. Using this as an estimate of the actual escape speed, find what fraction of a Maxwellian distribution of velocities lies above it.

(4) Consider a King model consisting of stars of mass m_1 with distribution function

$$f_1(E) = \begin{cases} \exp(-m_1 E) - 1 & \text{if } E < 0, \\ 0 & \text{if } E > 0, \end{cases}$$

and central potential $\phi_c < 0$. Consider adding a few stars of lower mass $m_2 < m_1$ with distribution function

$$f_2(E) = \begin{cases} \eta(\exp(-m_2 E) - 1) & \text{if } E < 0, \\ 0 & \text{if } E > 0, \end{cases}$$

where η is a small constant. Explain why (i) the two species are nearly in equipartition if $|m_2\phi_c|$ is large enough, and (ii) the heavier species dominates the mass density everywhere provided that η is small enough. (Though this appears to contradict Eq. (16.15), that equation depended on the assumption $M_1 \ll M_2$, which is broken here; cf. Merritt 1981.)

17

Gravothermal Instability

Can a million-body system be in equilibrium? More precisely, can a *model* of a million-body system exhibit equilibrium? The answer depends on the model and other conditions. But we already saw in Chapter 8 that equilibrium models of gravitational many-body systems can be constructed. Thus, if it is modelled as a self-gravitating perfect gas it will be in thermal equilibrium if its temperature is uniform. If it is modelled by a Fokker–Planck or Boltzmann equation, then the equivalent condition is that the single-particle distribution be Maxwellian. In both cases the system is required to have infinite mass and extent.

These isothermal models are, in a strict sense, artificial, but they are of great importance conceptually, and for other reasons. In order to make progress in understanding them we shall replace one form of artificiality with another. Instead of dealing with infinite systems, we shall enclose our isothermal system in a spherical enclosure, which at least has the merit of implying that our systems have finite mass and radius. We shall suppose that the enclosure is rigid and spherical; in the N-body model this means that stars bounce off it without loss of energy, while in the gas or phase-space models, it implies that the enclosure is adiabatic. We assume spherical symmetry, and that stars all have the same fixed individual mass. We work entirely, however, with a perfect gas model.

As we saw in Chapter 8, an infinite isothermal model may be specified by two parameters, and to specify a finite model we add a third, e.g. the radius of the enclosure, r_e. For thermodynamic purposes it is sensible to exchange the first two of these parameters for the total mass and energy, denoted by M and E; these are constant, because we are interested in systems within an impermeable, adiabatic enclosure. Thus, for fixed M and E there is a single-parameter family of solutions with parameter r_e.

All are in thermal equilibrium. Another useful way of parameterising the models is in terms of the one dimensionless way of combining these three parameters, or something equivalent. In Chapter 8 we showed that the isothermal model could be made dimensionless in terms of a scaled radius which we called z. As our dimensionless parameter, therefore, we may use z_e, i.e. the value of z corresponding to r_e.

So far it is only the existence and structure of the equilibria that we have been looking at. Now we take the obvious but rather far-reaching step of looking at the *stability* of these models, by which we mean stability in the thermodynamic sense. Before we do so, it is perhaps worthwhile to make some reassuring remarks about the meaning of entropy in this context, because one sometimes hears the assertion that 'thermodynamics doesn't apply to self-gravitating systems' (Chapter 2). In fact astrophysicists apply it all the time, in the context of stellar structure, for example. Indeed it is not hard to show that the above equilibria are determined by extremising the entropy, subject to the conditions of constant mass, energy and volume, and the hydrostatic condition (Box 17.1).

Box 17.1. Entropy extrema

We shall assume spherical symmetry, and so we may use the mass within radius r as an alternative radial coordinate. Since entropy is $S = (k/m) \int \ln(T^{3/2}/\rho)dM$, its first variation is

$$\delta S = \frac{k}{m} \int \left(\frac{3}{2} \frac{\delta T}{T} - \frac{\delta\rho}{\rho} \right) dM,$$

where δT and $\delta\rho$ are Lagrangian variations, i.e. following the mass. Similarly, the energy is

$$E = \frac{3}{2} \frac{k}{m} \int T dM - G \int \frac{M}{r} dM,$$

and its first variation is

$$0 = \delta E = \frac{3}{2} \frac{k}{m} \int \delta T dM + G \int \frac{M}{r^2} \delta r dM. \tag{17.1}$$

Now the hydrostatic condition tells us that $GM/r^2 = -4\pi r^2 \dfrac{dp}{dM}$, and so we can integrate the second term in Eq. (17.1) by parts to remove the dp/dM, provided that we know how to differentiate the rest of the integrand. In fact $r = \int dM/(4\pi\rho r^2)$, and so

$$\delta r = -(4\pi)^{-1} \int (\delta\rho/\rho + 2\delta r/r) dM/(\rho r^2).$$

Differentiation of this expression followed by a little rearrangement quickly leads to the relation

$$\frac{d}{dM}(r^2 \delta r) = -\frac{\delta\rho}{4\pi\rho^2},$$

which turns out to be just what we need. Using also the fact that $\delta r = 0$ at the outer boundary, we find that Eq. (17.1) may be rewritten as

$$0 = \delta E = \frac{3}{2}\frac{k}{m}\int \delta T \, dM - \frac{k}{m}\int T \frac{\delta \rho}{\rho} dM.$$

At this point we suddenly see that, if T is constant in the unperturbed system, then the expression on the right of this equation is just $T\delta S$. In other words, the first variation of S vanishes if the system is in thermal equilibrium; that is to say, the isothermal self-gravitating systems we have been studying extremise the entropy.

As in conventional thermodynamics, the appropriate criterion for thermodynamic stability (for the case of a rigid adiabatic boundary) is based on entropy considerations. As we have seen, equilibrium is characterised by the vanishing of the first variation of the entropy, δS. The series of models we have been discussing are precisely those determined by this condition. Stability depends on the second variation of S in the vicinity of an equilibrium. If this is negative definite, then the equilibrium is stable, since any dissipative process can only increase the entropy, and therefore no such process can cause the system to move indefinitely far from the initial equilibrium. If there are variations in the structure of the model for which the second variation $\delta^2 S$ is positive, then the equilibrium is unstable. Our aim, then will be to examine the thermal stability of the series of equilibria we have found.

There are some situations where the stability of an enclosed isothermal model is easily determined. Consider, for instance, the case $z_e \ll 1$. Figure 8.1 shows that the density is nearly uniform; the reason is that the effect of gravity is tiny in this case, and the gas behaves just like a perfect gas in a box. This is most certainly stable thermodynamically, and therefore the same must be true of the isothermal model in the weakly self-gravitating limit $z_e \to 0$.[1] For later reference we note that the energy of the model in this limit is

$$E = \frac{3}{2}\frac{M}{m}kT, \tag{17.2}$$

where m is the individual particle mass, and the other symbols have their usual thermodynamic meaning.

Starting from this stable limiting case, we now travel along the one-parameter family of equilibria looking for instability. In fact z_e will do very nicely as our parameter. As z_e increases we see that gravity plays an increasingly important part in determining the structure of the model. Its density becomes increasingly inhomogeneous (cf. Fig. 8.1), and we can expect the paradoxical thermodynamic behaviour of self-gravitating systems (Chapter 5) to make an appearance.

This is by now an old problem, and has been investigated quantitatively using several different techniques (Antonov 1962, Lynden-Bell & Wood 1968, Horwitz &

---

[1] Strictly, the system will evolve very slowly by the formation and hardening of a binary (Chapter 5). The time scale is very long, however, of order N relaxation times at least (cf. Problem 21.3).

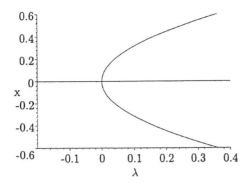

Fig. 17.1. Bifurcation diagram for Eq. (17.3) with $U = x^3/3 - \lambda x$.

Katz 1978, Chavanis 2002, Chavanis *et al.* 2001). A common, direct approach is to compute the second variation of S or some other appropriate variable, but this can be somewhat laborious, and here instead we employ what would nowadays be called bifurcation theory. Actually, the method we adopt, which loosely follows that of Katz (1978), has its origins in Poincaré's theory of linear series (cf. Jeans 1929). For a simple analogue, consider a one-dimensional conservative oscillator whose equation of motion is given by

$$\ddot{x} = -\frac{\partial}{\partial x}U(x, \lambda), \tag{17.3}$$

where the potential $U = x^3/3 - \lambda x$ and λ is a parameter. The analogy is that equilibria of our thermodynamic problem are given by a one-parameter family of extrema of the entropy, whereas in Eq. (17.3) the equilibria are given by a one-parameter family of extrema of U. The main difference is that the dynamical problem has one degree of freedom, whereas there are infinitely many in the problem of isothermal spheres. The equilibria for Eq. (17.3) occur at $x = \pm\sqrt{\lambda}$ if $\lambda \geq 0$, while there are none if $\lambda < 0$. This is depicted in the 'bifurcation diagram', Fig. 17.1, which shows that, as λ is decreased, the two equilibria approach, fuse and disappear. Such a feature is referred to as a 'saddle-node bifurcation'.

Now let us look at stability. In Fig. 17.1 the upper equilibrium (with $x > 0$) is stable, and the other is unstable. The fact to concentrate on is that stability changes precisely at the point where the equilibria disappear. Now it could easily be claimed that this is a special feature of this toy problem, but in fact it is 'generic', i.e. the systems for which it does *not* happen are the special ones.

In order to see this, consider Eq. (17.3) with a general potential about which we assume only that there is a sequence of equilibria $x(\lambda)$. Thus

$$(\partial/\partial x)U(x(\lambda), \lambda) = 0. \tag{17.4}$$

Differentiating with respect to λ, we see that

$$\frac{\partial^2}{\partial x^2}U(x(\lambda), \lambda)\frac{dx}{d\lambda} + \frac{\partial^2}{\partial\lambda\partial x}U(x(\lambda), \lambda) = 0. \tag{17.5}$$

Now suppose that there is a saddle-node bifurcation. At that point $dx/d\lambda$ is infinite, and it changes sign there as we pass along the series of equilibria. Provided, then, that

$$\frac{\partial^2}{\partial\lambda\partial x}U(x(\lambda),\lambda)\neq 0 \qquad (17.6)$$

there, it follows that $(\partial^2/\partial x^2)U$ changes sign there, and hence that the equilibrium changes stability. Now, without an example at hand, one might suspect that conditions at a saddle-node bifurcation are so special that Eq. (17.6) is never satisfied. But in fact our toy problem illustrates that this is not the case, and it is clear that it is systems in which Eq. (17.6) is *not* satisfied that are special. In general, then, we should expect a saddle-node bifurcation to be associated with a change of stability.

Now there are two more technicalities to deal with before all this machinery can be applied to the gravothermal problem, which has infinitely many degrees of freedom, while the toy problem has only one. In the first instance one would like a criterion which does not require plotting of x against λ, because there are infinitely many xs to choose from. One way of dealing with this is to look at the variation of U along the series of equilibria. In other words, define $W(\lambda)=U(x(\lambda),\lambda)$. Then it is easily shown, using Eq. (17.4), that

$$\frac{d^2W}{d\lambda^2}=\frac{\partial^2U}{\partial x\partial\lambda}\frac{dx}{d\lambda}+\frac{\partial^2U}{\partial\lambda^2}.$$

If this is combined with Eq. (17.5) it follows that

$$\frac{d}{d\lambda}\left(\frac{dW}{d\lambda}\right)=-\frac{\left(\dfrac{\partial^2U}{\partial x\partial\lambda}\right)^2}{\dfrac{\partial^2U}{\partial x^2}}+\frac{\partial^2U}{\partial\lambda^2},$$

and so the change of equilibrium can be detected by plotting $dW/d\lambda$ against λ and looking for a vertical tangent. If there are many degrees of freedom with generalised coordinates $\{x_i\}$, then the same conclusion follows, the calculation being most easily carried out in coordinates which diagonalise the second variation of $U(x,\lambda)$ at fixed λ.

Now how do we apply all this to the gravothermal problem? In fact almost everything is in place. Let us consider the series of equilibria corresponding to a fixed total mass (M) within an enclosure of fixed radius (and therefore fixed volume, V). For U we take the entropy S, and the series of equilibria can be parameterised as we please, but it is convenient to use the total energy E for the parameter λ. The reason for this is that the quantity to be plotted, i.e. $(\partial S/\partial E)_{M,V}$, then has a simple interpretation: it is the inverse of the temperature, T. (Recall our proof in Box 17.1 that $\delta S=0$ in thermal equilibrium, which followed from the result $\delta E=T\delta S$ if T is uniform.) Therefore we plot $1/T$ against E and look for vertical tangents in the resulting curve.

The result is shown in Fig. 17.2. Several parts of this result can be understood without numerical computation. In the almost uniform, weakly gravitating limit, it follows from Eq. (17.2) that $T^{-1}\simeq 3Mk/(2mE)$, which yields the part of the curve

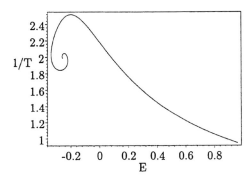

Fig. 17.2. The linear series of isothermal models of fixed mass and volume, in suitable scaled units. Compare with Chapter 5, Fig. 5.3.

going off to the right. Similarly, the singular isothermal solution and its asymptotic form (Problem 8.2) give rise to the spiral.

Now we proceed from the right-hand end of the series of equilibria, where the models are certainly stable, looking for a vertical tangent. It is already evident that such points exist, and we need numerical methods only to locate the first one. In fact it occurs when the scaled radius of the enclosure is $z_e = 34.2$, approximately. Stellar dynamicists remember more easily the corresponding value of $\exp(u_e) = 1/709$, cf. Eq. (8.4), which is the ratio of the space density at the edge to the central value. This is the last stable isothermal model inside a spherical, adiabatic container. Models with a larger density contrast are unstable thermally.

After all this heavy analysis and theory, it is a relief to realise that the physical origin of the instability is very easy to understand, provided that one reads the appropriate paper (Lynden-Bell 1968). The arguments have already essentially been deployed in Chapter 5. Imagine what happens if a little thermal energy is transferred from the inner part of an enclosed isothermal system to the outer part. The inner part should cool, but the gravitational field causes the stars there to drop into lower orbits, where they move faster than before. Just like a satellite spiralling slowly to Earth under atmospheric friction, the net effect is that the stars are moving faster, and the system is hotter, hotter even than at a time before the heat was transferred. The thermal energy transferred to the outer parts has a more familiar effect: these parts are held in not by gravity but by the wall of the enclosure. Just as in a non-self-gravitating gas in an enclosure, addition of thermal energy pushes up the temperature. How much the temperature goes up depends on the heat capacity of the outer part of the system. The larger and more extensive the outer halo, the smaller the rise in temperature. For a sufficiently large halo (sufficiently large r_e) the rise in temperature of the outer part is smaller than that of the inner part. Then a temperature gradient has been set up, more heat flows from the inside to the outside (provided that the gas is conducting), and the temperature gradient is enhanced. There is a runaway, the famous 'gravothermal catastrophe'.

This simple physical picture is suspicious, because it now seems as though the boundary condition, an adiabatic rigid enclosure, plays a crucial role. To show qualitatively that the boundary condition is not playing an essential role, let us consider an isothermal model with a very large value of r_e, i.e. very close to the centre of the spiral pattern in Fig. 17.2. There is a huge range of densities in this model. Therefore there is a huge range in the time scale for the convective transport of heat. On the time scale for heat transport in the inner parts, the outer parts act effectively as an insulating envelope. Also, the amount of mass in the inner parts is so small that adjustments in the structure there have little effect on the outer parts, which are also effectively a rigid barrier. In other words, the outer parts of the system have the properties of the boundary which we artificially inserted in order to make our model precise. It follows that the central parts of an isothermal system will be subject to the gravothermal instability, and that the boundary conditions far out in the system are irrelevant.

Precisely how the growth of the instability takes place is beyond the scope of the above analysis. In particular, it depends on the details of the mechanism by which heat is transported (Makino & Hut 1991). But various calculations confirm that the 'first' isothermal model in which thermal inhomogeneities can grow is one in which the density contrast is approximately the famous 709. This has been shown not only for models which treat the system as a self-gravitating, conducting, perfect gas (Nakada 1978), but also for Fokker–Planck calculations (Inagaki 1980). As we have said, however, the precise boundary conditions are not strictly relevant to the behaviour of self-gravitating systems, and the critical value of the density contrast is not the main issue.

There is another way of looking at the role of boundary conditions. So far we have imagined that energy can be exchanged only between one part of the system and another. Suppose now that the entire system may exchange energy with its surroundings. We imagine that the enclosure is still rigid and allows no mass in or out, but that the system can exchange energy with its surroundings, which are held at constant temperature. Then the heat capacity of the system is simply $\partial E/\partial T$. It is therefore subject to the kind of instability outlined in Chapter 5 as soon as the tangent to the curve in Fig. 17.2 becomes horizontal. Not surprisingly, this occurs for a smaller density contrast than in the previous case, because now the enclosure is taking on the thermodynamic role previously played by the outer parts of the system. The distinction between the two situations can also be described in terms of the appropriate ensemble, in the sense of statistical physics: the microcanonical ensemble for systems of fixed energy (the first case, in which the enclosure is adiabatic), and the canonical ensemble for systems of fixed temperature (the second case).

There is one final twist to this tale. We gave a physical explanation for the gravothermal catastrophe by contemplating a transfer of heat from the inside to the outside. But instabilities can grow in either direction; what happens if heat is transferred from the outside to the inside? It is not hard to see that both the inside and the outside cool down, but again the cooling of the outer parts is slight if they are sufficiently massive and extensive. Then heat starts to flow inwards, driving the gravothermal instability in reverse. We shall see in Chapter 28 that this is no mere thought experiment.

Problems

(1) A simple approximate model of an isothermal system in an enclosure of
radius r_e is defined by the density

$$
\rho(r) = \begin{cases} \rho_0, & \text{if } r < r_0; \\ \rho_0 \left(\dfrac{r_0}{r}\right)^2, & \text{if } r_0 < r < r_e. \end{cases}
$$

Compute $M(r)$, i.e. the mass within radius r, and deduce that the potential
energy is

$$
W = -\frac{x}{5(3x-2)^2}(45x - 30\ln x - 42)\frac{GM_e^2}{r_e},
$$

where M_e is the total mass within the radius r_e and $x = r_e/r_0 > 1$. Using the
appropriate form of virial theorem (Problem 9.1, where T denotes the *kinetic
energy*), compute the temperature and the total energy. Hence plot the diagram
analogous to Fig. 17.2. Show that the sequence includes models with positive
and negative heat capacity.

(2) In an enclosed isothermal model with enclosure radius r_e, the kinetic energy
is $\dfrac{3}{2}\dfrac{M}{m}kT$, where T is the temperature. Use the appropriate form of the virial
theorem, as in Problem 1, to show that the energy is given by

$$
\frac{Er_e}{GM^2} = \frac{1}{u'^2}\left(e^u + \frac{3}{2}\frac{u'}{z}\right),
$$

in the notation of the isothermal equation (Eq. (8.6)). By solving the latter
equation numerically, and locating the minimum of $Er_e/(GM^2)$ for varying
z, show that the first unstable model corresponds to a density contrast
$e^{-u} = 709$ approximately.

(3) Does a two-dimensional self-gravitating system, in which the interaction
potential between two masses m is $2Gm^2\ln r$, exhibit gravothermal instability? Think about this first from the point of view of the virial theorem
(Problem 9.2). Then use the analytical solution (Problem 8.3) to carry out
an analysis as in Fig. 17.2. See Katz & Lynden-Bell (1978).

18

Core Collapse Rate for Star Clusters

The cruel fate of a system forever striving to be what it can never be – in thermal equilibrium.

J. Goodman

The last two chapters have assembled most of the qualitative arguments by which the evolution of the core of a stellar system can be understood. In summary, the tendency towards equipartition drives the more massive stars to smaller radii. Unless their total mass is sufficiently small, equipartition cannot be reached by the time the heavier stars become essentially self-gravitating. When that happens they are eventually subject to the gravothermal instability. It is the purpose of the present chapter to flesh out this outline, but we shall do so in two passes, as it were. First we shall examine the time scales on which these processes act, and a number of factors which modify the simple picture; and we shall explain the qualitative nature of the resulting evolution. Then we turn to a more detailed description of one case which has been studied in great detail: self-similar collapse in systems with stars of equal mass.

The big picture

The time scale for equipartition, t_e, was discussed in Chapter 16 (see Eq. (16.14)). It is useful to compare it with the standard relaxation time t_r (Eq. (14.12)). For this purpose we evaluate the mean kinetic energy per unit mass for each species by $\langle E_i \rangle = v^2/2$, independent of mass, i.e. we assume equipartition of *velocities*. In Eq. (14.12) we write the factor $m^2 n$ as $\bar{m}\rho$, where $\bar{m}$ denotes the mean stellar mass, and ρ is the mass-density. We also identify the Coulomb logarithm in both time scales. For a two-component system it follows that $t_e/t_r = 1.25\bar{m}^2/(m_1 m_2)$.

In applications, we are often interested in models with many particles of low mass (m_1, say) and a few of high mass $m_2 \gg m_1$. In this case $\bar{m} \simeq m_1$, and so

$t_e/t_r \simeq m_1/m_2, \ll 1$. In other words, the fastest processes are associated with equipartition. Indeed, if one uses some detailed model (Chapter 9) to follow the evolution of a multi-component model (i.e. one in which the particles do not all have the same mass), the first thing one observes is that the heaviest particles drift towards the centre, increasing the central density and the central proportion of massive particles (Fig. 16.1). The greater the individual masses of the most massive particles, the faster the evolution.

There are factors which modify this picture in more realistic models of stellar systems. In the first place, the most massive stars also evolve (internally) most rapidly (Chapter 11). In the very rapid, late stages of their evolution, they lose most or all of their mass. Therefore the range of individual stellar masses shrinks, and this in turn affects the relevant time scale t_e. The second point to notice is that the most massive species may well consist of binaries with massive companions. Though the effects of close interactions between binaries and single stars are complicated (Parts VI and VII), in distant encounters a binary behaves essentially like a single star with a mass equal to the total mass of its components.

Eventually the tendency towards equipartition slows down drastically. As already mentioned, this does not mean that equipartition has been achieved; rather, the central parts of the system are now dominated by the most massive stars, and, as a first approximation, we may consider that the role of the less massive stars is unimportant. As a step towards understanding this new situation, we proceed to discuss the evolution of the core in a *single-component* system, in which all stars have the same mass. This problem is of obvious interest in its own right, and it has been rather well studied.

First let us consider the situation discussed in Chapter 17, i.e. an isothermal model enclosed in a spherical container. As we saw, this equilibrium is unstable only if the density contrast (between the centre and the enclosure radius) exceeds a factor of about 709. At the enclosure radius the relaxation time is $709t_{rc}$, where t_{rc} denotes the central value. Therefore we may expect that the time scale on which the instability acts is *at least* of this order. In fact it may be much longer, as the system is nearly isothermal, which reduces the thermal flux.

Now we should consider a more normal stellar system, still consisting of stars of equal mass, but bound only by their self-gravity, and not by any external wall. Such a system cannot be isothermal, and so the flux of heat which drives evolution of the core may be stronger than in the near-isothermal situation considered previously. Qualitatively, however, the result is the same: the centre contracts and becomes hotter. At the higher densities which result, the relaxation time decreases, and in time it becomes much shorter than at the larger radii where the heat flux drove the initial contraction. By the time this happens, the hot, central region of the cluster is effectively isolated (thermally) from the regions at larger radius. Thus there is a transition from a collapse which is driven by the heat flux at large radii to a collapse, analogous to that in an enclosed isothermal sphere, driven by the heat flux in a smaller, nearly isothermal region. As this region collapses further, the time scale becomes

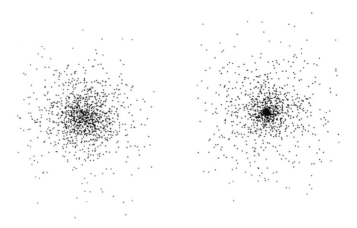

Fig. 18.1. Core collapse from a King model with $W_0 = 3$, $N = 1285$. The initial model is shown on the left, and the view on the right shows the same model to the same scale after core collapse. The model has a tidal cutoff.

shorter still, and there is a 'runaway' evolution towards high central densities and temperatures (Fig. 18.1).

The phenomena we have been describing are referred to collectively as 'core collapse'. Its study dominated much theoretical work on the evolution of dense stellar systems in the 1970s. The time it takes to reach completion is determined mainly by the early phase, in which the central parts are not yet thermally isolated from the outer parts of the system, and when the central relaxation time scale is longest. In fact, detailed numerical experiments of various kinds, using Plummer's model for initial conditions, and equal masses, show that the time taken is about $15t_{rh}$, where t_{rh} is the initial half-mass relaxation time (Chapter 14). For a system with a distribution of individual stellar masses, the time to completion of core collapse is determined by the time scale of mass segregation, and we have already seen that this is smaller than the relaxation time scale of an equivalent equal-mass system. Detailed experiments confirm this. Consider, for example, a system in which the ratio of the largest to the smallest mass is 15:0.4 (values which do not look so strange to an astrophysicist who interprets the unit of mass as the mass of the Sun), with a distribution of masses $f(m) \propto m^{-2.5}$ in this range. Then the time to the end of core collapse is as short as about $0.9t_{rh}$ (Chernoff & Weinberg 1990). The time scale of core collapse is also altered if the system is rotating (Lagoute & Longaretti 1996, Einsel & Spurzem 1999, Kim *et al.* 2002).

Self-similar core collapse with equal masses

So far, our discussion of core collapse has been short of quantitative detail on anything except time scales. We have arrived at a broad-brush picture in which the core of the cluster becomes smaller, hotter and denser. In addition, when a mixture of stellar masses is present, the core is increasingly dominated by the heaviest species. But it is not yet at all clear what this core will look like. In the remainder of this chapter we shall try to flesh out the portrait. We concentrate on systems in which all stars have the same mass, and analyse the late evolution, i.e. when the core is no longer in thermal contact with the outer parts of the system. The approach we adopt was pioneered by Lynden-Bell & Eggleton (1980).

It is clear that the central parts of the collapsing core must be nearly isothermal. If not, relaxation would quickly thermalise the largest region around the centre which is gravothermally stable. Therefore the core can be nearly characterised, like an isothermal model, by two parameters, e.g. the central density ρ_c and the central one-dimensional velocity dispersion σ_c (cf. Chapter 17).

We now make the assumption that the structure around the core, while not being precisely isothermal, is nevertheless still determined by ρ_c and σ_c. In other words, at each time during the collapse of the core the structure may be obtained from that at any other time by a simple rescaling. Often such evolution is referred to as 'homologous' or 'self-similar'. This apparently drastic assumption has been of great benefit in stellar dynamics. Indeed, we shall see in Chapter 33, where we summarise the whole of the evolution of a star cluster, that much of it can be described by patching together various kinds of self-similar evolution.

There are other areas of applied mathematics where such behaviour is common. In particular, in problems of 'finite-time blow-up', i.e. solutions of partial differential equations which tend to infinity at some point in space as t approaches some finite value, it is often found that the form of solution at late times, and in the vicinity of the singularity, is self-similar. A familiar example is the 'backward heat equation' $\partial u/\partial t = -\partial^2 u/\partial x^2$, which has the self-similar solution

$$u = \frac{1}{\sqrt{-4\pi t}} \exp\left(\frac{x^2}{4t}\right)$$

for $t < 0$. It is perhaps not surprising that such behaviour should arise also in core collapse in stellar dynamics. This is governed, as we saw in Chapter 17, by a property of stellar systems analogous to a negative heat capacity. If we were to derive the standard heat equation for such a material, we would obtain the backward heat equation.

For the star cluster problem, derivation of the self-similar solution requires a modicum of numerical integration, and our treatment will confine itself to some basic observations and a brief description of the results. In the first place let us note that the time scale on which core parameters such as ρ_c vary is t_{rc}, the central relaxation time. In self-similar evolution it follows that $t_{rc}\dot{\rho}_c/\rho_c$ is constant, and the same is true for other core parameters. It follows that these parameters will vary with each other as power laws (the powers being determined by the various constants) and, if we ignore variations of the Coulomb logarithm, that t_{rc} is a linear function of time.

Let us suppose, then, that

$$\rho_c \propto r_c^{-\alpha},\tag{18.1}$$

where α is some constant. Since we are supposing that the evolution is self-similar, the density profile at any time (i.e. the function $\rho(r, t)$ at some t) may be obtained from the profile at any other time by scaling. In other words,

$$\rho(r, t_1) = \frac{\rho_c(t_1)}{\rho_c(t_0)} \rho\left(\frac{r_c(t_0)}{r_c(t_1)} r, t_0\right).\tag{18.2}$$

Now at large radii, where the relaxation time scale is huge compared with that in the core (or even with the evolution time scale), the density profile must be almost stationary. By substituting (18.1) into Eq. (18.2) it follows that, at large radii,

$$\rho(r) \propto r^{-\alpha}.\tag{18.3}$$

This is the density profile that must be left behind by the outlying material of the contracting core.

Now we are in a position to say something about α on physical grounds. From Eq. (18.3) it follows that the mass at large radii varies as $r^{3-\alpha}$, and so the gravitational acceleration F varies as $r^{1-\alpha}$. By the condition of hydrostatic balance, i.e.

$$\frac{\partial}{\partial r}(\rho\sigma^2) = \rho F,$$

it is easy to deduce that the mean-square velocity must vary as $r^{2-\alpha}$. Since the collapse of the core is driven by a negative temperature gradient, it follows that $\alpha > 2$.

Now the velocity dispersion in the core is related to ρ_c and r_c by the definition of the core radius (Eq. (8.9)), and so it follows that

$$\sigma_c^2 \propto r_c^{2-\alpha}.\tag{18.4}$$

Also, the energy contained within the core is of order $\rho_c r_c^3 \sigma_c^2$, and by Eqs. (18.1) and (18.4) we see that this varies as $r_c^{5-2\alpha}$. It is physically reasonable to suppose that this must remain finite as the core radius contracts, and so $\alpha < 5/2$.

It has been noted (Lancellotti & Kiessling 2001) that such a solution would not be a self-similar solution of the full Fokker–Planck equation. But this requirement is too strong, for it requires that the orbital motions, as well as the collapse of the density profile, take place on the same time scale. Here we are studying collapse, on a relaxation time scale, of a system where orbital motions (on the crossing time scale) are much faster. Then it is appropriate to solve only the orbit-averaged Fokker–Planck equation, for which a wider class of self-similar solutions are possible.

Since collapse is driven by relaxation, we may estimate that the time to collapse, τ, is proportional to the relaxation time, i.e. $\tau \propto \sigma_c^3/\rho_c$ (Eq. (14.12), except that we ignore the Coulomb logarithm). By Eqs. (18.1) and (18.4) we deduce that $r_c \propto \tau^{2/(6-\alpha)}$. For our range of estimates of α the power here lies between 0.5 and about 0.57.

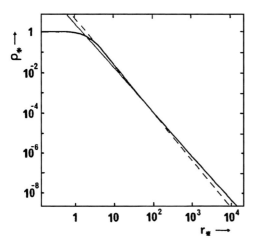

Fig. 18.2. The self-similar model of Lynden-Bell & Eggleton (1980). The dashed and solid straight lines correspond to the asymptotic relation $\rho \propto r^{-2.4}$ and the asymptotic profile of the collapsing model, $\rho \propto r^{-\alpha}$. The variables plotted are the scaled radius and density, r/r_c and ρ/ρ_c, respectively.

It does not appear to be easy to sharpen the possible range of values of α further without numerical methods. Indeed, more detailed models show that $\alpha \simeq 2.2$ or 2.3, and Fig. 18.2 shows the numerically computed density profile for a self-similar gaseous model. For the corresponding Fokker–Planck model the e-folding time scale for the central density is approximately $280t_{rc}$, which is not so far from the estimate of $709t_{rc}$ obtained previously on the basis of physical arguments. It should be pointed out, though, that this homologous phase of core collapse is a good approximation only very late on. It is preceded by a much longer phase where the collapse time scale is a smaller multiple of the central relaxation time.

Finally, it should be stressed that this process of core collapse is not quite the catastrophic process which the words tend to convey. This is no headlong rush like the collapse to a black hole. It takes place on a time scale which is very long even relative to the central relaxation time. It also involves a rapidly decreasing fraction of the total mass of a cluster. But it does imply that the central density reaches arbitrarily high values within a finite time. For many Galactic globular clusters it is a time considerably smaller than their age. Even at the present day, a couple of clusters approach this endpoint of core collapse every billion years in our galaxy (Hut & Djorgovski 1992). It is accompanied by a marked increase in the rate at which stars escape (cf. Fig. 5.2), and in the rate at which interesting encounters take place (Chapter 31). What happens next is the point at which Chapter 27 takes up the story.

Problems

(1)   Show from the isotropic Fokker–Planck equation (Eq. (14.18)) that the rate of change of f at the bottom of the potential well (energy $\phi(0)$) is given by

$$\dot{f} = 16\pi^2 G^2 m^2 \ln \Lambda \left(f' \int_{\phi(0)}^{\infty} f(E)dE + f^2 \right),$$

where f and f' are evaluated at $\phi(0)$, except in the integral. Hence show that the time scale of the initial collapse of a Plummer model (Table 8.1) is of order $0.88 t_{rh}$.

(2) Globular star clusters in orbit around our Galaxy contain only older stars, with an age of order ten billion years. This implies that nearly all single objects in the cluster are less massive than the Sun, with the exception of neutron stars and black holes. During intermediate stages of core collapse, the lighter stars tend to leave the core, through mass segregation, with the result that the typical mass of stars in the core will lie in a relatively narrow range, somewhere between 0.5 and $0.8 M_\odot$, with neutron stars with a mass around $1.4 M_\odot$ only slightly heavier. In contrast, black holes can easily have masses of $10 M_\odot$ or more. Let us investigate what happens with a population of black holes, remnants of the heaviest stars that burned up early on in the history of a globular cluster.

(a) Argue that the black holes will form a 'core within the core' of the globular cluster, on a time scale of a half-mass relaxation time or shorter, and show that the radius of that inner core is at least three times smaller than that of the core of the cluster. (Assume that the total number of black holes is small, so that the other stars are hardly affected by the presence of the black holes, initially.)

(b) Argue that the central density of the black holes will increase relative to the central density of the other stars, and make an estimate of the number of stars left in the core by the time the densities between holes and stars become comparable.

(c) From this point on, the Spitzer mass segregation instability will set in, if the number of black holes is sufficiently large. Argue that, subsequently, the black holes will undergo core collapse on a shorter time scale than the normal stars in the core. In particular, show that the ratio of time scales is of order $\mu^{5/2}$, where μ is the mass ratio between a black hole and a typical single star.

(This problem continues in Problem 23.3.)

(3) Using the estimate in Chapter 13 for the e-folding time for the growth of errors, show that the number of decimal places required for an accurate numerical integration to core collapse is roughly proportional to N.

Part VI

Gravitational Scattering: $N = 3$

Core collapse leads to high stellar densities, where interactions may involve more than just two stars at a time. The chapters in this section are therefore devoted to three-body interactions, especially interactions between a binary star and a single star. One of our aims in these chapters is to show that important aspects of the three-body problem can be understood from various points of view, even though the problem itself lacks a general mathematical solution.

Chapter 19 takes a phenomenological approach, applying notions of equipartition and energy conservation. This already classifies encounters according to whether the binary is *hard* or *soft*. In some interactions with hard binaries the result (temporarily) is like a miniature star cluster of three stars, and our previous knowledge of the behaviour of star clusters can suggest how this evolves.

Chapter 20 takes an informal mathematical view of the same phenomena. We see that the breakup of triple star clusters exhibits a sensitive dependence on initial conditions, partly justifying a statistical treatment. One of the standard examples in which this is most readily understood is *Sitnikov's problem*, which we use to introduce the Smale horseshoe. Finally we prove informally a theorem which shows that permanent capture into a triple configuration is (practically) impossible, and end with some recent surprising discoveries about permanently bound triple systems.

Chapter 21 takes a course half way between the previous ones, exploiting a mixture of approximate analytical tools and physical arguments to develop theoretical results on the outcome of three-body interactions. This is expressed in the language of *cross sections*, in analogy with similar problems in atomic scattering theory. Here the analogue of Coulomb repulsion is *gravitational focusing*. We estimate the rate at which binaries form within a million-body system. We treat in some detail the case of *adiabatic* encounters between a hard binary and a single star, when the single star remains 'outside' the binary and moves on a time scale long compared with the period of the binary. In the opposite extreme, we analyse cases in which all three stars come within a comparable small distance from each other, which produces results akin to those of 'threshold' scattering in the atomic context.

Far as these analytical techniques go, it is hard to make them completely quantitative and sufficiently accurate, which may be accomplished by computer simulation (Chapter 22). We describe some of the issues involved in integrating few-body systems efficiently (i.e. accurately and quickly), and the manner in which the calculations may be organised and automated.

Chapter 23 takes the first steps in applying these results to the evolution of the million-body problem. The cross sections are turned into quantities representing the rate of occurrence of three-body interactions, and the mean rate of change of energy of a binary in three-body interactions. One effect of these interactions is that binaries are driven from the dense central core of an N-body system into the surrounding halo. The most important results concern the rate at which energy is imparted to the single stars, as it is this process which halts core collapse and allows the system to evolve after that.

19

Thought Experiments

As has been seen in Chapter 18, two-body relaxation predicts its own downfall. It leads to the collapse of the core and, at the level of simplified models, infinite central density. Clearly, some new dynamical processes, beyond two-body encounters, must come into play. The very high density is the clue, for it suggests that a third body may, with increasing probability, intervene in the two-body encounters which mediate relaxation. In Chapter 27 it will be seen that three-body encounters do indeed act on a sufficiently short time scale, late in core collapse, to have a decisive influence on events. As we note there, this is not the only mechanism that can work, but we concentrate on it for the time being.

The mechanism is a two-stage one, both stages involving three-body encounters.[1] In the first stage, which we consider in Chapter 21, a three-body interaction leads to the formation of a *binary star* and a single third body (which acts as a kind of catalyst). In the second stage, this binary interacts with other single stars (again in three-body reactions). In this chapter we shall study three-body encounters in isolation, in order to uncover those properties which allow them to play their crucial role in rescuing the cluster from collapse. Clusters get into this difficulty because of their negative heat capacity, and in fact it is the negative heat capacity of binaries which comes to their rescue.

---

[1] We have in mind systems which initially have no binaries. If 'primordial binaries' (Part VII) are present, the first stage is not needed.

Energetics of three-body scattering

The energy of a binary star is

$$E_b = \left(m_1 v_1^2 + m_2 v_2^2\right)/2 - Gm_1 m_2/r,$$

where m_i and v_i are, respectively, the mass and speed of the ith component ($i = 1, 2$), and r is their separation. It is more convenient to write this as

$$E_b = \varepsilon + (m_1 + m_2)V^2/2, \tag{19.1}$$

where V is the speed of the centre of mass of the binary and

$$\varepsilon = \frac{1}{2}\frac{m_1 m_2}{m_1 + m_2}v^2 - Gm_1 m_2/r; \tag{19.2}$$

here v is the relative speed of the components. The 'binding energy' of the binary is the positive quantity $-\varepsilon$; ε, which is the quantity we prefer to work with, is referred to as the 'internal energy' of the binary. The second term on the right of Eq. (19.1) can be thought of as the kinetic energy of the barycentre. It is often called the 'external energy' of the binary, though if the binary is immersed in an N-body system the potential due to the other members would also be included. Note that the kinetic energy of internal motion, when expressed in terms of the relative speed v, involves the *reduced mass* $m_1 m_2/(m_1 + m_2)$ of the two components.

Now we add a third body, of mass m_3, and at first we suppose that it is well separated from the binary. Then the energy of this triple system is obtained from E_b by adding the kinetic energy of m_3. The result takes its simplest form in the frame of the barycentre *of the triple system*, and is then

$$E_3 = \varepsilon + \frac{1}{2}\frac{(m_1 + m_2)m_3}{m_1 + m_2 + m_3}V^2, \tag{19.3}$$

where V is the speed of m_3 relative to the barycentre of the binary. The second term on the right is the combined kinetic energy of the third body and the barycentre of the binary.

Now we suppose that the third body approaches the binary, interacts with it briefly, and then departs once again to a great distance. (This is what physicists refer to as a 'scattering problem'.) Suppose that the energy of the binary and the relative speed at a time long after the encounter are, respectively, ε' and V'. Then it is convenient to gauge the effect of the interaction by the relative change in the internal energy of the binary, i.e. $\Delta = (\varepsilon' - \varepsilon)/\varepsilon$. If $\Delta > 0$ the internal energy of the binary has become more negative, i.e. it has become tighter, or more bound. By conservation of energy E_3 it follows that $V' > V$ in this case. It is this possibility which is all-important in the termination of core collapse, and so in this chapter we place considerable emphasis on deciding whether Δ is greater or less than zero on average.

There are three related qualitative arguments which suggest what the net effect of such encounters may be. One is based on the negative specific heat of self-gravitating systems (Chapter 5). We already saw in the last chapter that this property leads to increasing density in the most usual situation. More precisely, this is what happens to a core in thermal contact with a cooler halo, and we also saw that, if the surroundings are hot, the evolution could run in reverse. One might guess that the same is true of binaries, i.e. that binaries get tighter (more bound) as they react with other, slow moving stars in three-body interactions, and looser (i.e. less bound) if the stars they encounter move fast.

The second argument, based on the tendency to equipartition of energies, strengthens and clarifies these conclusions, and runs as follows. We begin by noticing that there are degrees of freedom associated with both terms on the right of Eq. (19.3). If the second term is much bigger than the first (i.e. we consider fast encounters, or binaries of low binding energy), the kinetic energy of relative motion will tend to diminish and the internal energy of the binary will tend to increase (i.e. it will tend to become unbound, with $\Delta < 0$). Similarly, a binary with high binding energy will tend to become even more bound, and $V' > V$ on average.

For the third argument we assume that the interaction can be broken up into an encounter between the third body and each of the binary components separately. If the binary is hard, its components move fast, and impart some of their energy to the incomer, rather like a stone deflected by a rotary lawnmower. In this case energy is extracted from the binary, which hardens. Similarly, if the incomer is moving fast, it transmits some of this energy to the binary components, and the energy of the binary increases.

The conclusions of all three arguments suggest that there is a watershed binding energy, near

$$|\varepsilon| \sim (m_1 + m_2)m_3 V^2/(m_1 + m_2 + m_3), \tag{19.4}$$

above which binaries tend to become more bound, and below which they tend to become less bound. Though we shall mention some refinements shortly, we note here that this kind of result is amply supported by numerical experiments (Chapter 23; see, for example, Fig. 23.1). The distinction is often expressed by saying that binaries with $|\varepsilon| < (m_1 + m_2)m_3 V^2/(m_1 + m_2 + m_3)$ are 'soft', which indicates their relative fragility to encounters, while those with $|\varepsilon| > (m_1 + m_2)m_3 V^2/(m_1 + m_2 + m_3)$ are called 'hard'. In practice the distinction is not quite that simple. The important point is to distinguish those binaries which harden on average from those which tend to soften. For equal masses, it turns out that our expression for the threshold energy has the correct form, and there is a numerical coefficient near unity. If the masses are equal and there is a distribution of velocities V, then Eq. (19.4) is correct if we add a different numerical coefficient and replace V^2 by its mean square value.

When the masses are different, rather less is known about the correct threshold energy. When m_3 is tiny, i.e. the incomer is a test particle, it is clear that the critical value is independent of m_3, unlike an estimate based on Eq. (19.4). For this case it has been shown that, if the two members of the binary have equal mass, then the critical value of V is of order the orbital speed of the binary components about each other (Hills 1990; see also Fullerton & Hills 1982). Sigurdsson & Phinney (1993) have given useful data for masses relevant to globular clusters.

Soft binaries

The equipartition argument is attractively simple, provided one comes to it with a sufficiently developed physical intuition. But it is not hard to confirm its prediction with somewhat more concrete considerations, as we now show. We first consider the case of a fast encounter, i.e. a very soft binary, in which the first term on the right of Eq. (19.3) is much smaller than the second. Then the encounter is nearly impulsive, the relative position of the two components of the binary hardly alters, and the main effect of the encounter is on their relative velocity $\mathbf{v}$. If this changes by an amount $\delta\mathbf{v}$, then Eq. (19.2) shows that ε changes by

$$\delta\varepsilon \simeq \frac{m_1 m_2}{m_1 + m_2}\left(\mathbf{v}\cdot\delta\mathbf{v} + \frac{1}{2}\delta\mathbf{v}^2\right). \tag{19.5}$$

In an impulsive encounter $\delta\mathbf{v}$ depends only on the position of the components of the binary and not on $\mathbf{v}$. Furthermore, in any reasonable statistical distribution of binaries, velocities $\pm\mathbf{v}$ will be present with equal probabilities, and so the mean value of $\mathbf{v}\cdot\delta\mathbf{v}$ will be zero. It follows that the mean value of $\delta\varepsilon$ will be positive, i.e. that soft binaries soften on average, as expected.

Qualitatively different encounters are possible. If m_3 comes sufficiently close to one component of the binary, then the velocity of that component may change by so much that its new velocity exceeds the velocity of escape from its companion. Such an encounter is referred to as an *ionisation*, by analogy with atomic scattering. In the usual case, when the binary survives, we may refer to the encounter as a 'fly-by' (which is a name with definite space-age connotations). There is a third possibility for these fast encounters, and it is most easily understood in the case of equal masses. In a very close *two*-body encounter between equal masses the two masses almost exactly exchange velocities, just as when a billiard ball hits another stationary one. If, therefore, in a fast encounter between m_3 and a binary, the intruder scores almost a direct hit on one component, that component will leave the binary with almost the speed of approach of m_3, and the latter takes its place as one component. Such an encounter is an example of an *exchange*.

As an aside, it is worth remarking that the theory of exchange encounters has a curious history. The earliest numerical examples (Fig. 19.1) were given by Becker (1920) (working at Glasgow University, incidentally), but there was a community of theorists who preferred to settle such questions by pure thought. One of them

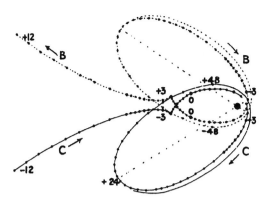

Fig. 19.1. One of the exchange orbits computed by Becker in 1920. The position of one particle is used as the origin of co-ordinates, and the initial conditions were chosen so that particles *B* and *C* are symmetric under time reversal and a reflection.

(Chazy) conjectured that exchange was impossible, and such was the suspicion of numerical methods among this community, and such was Chazy's influence, that his view held sway for over half a century. According to Marchal (1990), who gives a lively account of all this history, it was only in 1975 that the possibility of exchange was convincingly established. Meanwhile, astrophysicists, largely ignorant of these niceties, had already spent some years estimating the first cross sections for exchange, both numerically and by approximate analysis. As it happens these were published in the same year (Hills 1975, Heggie 1975).

Hard binaries

Now let us consider the opposite extreme, i.e. encounters in which the second term on the right of Eq. (19.3) is small. We shall also suppose that the encounter takes place in such a way that the minimum distance of the intruder (m_3) from the binary is not much greater than the size of the binary (technically, the length of its semi-major axis). Typical energies in the interaction are therefore of order ε. If a positive energy of order $|\varepsilon|$ is donated to the third body (or, more precisely, to the kinetic energy of its motion relative to the binary), then the third body will recede with much higher speed than its original speed of approach. (This follows from our assumption that $|\varepsilon|$ much exceeds the second term on the right of Eq. (19.3).) Also, by energy conservation, the binding energy of the binary has increased by an amount of order $|\varepsilon|$, i.e. it has become harder, as expected. What if, however, an energy of order $|\varepsilon|$ is *extracted* from the kinetic energy of the third body? Since $|\varepsilon|$ much exceeds the kinetic energy of the third body, the third body cannot escape, at least not right away. What happens is that it must return to the vicinity of the binary and there exchange another packet of energy. Again there are two possibilities: either m_3 can now escape or it cannot. It can be seen that, in general, the only way in which the encounter can end is with the eventual escape, after some large or small number of interactions, of one of the participating stars (Fig. 19.2). When it escapes, it will do so with a kinetic energy of

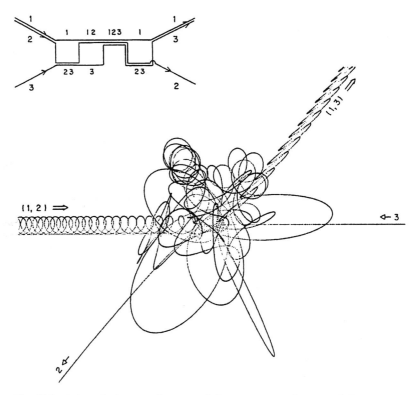

Fig. 19.2. A three-body scattering event. It is a resonant exchange, and the sequence of interactions is summarised in the pictogram at upper left. From Hut & Bahcall (1983); the authors had to look quite hard to find an example as *simple* as this!

order ε, which much exceeds its initial kinetic energy, and so again the net effect of the interaction will have been to harden the binary.

From the above discussion it can be seen that the encounter may be over quite promptly. If, on the other hand, there is a complicated interplay of the three participants, then the encounter is described as a 'resonance'. The name again comes from atomic or nuclear scattering, where it refers to the unusually large probability of interactions which can occur if the participants can form a temporary bound state. In the gravitational case what we see is a temporary, bound triple system. As we shall see in the next chapter, such a configuration cannot last for ever, but while it lasts it behaves very much like a star cluster with $N = 3$. In particular, the motion exhibits sensitive dependence on initial conditions (Chapter 13), which suggests that the behaviour of an ensemble of such systems may exhibit some statistical features. Indeed, the distribution of lifetimes of these systems is nearly exponential (Agekyan *et al.* 1983), like the random decay of an atomic nucleus. It turns out that this statistical approach is a fruitful way of attempting a theoretical prediction of the distribution of the energy of the eventual escaper (Heggie 1975, Nash & Monaghan 1978). More obviously, it suggests that, in the case of equal masses, the probability that any of the three participants is the escaper is $1/3$; thus the probability that the resonance leads to exchange should be $2/3$. These resonances in which all three particles interact closely are therefore sometimes called 'democratic', even in the case of unequal masses.

There is another interesting result which follows from thinking of a resonance as a small star cluster. In the case of unequal masses our knowledge of the escape rate in N-body systems (Chapter 16) suggests that the body of lowest mass is most likely to escape. It follows that a resonant encounter of a binary with a star whose mass considerably exceeds that of one of the components is liable to lead to exchange of that component. In other words, such encounters tend to increase the mass of the components of a hard binary.

Finally we must turn briefly to the case in which the distance of closest approach of the third body, denoted by r_p, remains considerably greater than a, the semi-major axis of the binary. Now the energy exchanged between the two terms in Eq. (19.3) may be quite small. In fact, it may be exponentially small, so that $-\ln|\Delta|$ varies roughly as a positive power of r_p/a. The reason for this is that, unless the masses are very disparate, the time scale on which the third binary is within a distance of a few r_p of the binary much exceeds the period of the binary. Then the perturbation by the intruder is *adiabatic*, and an exponentially small net perturbation is to be expected (Chapter 21). Still, the encounter may result in a fly-by or in formation of a resonance. Now, however, the third body has too much angular momentum relative to the barycentre of the binary for a close interaction, and the resonance is very unlikely to lead to exchange. Such resonances are referred to as 'hierarchical'. They usually resolve, sooner or later, with the ejection of the original intruder.

When we contemplate Fig. 19.2 it is not hard to see why the three-body problem is not analytically soluble in general. Yet it has simple analytical properties which can be very useful. We have already used energy conservation effectively in the above discussion, and another valuable property is the invariance of the equations of motion under time-reversal. For example, time-reversal of an ionisation encounter yields an encounter in which three single stars interact to yield one binary and a single star which carries off the excess energy. The same argument applied statistically also tells us how the rate at which binary stars are formed is related to the rate at which they are destroyed. The general three-body problem may well be insoluble from any strict point of view, but there is a large amount that may be understood from quite simple considerations. And our study of the three-body problem in this book has just begun.

Problems

(1) Answer the following riddle: How do you make a binary star with a grain of sand? What this means is the following. Two stars approach on a hyperbolic relative orbit, and your task is to throw in a grain of sand so as to extract enough of their kinetic energy to bind them into a binary. (This is classical mechanics of point masses.)

(2)   For an N-body system of energy E in virial equilibrium, show that the
       threshold energy between hard and soft binaries is of order E/N.

(3)   Suppose single stars of mass m are distributed uniformly in space with space
       density n and have a Maxwellian distribution of velocities. Assuming that the
       two stars are uncorrelated, show that the number density of very soft binaries
       of energy ε is

$$n_b \simeq \frac{\pi^{3/2}G^3m^6n^2}{4(kT)^{3/2}}|\varepsilon|^{-5/2}.$$

(4)   In quantum statistical mechanics, the equilibrium distribution of energies of
       hydrogen atoms is given by

$$f(E) = w(E)\rho(E)\exp(-E/kT),$$

where the first factor on the right is the statistical weight (multiplicity) of one
energy level, the second counts the density of levels per unit energy, and the
third is the usual Boltzmann factor. By applying the correspondence principle
to high-energy states, obtain the corresponding result for the distribution of
the binding energy of binaries in statistical equilibrium, i.e.

$$f(\varepsilon) \propto |\varepsilon|^{-5/2}\exp(-\varepsilon/kT).$$

Mathematical Three-Body Scattering

While the last chapter roughed out a picture of what happens in an interaction between a binary star and a third body, there are three very different ways in which the picture can be sharpened. One is to develop approximate analytical results on the outcome of an encounter, and that is successful in various limiting cases, e.g. very distant encounters, very hard binaries, and so on. This is the approach of Chapter 21. To cover the middle ground between these extremes, there is no substitute for computational studies (Chapter 22). In the present chapter, however, we push the analytical methods in the opposite direction, and examine minute corners of parameter space which may be of no conceivable value in applications. The merit of this approach is that rigorous statements become possible, at least in expert hands, and the resulting ideas help to develop our intuition of what can happen in more realistic situations. Our approach is quite informal, and places emphasis on the ideas behind the proofs, without *any* technical details.

Fractals and chaos

We first turn to resonances, those long-lived but temporarily bound triple systems that often arise in scattering events. Our first aim here is to discuss one situation which makes it particularly clear why the outcome of a resonance depends sensitively on the initial data. We can then argue, at least physically, that the system forgets details of the initial conditions of its formation, and that the breakup is determined (in a statistical sense) only by the quantities which are preserved in the evolution, i.e. energy and angular momentum.

It is not hard to see how sensitive dependence can arise. Consider first a resonant encounter in which the eventual escaper only just escapes. Now let us vary one of the initial conditions, which we shall call x, and suppose the condition that the escape occurs is that $x > x_*$, where x_* is some critical value. If x is just below x_*, the would-be escaper moves to an enormous distance from the binary before returning, and then what happens depends on the phase of the binary at the moment when the would-be escaper returns to its vicinity. But as $x \rightarrow x_*$, the phase of the binary changes by 2π infinitely often, and in each corresponding interval of x (i.e. corresponding to a given interval of phase of the binary, modulo 2π) virtually the same sequence of eventual outcomes will be observed (Hut 1983b). Furthermore, if we look at one such sequence, we may be able to find another critical value of x, say x_{**}, where another near-escape occurs. In its vicinity there will be a similar accumulation of patterns of outcome, and furthermore this accumulation will occur infinitely often as $x \rightarrow x_*$. Evidently this complicated dependence on the initial parameter x may be visible at arbitrarily many levels of refinement, as in a fractal pattern. Contemplation of this intricate structure gives a vivid impression of how sensitively the outcome of resonant three-body encounters depends on initial conditions (Fig. 20.1).

The foregoing discussion refers, as always in this chapter, to three-body inter-actions in isolation. In the context of the million-body problem, however, distant excursions of one member of the triple would be disturbed by passing stars. The distant third body makes a very soft binary with the centre of mass of the other two stars, and this would tend to be disrupted, by analogy with 'pressure ionisation' in gases.

In order to lend some precision to these ideas we shall consider one of those very simple but amazingly rich problems with which dynamical astronomy abounds. It is a special case of the restricted problem of three bodies (i.e. the problem in which one of the stars is massless), and is usually called 'Sitnikov's problem' (see also Chapters 4 and 13). The two massive stars have equal masses m, and orbit in the x, y plane (Fig. 20.2) with position vectors $\mathbf{r}$ and $-\mathbf{r}$, where $\mathbf{r}$ is given by Eq. (4.3). The third body moves on the z-axis, and the symmetry of the positions of the massive stars ensures that it stays there, though the problem is unstable to off-axis displacements. We shall use units in which the period of the binary is 2π. Our treatment follows Moser (1973) but omits many details.

Now, following Poincaré, we construct a *surface of section* for this problem, which may be done as follows. Each time the third body crosses the plane of the binary ($z = 0$) we note the time t and its speed $v = |\dot{z}|$. Next, using v and t as radial and angular polar coordinates, respectively, we plot the corresponding point in a plane.[1]

---

[1] Note that we seem to have lost information because values of t differing by a multiple of 2π occupy the same place, but in fact this does not matter: if two test particles arrive at $z = 0$ with the same value $\dot{z}$ but at times differing by a multiple of 2π, they follow the same subsequent motion with a constant time delay, because the perturbing effect of the binary is 2π-periodic. For a similar reason it does not matter that we have dropped the sign of $\dot{z}$ by using v instead.

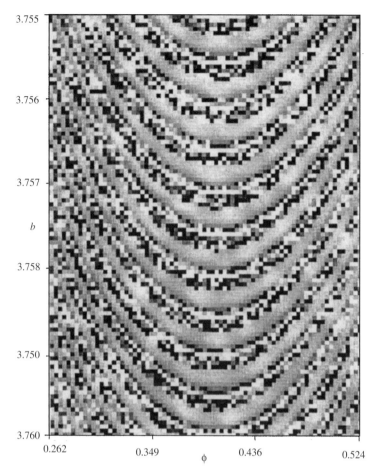

Fig. 20.1. Angle of deflection of the third body in a set of scattering events. The system is coplanar, the masses are equal, the initial orbit of the binary is circular and the initial speed of the third body is fixed. The initial conditions are then determined by the two parameters b (which is the closest distance to which the third body would approach the centre of the binary if it were undeflected) and ϕ, which is the phase of the binary. For each pair of values the angle of deflection of the third body is characterised by the shade of grey. There is one episode in which the third body is almost ejected to infinity, and the 'parallel' stripes arise because the subsequent outcome is similar for motions in which the third body arrives back at the binary after times which differ by a multiple of the binary period. (From Boyd & McMillan 1992, copyright 1992 by the American Physical Society.)

Thus any point on this diagram specifies the time t at which the third body crosses the plane $z = 0$, and its speed, for some solution of the problem. Following this solution, we wait until it crosses the same plane the next time, again measure t and v, and plot the corresponding point. In this way we have constructed a *map* of the plane, called the *Poincaré map*, which takes the initial point to the new point. We shall call this map ϕ, and say that it takes a point (v_0, t_0) to the point $(v_1, t_1) = \phi(v_0, t_0)$. Actually

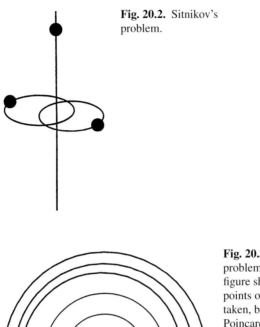

Fig. 20.2. Sitnikov's problem.

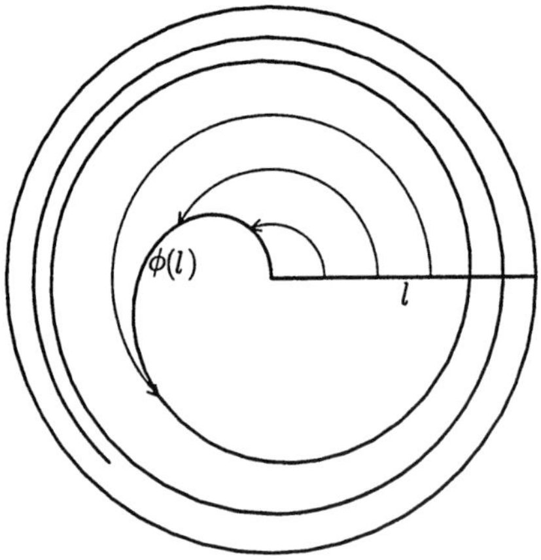

Fig. 20.3. Sitnikov's problem for $e = 0$. The figure shows how several points on a radial line l are taken, by the action of the Poincaré map ϕ (thin arcs with arrowheads) to corresponding points on a spiral curve, which is $\phi(l)$.

it is defined only on a subset of the plane, consisting of those points which lead to motions which do eventually recross the plane $z = 0$.

If $e = 0$ it is easy to visualise this map. In this case the two massive bodies stay at a constant distance from the z-axis and the force experienced by m_3 is time-independent, even though the massive stars are in motion. Hence it is clear that $v_1 = v_0$. Also, the larger the value of v_0 (below the escape speed) the longer the time $t_1 - t_0$ until m_3 returns. The result of these two facts is that ϕ rotates points along circles, the angle of rotation being larger for larger circles (Fig. 20.3). In fact in this case ϕ is an example of a 'twist' map, as the figure suggests. It can be used to illustrate a number of concepts, such as 'phase mixing', and the distinction between 'isolating' and 'non-isolating' integrals, which were introduced in Chapter 10.

Now we go on to consider the case in which e is non-zero but small. In this case the force experienced at a fixed point on the z-axis is time-dependent. Therefore the energy of the massless particle is no longer constant, and it is rather clear that points corresponding to a single orbit no longer lie on a circle in the surface of section.

One might think they simply lie on a slightly distorted curve, especially if e is small. Indeed this is the kind of result that can often be established by using the famous KAM Theorem of Hamiltonian dynamics. In the present case, however, it would only be applicable usefully to a proportion of initial conditions in the vicinity of the origin, i.e. to orbits of low energy. For other orbits the behaviour of the points on the surface of section can be much more complicated. In fact it can be shown for certain orbits that the points sit on a *fractal* set (Box 20.1). Not surprisingly, the theory turns on the neighbourhood of a point (P in Fig. 20.4) where an escaper just escapes.

Box 20.1. Sitnikov's problem in general

We outline some theory of the Poincaré map for Sitnikov's problem when $e > 0$. Because the potential is time-dependent, the escape speed from the plane $z = 0$ depends on the time at which a particle crosses this plane. It follows that the set of points on which ϕ can be defined is not a circular disk (as in the case $e = 0$), and we shall denote it by D_0 (Fig. 20.4). Let its image under ϕ be denoted by $D_1 = \phi(D_0)$. Points in D_1 which are outside D_0 give motions which escape, never returning to the plane $z = 0$.

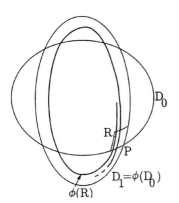

Fig. 20.4. The domain of definition of the Poincaré map in the eccentric Sitnikov problem, and its image.

Now draw a little curvilinear rectangle R in the corner beside P, where the boundaries of D_0 and D_1 intersect, and consider what happens to R under ϕ. As in the case $e = 0$, the image of R spirals infinitely many times just inside the boundary of D_1. Therefore R itself contains infinitely many pieces of $\phi(R)$. Different pieces correspond to different times at which m_3 returns to the plane $z = 0$; for neighbouring pieces the times differ by nearly one revolution of the binary.

Now a wonderful thing becomes apparent if we apply ϕ to R, with its pattern of stripes, once again. Then it is apparent that, not only does $\phi(R)$ cross R in these stripes, but *within each stripe* is a complete replica of the infinite pattern of stripes. If we choose a point within a tiny stripe inside a large stripe, we can then tell how many binary periods elapsed between each pair of *three* successive crossings of the plane $z = 0$ (Fig. 20.5). Once this point is understood it is also clear that, if these numbers of periods are large enough, they can be chosen arbitrarily, simply by choosing an arbitrary stripe within an arbitrary stripe. And we need not stop at three successive crossings. We can repeat the process of applying ϕ indefinitely, obtaining a *fractal* pattern of stripes within stripes within stripes...*ad infinitum*. And we can choose initial conditions leading to an arbitrary sequence of times between successive crossings (as measured, say, in the nearest whole number of binary periods).

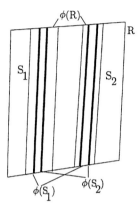

Fig. 20.5. A piece of the surface of section and parts of its image under ϕ and ϕ^2 (i.e. ϕ applied twice). Only two of the infinitely many stripes are shown.

Now it is also possible to extend the argument into the past. This time, however, we get a fractal set of stripes which cross the first set transversely. There is a very important set of points, called I, which are obtained by taking the points which lie on both sets of stripes at all levels of the fractal structure. These correspond to orbits which never escape, no matter how far forward or back in time we go. Now we can construct orbits which exhibit any sequence of times between successive crossings, not only into the future, but into the past as well. Examples are given in the text.

Now consider two points which lie on the same component of the fractal of horizontal stripes, and on the same component of $\phi^n(R)$, for some largest positive n. These points both lie on I, and are very close together. On the other hand, they lie on different components of $\phi^{n+1}(R)$, by assumption. Therefore after n crossings of the plane $z = 0$ the two orbits differ greatly in the number of orbital periods before the next crossing. This illustrates sensitive dependence on initial data.

This structure has some remarkable consequences. Thus: suppose we characterise an orbit by the numbers of binary periods which elapse between each successive crossing of the plane of the binary, i.e. by a sequence such as ... 2, 5, 314 159, 27 182 818, 5 .... Then for each such sequence[2] there is a corresponding orbit! This result is very powerful, and produces a whole zoo of fascinating orbits. You can construct periodic orbits using a constant sequence, capture or escape orbits by terminating the sequence with ∞ at one end or the other, and orbits which are captured and then escape again (by terminating it at *both* ends). You can even construct orbits which stay together for arbitrary long times and then go their own separate ways. This is what we mean by sensitive dependence on initial conditions. And similar examples can be used to

---

[2] For technical reasons the numbers must be chosen larger than some fixed number. This is understood in all such statements hereafter.

show that, when $e \neq 0$, not only is the energy not conserved, there is no smooth function which is; in other words, the problem is non-integrable.

Capture

At several points in the previous section we tacitly assumed something about the eventual fate of triple interactions, which is that they eventually result in escape of one of the particles. This is what is referred to as 'hyperbolic–elliptic' motion, as the escaper moves on a nearly hyperbolic orbit relative to the centre of mass of the binary, while the orbit of the binary is nearly elliptic. But there are other possibilities (see Alekseev 1981, or Arnold *et al.* 1997), and we even mentioned one: elliptic–parabolic motion, in which the escaper only just escapes. But why should escape, of whatever kind, be regarded as inevitable? Couldn't an approaching third body be permanently captured by a binary?

We already touched on this question in the last chapter with a physical argument. But it is not hard to present, at least in outline (Box 20.2), a mathematical proof which shows that, while permanent capture is not impossible, its probability is zero: among all initial conditions, the set leading to permanent capture 'has measure zero', provided that the interaction between the stars is non-dissipative.[3] The same is true of hyperbolic–parabolic motion: almost any change in initial conditions will tip the balance from parabolic escape to either hyperbolic escape or a return to the vicinity of the binary.

The content of the mathematical proof bears some resemblance to the fact that, if you keep filling a bottle, it eventually overflows. In both situations volumes and incompressibility lie at the heart of the argument. The other critical factor in the mathematical proof is the invariance of the equations of motion under time translation.

Invariance under time reversal does not have a role in the proof, but can be used to extend it. By reversing time, the corresponding result on capture shows that orbits which were bound[4] forever in the past, but eventually escape, are also rare. The only orbits which one expects to find are of three sorts: (i) triples which are permanently bound, (ii) triples which never become bound, and (iii) triples which become temporarily bound (perhaps more than once, perhaps many times).

Of these three types, the last two have become so familiar that many people must have concluded that nothing else can happen. But stable permanently bound triples do exist. Despite an earlier example investigated by Hénon (1976), one of the astonishing developments in this subject took place around the start of 2000, when Chenciner & Montgomery (2000) discovered, more or less by pure thought, a remarkable stable

---

[3] Capture even by a single star is quite possible if the interaction raises tides on either star; cf. Chapter 31.

[4] Remember that we are here defining 'bound' in a particular sense, based on the locality of the three bodies.

Box 20.2. **The improbability of permanent capture**

Following J.E. Littlewood (see Littlewood 1986), we shall say that the third body has been captured if, after some time, all three bodies remain forever within some sphere S (in space) centred on the barycentre of the three bodies. Now consider the phase space of the three-body problem, and let $x(t)$ denote an orbit in this space, so that x is really a vector of positions and momenta. Let V_0 be the set of initial conditions $x(0)$ such that all three bodies lie within S for all positive times $t \geq 0$ (Fig. 20.6). Next, let V_1 be the set of points $x(1)$ which these orbits, originating in V_0 at time $t = 0$, occupy at time $t = 1$. It follows from Liouville's Theorem that the volumes (or measures) of V_0 and V_1 are equal. It is also easy to see that V_1 lies within V_0: for any point P in V_1 lies on an orbit which started at a point in V_0, and on the resulting orbit all three bodies subsequently lie within S; therefore P could be used as an initial condition for an orbit lying entirely within S; therefore P lies in V_0. To summarise so far, what we have now established is that V_1 is contained within V_0 and has the same volume.

Now we concentrate on the set of points which are in V_0 but not in V_1. By what has just been proved this set has zero volume. Let us use one of these points as initial conditions. On this orbit, all three bodies lie within S for $t \geq 0$, since $x(0)$ lies within V_0. If, in addition, all three bodies lay within S during the interval from $t = -1$ to $t = 0$, then the point $x(-1)$ could be used as initial conditions for an orbit with the bodies inside S for $t \geq 0$; in other words, $x(-1)$ would lie in V_0, and so $x(0)$ would lie within V_1. This contradiction establishes that the points in V_0 which are not in V_1 include those orbits which are captured during the interval between $t = -1$ and $t = 0$. (There may be others, including orbits in which one of the bodies is temporarily ejected from S.) Therefore the measure of capture orbits is zero.

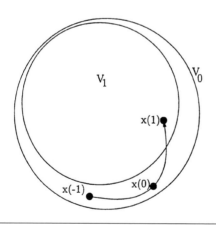

Fig. 20.6. V_0 is the set of initial conditions leading to capture for $t \geq 0$, and V_1 is where these points are at time $t = 1$. The curve is an impossible trajectory (see text).

periodic solution of the three-body problem with equal masses.[5] The three particles chase each other round an orbit resembling a figure 8. Even more wonderful stable periodic motions are possible (Fig. 20.7; see also Montgomery 2001, and other papers and web references therein).

[5] It had been found previously by Moore (1993) using a numerical technique, but few had paid much attention at that time.

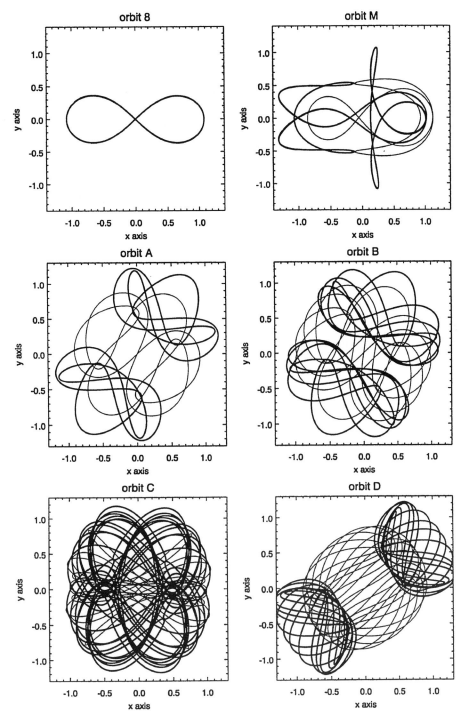

Fig. 20.7. Some periodic motions of the three-body problem (M. Ruffert). Top left is the orbit of Chenciner & Montgomery, with numerical initial conditions found by C. Simó. All three particles traverse the same orbit, and the motion is periodic in an inertial frame. The other orbits are periodic in a rotating frame. At top right, two bodies traverse the thicker curve, while one traverses the thinner curve. In the remaining four frames the three bodies traverse different curves. All motions are stable for equal masses, and have zero angular momentum.

Problems

(1)   In a hierarchical triple the third body orbits well 'outside' the inner binary. Suppose the inner binary has period 2π and that the orbit of the third body takes it close to the inner binary at times $t_0 < t_1 < t_2 < \ldots$. We suppose that the energy of the outer binary remains constant between close approaches, and that it changes at each close approach by an amount that depends periodically on the phase of the inner binary at that instant, i.e. that $E_{n+1} = E_n + \epsilon \cos t_n$, where E_n is the energy of the outer binary between times t_{n-1} and t_n. Show, in a suitable scaling, that we may take $t_{n+1} = t_n + (-E_{n+1})^{-3/2}$ approximately. Show that the map $(E_n, t_n) \to (E_{n+1}, t_{n+1})$ is area preserving. Write a computer program to iterate this map, given ϵ, t_0 and E_0, to find the time t_n at which the third body escapes, i.e. $E_{n+1} > 0$. (You may have to insert a line to stop the program if E becomes too small or t becomes too large.) By plotting the escape time against E_0 for fixed ϵ and t_0 at various resolutions, try to find fractal structure in the plot.

(2)   Parabolic escape is improbable, and so parabolic capture followed by parabolic escape seems doubly improbable. But is it possible?

(3)   'Oscillatory motion' as $t \to \infty$ is a class of motions in which, roughly speaking, the size of the system eventually exceeds any bound, but there are always times in the future at which the size of the system lies within some fixed bound. Using Sitnikov's example, make plausible the existence of a motion with this property both in the future and in the past.

(4)   In the collinear parabolic Sitnikov problem the two masses lie at $x_1(t) = -x_2(t) = (3t)^{2/3}/2$ in suitable units, and the equation of motion of the test particle is $\ddot{z} = -\dfrac{2z}{(z^2 + x_1^2)^{3/2}}$ (Chernin & Valtonen 1998). By introducing a new time variable $\tau = \ln t$ and scaling z by $t^{2/3}$, show that the equation of motion becomes that of a damped oscillator. Hence determine and interpret the asymptotic motion of the body as $t \to \infty$.

21

Analytical Approximations

The previous two chapters were intended to develop a qualitative understanding of the nature of the interactions between binary and single stars, with no more than order-of-magnitude estimates. The present chapter attempts to sharpen these ideas with some approximate quantitative results. We imagine that binaries and single stars are distributed throughout some region of space, such as a part of a star cluster, and we want to know how frequently three-body interactions are taking place.

Cross sections

What is important in applications (Chapters 23f) is the energetics of these interactions, and that is why such stress was laid on the distinction between soft and hard pairs in Chapter 19. In the present chapter this consideration implies that we may be interested in interactions with binaries of a given energy. Encounters with such binaries, however, are taking place all the time with stars which approach from random directions and random distances. Therefore, besides the energy of the binary, we usually do not know or care about the other properties of the participants, except for their statistical distribution. This is true of the approach path of the third body, and also usually it is true of the other five parameters (besides the energy) which determine the relative motion of the binary components.[1]

An analogy with atomic scattering may now be clear. There also we may be concerned with binaries (atoms) occupying a specific energy level and orbital angular

---

[1] One exception may be the eccentricity; an important special case is that of binaries in initially circular orbits.

momentum (analogous to the energy and eccentricity of a binary). In the atomic context interactions are characterised by the *cross section* of a specific type of interaction, and in stellar scattering theory we do the same.

Consider a third body, of mass m_3, approaching a binary with relative velocity $\mathbf{V}$. The initial orbit of the intruder is characterised by the point at which its orbit, had it been unaffected by the binary, would have intersected an imaginary plane Π passing through the barycentre of the binary and orthogonal to $\mathbf{V}$. Suppose now we wish to calculate the cross section for exchange reactions. We simply measure the area, in the plane Π, corresponding to all encounters which lead to the required outcome. Having done this for a given initial configuration of the binary, we now repeat the process for the appropriate distribution of its parameters (such as the orientation of its orbit). The resulting averaged area is the cross section for the required process.

In applications we next have to convert the cross section to a reaction rate, i.e. use expressions for the rate at which incomers strike the special region on the plane Π. In time δt this is just the number of stars in a volume $\sigma V \delta t$. The relative velocity $\mathbf{V}$ itself has some probability density $f(\mathbf{V})$, which can be computed from the distributions of velocity of binaries and single stars, and so the reaction rate is

$$R = n \int V \sigma f(\mathbf{V}) d^3 \mathbf{V}, \tag{21.1}$$

where n is the space density of single stars, and f is normalised with respect to unity. In order of magnitude work R is approximated by $n \sigma v$, where σ is the cross section for some typical velocity v – what is called literally an 'n-sigma-v' estimate (cf. Box 13.1). For the time being, however, we concentrate on σ.

As an illustration we shall estimate the cross section for *close* encounters, defined to be those in which the minimum distance of the intruder from the barycentre of the binary, R_p, is equal to a, the initial semi-major axis of the binary. We use a Keplerian approximation for the relative motion of the intruder and barycentre, and conservation of energy and angular momentum lead to

$$pV = R_p V_p \tag{21.2}$$

and

$$\frac{1}{2} V^2 = \frac{1}{2} V_p^2 - \frac{GM_{123}}{R_p}, \tag{21.3}$$

where V_p is the relative speed at closest approach and M_{123} is the total mass of all three stars. Setting $R_p = a$ and solving for p yields the following estimate for the cross section $\sigma = \pi p^2$:

$$\sigma = \pi a^2 \left(1 + \frac{2GM_{123}}{aV^2} \right).$$

This formula has a simple interpretation. The geometrical cross section of the relative orbit of the binary components is of order πa^2, but their gravitational action on the intruder causes the intruder to be drawn towards the binary, and this gives rise to the factor in brackets. It is often called the 'gravitational focusing' factor, where we are thinking of an imaginary 'beam' of intruders.[2] In the hard binary limit this term dominates, and

$$\sigma \simeq \frac{2\pi G M_{123} a}{V^2}. \tag{21.4}$$

As an application of these formulae, let us estimate the rate at which binary stars are formed in a stellar system, ignoring all numerical factors. If a new binary is to survive it must be hard, but very hard binaries can only be formed if three stars enter a very tiny volume, for which the probability is tiny. Therefore we assume that a new binary is only slightly hard. For its energy we take $|\varepsilon| \sim mv^2$ in the case of equal masses, where v is the root mean square speed (cf. Eq. (19.4)). This energy corresponds to a semi-major axis $a \sim Gm/v^2$. Using Eqs. (21.1) and (21.4), we see that the rate at which a given star encounters another at this distance is of order $G^2 m^2 n/v^3$, where n is the number of stars per unit volume. The probability that a third star is within this volume (of order a^3) at the same time is of order na^3, and so the rate of formation of triple systems involving a given star is of order $G^5 m^5 n^2/v^9$. Since there are n stars per unit volume, the rate is of order $G^5 m^5 n^3/v^9$ per unit volume. In fact much more elaborate calculations yield a value

$$\dot{N}_b = 0.75 \frac{G^5 m^5 n^3}{\sigma_v^9}, \tag{21.5}$$

where $v^2 = 3\sigma_v^2$ (Goodman & Hut 1993). (The subscript should help avoid confusion with a cross section.)

It has already been said that we can define cross sections for encounters of a given type. For example, if we are interested in the amount by which the binding energy of the binary changes we may proceed as follows. Let Δ be the relative change in the binding energy, and let $\sigma(\Delta)$ be the cross section of encounters in which the relative change exceeds Δ. In fact we already know (from a discussion in Chapter 19) that a substantial fraction of 'close' encounters changes the energy of the binary by an amount comparable with its original value, and so Eq. (21.4) is an order-of-magnitude estimate of $\sigma(1)$. Equivalent information on the behaviour of interactions is contained in the *differential cross section* $d\sigma/d\Delta$. This is especially useful for the interpretation of numerical data on cross sections (Chapter 22).

There are several ways in which the determination of cross sections may be approached analytically. In the first place the nature of the equations of three-body motion places some non-trivial constraints on the cross sections. We have in mind here the principle of *detailed balance*, which relates the cross sections of inverse processes, i.e. reactions which are related to each other by time-reversal. A simple physical way of understanding detailed balance is to imagine a stellar system of

---

[2] Fig. 14.2 gives a picture of what this means.

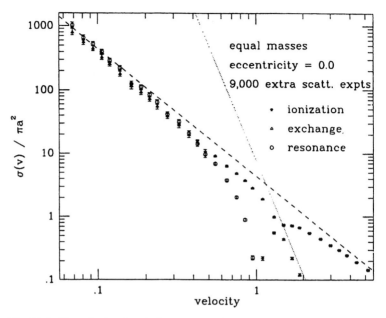

Fig. 21.1. Numerical and analytic estimates of cross sections for ionisation, exchange and resonance. The initial eccentricity of the binary is zero, and the relative velocity of the incoming star is given in units of the minimum speed required energetically to ionise the binary. Masses are equal. The dotted line gives the theoretical estimate for exchange in the soft limit ($v \gg 1$). The dashed line is the theoretical estimate for ionisation, which occurs only if $v \geq 1$. For equal masses it happens to coincide very nearly with Eq. (21.4), the cross section for 'close' encounters with a hard binary. This is why the numerical points follow this line approximately for $v \ll 1$. From Hut & Bahcall (1983).

single and binary stars in dynamic equilibrium, when the rates of inverse processes must balance exactly. This allows us to relate the cross sections of inverse processes; these cross sections do not depend on the assumption of dynamic equilibrium, and so these relations are valid more generally (Problem 2).

The second analytical approach to the determination of cross sections is to use approximate solutions of the equations of three-body motion. Different approximations are applicable in different regimes (Fig. 21.1). Some involve relatively routine calculations; high-speed encounters, for example, can be handled by an impulsive approximation. This is relevant to soft binaries, and these are energetically of little importance. We shall now consider two other kinds of analytical approximation which are less straightforward but rather more significant.

Adiabatic encounters

Let us consider first a 'wide' encounter with a hard binary, i.e. one in which R_p considerably exceeds a. Then the angular velocity of the third body about the binary,

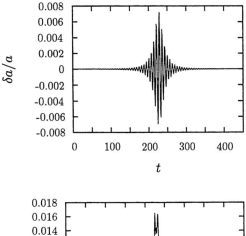

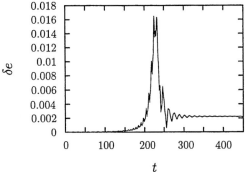

Fig. 21.2. Relative change in the semi-major axis a (top) and the eccentricity e in an adiabatic encounter, plotted against time. The binary is circular initially.

when at closest approach, is V_p/R_p. Now the mean angular velocity of the binary components (total mass M_{12}) is given by Kepler's Third Law in the form

$$n = \sqrt{GM_{12}/a^3}. \tag{21.6}$$

(Note the change of meaning of n.) By use of Eq. (21.3), it follows that the ratio of these frequencies is

$$\frac{nR_p}{V_p} = \left(\frac{R_p}{a}\right)^{3/2} \left(\frac{2M_{123}}{M_{12}} + \frac{R_p}{a}\frac{aV^2}{GM_{12}}\right)^{-1/2}. \tag{21.7}$$

We shall suppose that the masses are not very disparate. Since the binary is hard, $aV^2/(GM_{12}) \ll 1$. It follows that, if R_p considerably exceeds a, the time scale of the encounter will be long compared with the period of the binary. In this case the encounter is 'adiabatic'. It does not follow that the binary exhibits adiabatic invariance, with exponentially small changes as in systems with one degree of freedom, because it is a degenerate system with several (Arnold 1983, Goldstein 1980). For example, it is easily shown that, if the masses are comparable, the change in eccentricity is of order $(a/R_p)^{3/2}$, i.e. of order the ratio of time scales (Fig. 21.2), and not exponentially small. But it *is* known that the change in *energy* of the binary vanishes (as $R_p/a \to \infty$) faster than any power of this ratio (Heggie 1975).

It is not hard to give a calculation which shows (non-rigorously) that the change in energy in fact declines exponentially (Box 21.1). To lowest order the change in energy is given by an integral of rapidly oscillating terms with slowly varying coefficients. (These two features arise because of the two time scales in the problem.) Such integrals may be handled by the method of steepest descents, which involves deforming the contour of integration through a saddle point. It is an interesting feature of this particular problem that the saddle is located (in the complex plane) at the point where the third body would experience a collision with the centre of mass of the binary, another instance of the curious role of singularities in this subject. At this point the oscillatory terms of the integrand turn into exponentials, and these determine the dependence of the energy change (see Box 21.1, Eq. (21.14)). Not surprisingly, the exponent is essentially the ratio of the time scales in the problem: the short period of the binary, and the much longer time scale of the encounter. This is rather common behaviour for the effect of adiabatic perturbations.

Box 21.1. Theory of adiabatic encounters

Let ε be the energy of a binary, and let $\mathbf{r}$ denote the position of one component relative to the other. Thus

$$\varepsilon = \frac{m_1 m_2}{2M_{12}}\dot{\mathbf{r}}^2 - \frac{Gm_1 m_2}{r},$$

where $M_{12} = m_1 + m_2$. Also,

$$\dot{\varepsilon} = \frac{m_1 m_2}{M_{12}}\dot{\mathbf{r}}\cdot\mathbf{f}, \tag{21.8}$$

where $\mathbf{f}$ is the perturbative acceleration due to m_3, i.e. its contribution to the equation for $\ddot{\mathbf{r}}$. The exact expression for this is

$$\mathbf{f} = Gm_3\left(\frac{\mathbf{R} - \mu_2\mathbf{r}}{|\mathbf{R} - \mu_2\mathbf{r}|^3} - \frac{\mathbf{R} + \mu_1\mathbf{r}}{|\mathbf{R} + \mu_1\mathbf{r}|^3}\right),$$

where $\mu_i = m_i/M_{12}$ and $\mathbf{R}$ is the position of m_3 relative to the barycentre of the binary. For encounters which are not too close we approximate this at lowest (quadrupole) order in a multipole expansion, i.e. write

$$\mathbf{f} \simeq -\frac{Gm_3}{R^3}\left(\mathbf{r} - \frac{3\mathbf{r}\cdot\mathbf{R}}{R^2}\mathbf{R}\right). \tag{21.9}$$

In first order perturbation theory we substitute unperturbed motion for $\mathbf{r}$ and $\mathbf{R}$ in Eqs. (21.8) and (21.9). Terms involving $\mathbf{r}$ are periodic with period $2\pi/n$, and so the right-hand side of Eq. (21.8) may be written as a Fourier series (with $\mathbf{R}$-dependent coefficients). By averaging Eq. (21.8) at fixed $\mathbf{R}$, we see that this series has no constant term, and the harmonics of lowest order involve a factor proportional to $\exp(\pm int)$. (Harmonics of higher order oscillate more rapidly, and from what follows it would be seen that they would yield a negligible contribution to our result.) We use a parabolic approximation for the motion of the third body (i.e. $\mathbf{R}$). The solution is standard, if relatively unfamiliar:

$$R = q(1 + \sigma^2), \tag{21.10}$$

where $q\ (=R_p)$ is the distance of closest approach, and σ (not a cross section!) is related to the time via the equation

$$\sigma + \frac{1}{3}\sigma^3 = \sqrt{\frac{GM_{123}}{2q^3}}\,t. \tag{21.11}$$

Here we have taken $t = 0$ at the instant of closest approach of the third body, and $M_{123} = m_1 + m_2 + m_3$. (See Problem 4.2.)

The relative change in energy is given by integrating Eq. (21.8). This yields several terms, of which we take the integral

$$I = \int_{-\infty}^{\infty} R^{-3} \exp(int)dt \tag{21.12}$$

as a prototype. The above discussion of time scales implies that the exponential is rapidly oscillating compared with the remainder of the integrand. It is therefore suitable for evaluation by the method of steepest descents. In preparation for this we first transform from t to σ via Eq. (21.11), which, with the aid of Eqs. (21.3) and (21.10), yields

$$I = \frac{2}{V_p R_p^2} \int_{-\infty}^{\infty} \frac{\exp\left(\dfrac{2inR_p}{V_p}[\sigma + \sigma^3/3]\right)}{(1 + \sigma^2)^2}\,d\sigma. \tag{21.13}$$

The method proceeds by deforming the contour of integration from the real axis to one passing through a suitable saddle of the exponent. Differentiating the exponent, we find that the appropriate saddle is located at $\sigma = i$. The fact that this coincides with a singularity of the integrand complicates the asymptotic evaluation of the integral, but does not alter the exponential dependence of the result. Evaluating the exponent at the saddle point we find, in fact, that

$$I \propto \exp\left(-\frac{4}{3}\frac{nR_p}{V_p}\right), \tag{21.14}$$

though the 'proportionality' involves a further power-law dependence on the orbital parameters (see Problem 6). As we have seen, the exponent here has a simple interpretation in terms of the two time scales of the problem.

Finally, we express the result in terms of the impact parameter p and the relative speed at infinity, V. We also assume that the encounter is not too distant, compared with the hardness of the binary, and neglect the second term in the bracket in Eq. (21.7). We can connect with the parabolic approximation (for which p would be infinite) by equating the specific angular momentum (Eq. (21.2)), which yields $pV = \sqrt{2GM_{123}R_p}$, and so

$$I \propto \exp\left(-\frac{1}{3}\frac{(pV)^3}{(GM_{123}a)^{3/2}}\left(\frac{M_{12}}{M_{123}}\right)^{1/2}\right). \tag{21.15}$$

Now we show how to estimate the corresponding cross section. We use Eq. (21.15) of Box 21.1 as an estimate of the relative energy change Δ, though it is pointed out in the accompanying text that the coefficient of the exponential also depends on the orbital parameters. When we solve for p, however, this coefficient enters as a logarithm, and the resulting effect on the cross section is weak. Aside from this logarithm, then, we estimate that

$$\sigma = \pi p^2 \sim \frac{\pi G M_{123}^{4/3} a}{M_{12}^{1/3} V^2}(3 \ln |\Delta|)^{2/3}. \tag{21.16}$$

The differential cross section $d\sigma/d\Delta$ varies essentially as $1/|\Delta|$. Note the similarity to Eq. (21.4).

Close triple encounters

Now we consider the opposite extreme, i.e. encounters which can lead to *very* large (positive) values of Δ. The semi-major axis of the final binary, a_1, must be much smaller than the initial value a, and this requires the interaction of all three stars in a region of space whose size is of order a_1.

We first outline an argument which attempts to estimate the cross section for such processes, and then we discuss a refinement which introduces some interesting fresh links with the theory of the three-body problem and atomic scattering theory.

First we estimate the probability that the components of the binary lie within a distance $a_1 \ll a$ at the time of the encounter. The phase space for their relative motion has coordinates $\mathbf{r}$ and $\mathbf{v}$, and the volume of phase space accessible by a binary of energy $\varepsilon = -\dfrac{Gm_1m_2}{2a}$ is $\displaystyle\int \delta(\varepsilon - \dfrac{m_1m_2}{2M_{12}}\mathbf{v}^2 + \dfrac{Gm_1m_2}{r})d^3\mathbf{r}d^3\mathbf{v}$. After integration with respect to $\mathbf{v}$ we see that this is

$$4\pi \frac{G^{1/2}M_{12}^{3/2}}{m_1m_2} \int_{r<2a} \sqrt{2/r - 1/a}d^3\mathbf{r}.$$

The fraction of this volume for which $r < a_1 \ll a$ is easily seen to be of order $(a_1/a)^{5/2}$, and this is an estimate of the required probability. Note, however, that it is to be interpreted as applying to an ensemble of binaries. A single binary has a specific pericentre distance, and smaller values of r *never* occur.

If we assume that the binary is hard, we may easily estimate the cross section of this tiny target by replacing a by a_1 in Eq. (21.4). Multiplying by the foregoing estimate of the probability that the binary components are sufficiently close together, we find that the cross section for encounters leading to such a small value of the semi-major axis is of order

$$\sigma \sim \frac{GM_{123}a}{V^2}\left(\frac{a_1}{a}\right)^{7/2}, \tag{21.17}$$

i.e. $\sigma \sim GM_{123}a(1 + \Delta)^{-7/2}/V^2$. Hence the differential cross section varies as $d\sigma/d\Delta \sim GM_{123}a(1 + \Delta)^{-9/2}/V^2$.

Now for the complication. Our previous treatment is really too rough, because all three stars must arrive simultaneously within the small interaction region of size of order a_1, and as they approach they interact with each other in a way that cannot, as we have implicitly assumed, be broken down into two-body approximations. Another way of saying the same thing is that three-body gravitational focusing is different from the two-body kind that we introduced earlier in the chapter.

A better way to handle the simultaneous close approach of three bodies is to treat it as a perturbation from an exact triple collision. Solutions for a triple collision can be written down exactly, and then linearisation about one of these solutions can be adapted to provide the required information. In the context of atomic scattering theory, and threshold scattering in particular, this approach was initiated by G.H. Wannier (1953). It is equally successful in the gravitational problem under consideration here.

It would take too long to analyse three-body gravitational scattering in detail from this point of view, but we can illustrate the ideas involved by studying close *two*-body encounters. In particular, we rederive a consequence of Eq. (21.4) by looking at the relative motion of the third star and the barycentre of the binary as a perturbation of a two-body collision orbit (Box 21.2). The trick is to find the time at which the deviation from the collision orbit grows to become of the same order as the spatial size of the collision solution itself. At this time the collision solution breaks down as an approximation, and one can interpret the corresponding length scale as the distance of closest approach.

It turns out that a similar calculation is possible for three-body motions (Simonovic & Grujic 1987, Heggie & Sweatman 1991), and it leads to a similar relation which shows how the cross section for a close *three*-body interaction within a small distance r depends on r. The calculation is complicated because there are several configurations which yield exact collision orbits, and because the resulting eigenvalue problems are somewhat more involved. The kind of result obtained may be illustrated by the case of equal masses, which is quite easily worked out. In this case it turns out that the cross section given in Eq. (21.17) does not have quite the correct dependence on a_1: the exponent should be $2 + \sqrt{13}/2, \simeq 3.802\ldots$.

While this discussion dealt with the case in which a binary is greatly hardened by an encounter ($\Delta \gg 1$), similar considerations occur in encounters in which a binary is almost destroyed. In this case all three participants must emerge from the encounter on orbits close to an exact triple expansion solution. When the size of the configuration has enlarged sufficiently, one pair 'peel off' to reform a binary, leaving the single star to escape. The way in which the cross section for such events depends on Δ can be obtained either by similar calculations, or by detailed balance. Therefore similar curious exponents occur. They also occur in 'ionisation' events, in which the binary is just destroyed. Analytically, the cross section varies with E, the energy of the triple system in a barycentric frame, as $E^{(\sqrt{13}-1)/2}$ when E is small and positive. In fact it had been noticed by Hut and Bahcall (1983), on the basis of

Box 21.2. Close two-body encounters as perturbed collisions

Consider the planar two-body equations

$$\ddot{x} = -\frac{x}{r^3}$$

$$\ddot{y} = -\frac{y}{r^3},$$

where $r^2 = x^2 + y^2$. An example of an exact collision orbit is given by $x = C(-t)^{2/3}$, $y = 0$, where $C = (9/2)^{1/3}$. Linearisation about this motion gives the following equations for the small deviations δx and δy:

$$\ddot{\delta x} = \frac{4}{9(-t)^2}\delta x,$$

$$\ddot{\delta y} = -\frac{2}{9(-t)^2}\delta y.$$

Searching for a power-law solution proportional to $(-t)^\lambda$ leads to a simple eigenvalue problem, and the general solution is

$$\delta x = C_1(-t)^{-1/3} + C_2(-t)^{4/3}$$

$$\delta y = C_3(-t)^{1/3} + C_4(-t)^{2/3}, \tag{21.18}$$

where the C_i are constants.

   These terms have simple interpretations. The term in C_1 corresponds to a slight shift in the origin of time in the collision solution $x = C(-t)^{2/3}$, and the term with C_4 corresponds to a small rotation of the collision orbit off the x-axis. The term in C_3 results from adding some angular momentum to the rectilinear collision solution, as can be seen by computing the angular momentum, and the remaining term (in C_2) results from changing its energy. We concentrate on the term in C_3. It determines both the impact parameter p and the distance of closest approach. To make the discussion as simple as possible we take $C_1 = C_2 = C_4 = 0$.

   At fixed large negative t, we see that the term in $C_3 \propto p$. As t increases towards the time of collision ($t = 0$) this term becomes comparable with the value of x on the unperturbed (exact collision) orbit when $C_3(-t)^{1/3} \sim (-t)^{2/3}$, i.e. when $(-t)^{1/3} \sim C_3$. At this time $r \simeq C(-t)^{2/3} \sim C_3^2 \propto p^2$. In fact this relation shows how the minimum distance in the exact two-body motion is related to the impact parameter. It follows that, for encounters in three dimensions, the cross section for encounters at distance r is proportional to r, which we have already established in Eq. (21.4).

about 10^4 numerical experiments, that the cross section varies as about $E^{1.3}$. This was surely one of the most bizarre numerical algorithms for computing the square root of 13.

   These numbers are closely related to the 'Siegel exponents' of the three-body problem, which are named after the mathematician who first exhaustively analysed triple collisions (Siegel & Moser 1971). In the neighbourhood of triple collisions an analysis like that in Box 21.2 gives solutions proportional to $(-t)^\lambda$, and the possible values of λ (the Siegel exponents) include $\lambda = (-1 \pm \sqrt{13})/6$ (Problem 7). The fact that some of the exponents in this case are irrational points to the non-existence of any technique for regularising triple collisions in general. This hinges on whether the powers of t can be evaluated for both positive and negative t. The contrast with two-body collisions, which are regularisable (Sundman 1913), is made clear by noting the form of the near-collision solution in Eq. (21.18). Those forms can be evaluated (with a real result) for both positive and negative t. This would not be true for irrational exponents like $(-1 \pm \sqrt{13})/6$, which is the typical situation in the three-body problem.

This completes our exploration of the analytical approximations that may be employed in gravitational three-body scattering. Much has been omitted, including the important question of *exchange* cross sections (e.g. Hut 1983a). Enough has been done, however, to arrive at a picture of the differential scattering cross section $d\sigma/d\Delta$ for hard binaries, which is the most fundamental problem. If the masses are not too different, Eq. (21.16) leads approximately to the relation $d\sigma/d\Delta \sim Gma/(V^2|\Delta|)$ in the limit $|\Delta| \ll 1$, while the result we have just discussed leads to the complementary result $d\sigma/d\Delta \sim Gma/(V^2\Delta^{4.802\cdots})$ for equal masses in the limit that $\Delta \gg 1$.

Problems

(1)   A rapidly moving third body approaches one component of a soft binary much more closely than the other. Use Eqs. (14.3) and (14.4) in Box 14.1 to estimate the change in velocity of this component. Substitute this result into Eq. (19.5), using the statistical argument in the associated text, and hence estimate the rate of softening of the binary. (Impose a cutoff in impact parameter p at the semi-major axis (a) and at the $90°$-deflection distance.) Hence estimate the lifetime of a soft binary, in terms of the relaxation time.

(2)   *Detailed balance*
Let E_3 be the energy of three particles (a binary and a single star involved in an interaction), in their barycentric frame. Using Eq. (21.1), show that the rate of encounters (per binary) in which E_3 lies in a range of width δE_3 is $4\pi n V^2 \sigma f(V)\delta E_3/\mu$, where μ is the reduced mass of the binary and single star. If binaries have energy distribution $f(\varepsilon)$, i.e. the spatial number density of binaries with binding energy in a range of width $\delta\varepsilon$ is $f(\varepsilon)\delta\varepsilon$, deduce that the rate of encounters in which binaries in this energy range are converted into binaries of energy ε' in a range $\delta\varepsilon'$, and the triples involved have energy in the stated range, is $\dfrac{4\pi n V^2}{\mu\varepsilon}\dfrac{d}{d\Delta}\sigma(\varepsilon, \Delta)f(V)f(\varepsilon)\delta E_3\delta\varepsilon\delta\varepsilon'$, where the arguments of σ remind us of the parameters of the interaction, and $\Delta = (\varepsilon' - \varepsilon)/\varepsilon$.

In dynamic equilibrium this is balanced by the rate of the inverse process. Deduce that

$$\frac{4\pi n V^2}{\mu\varepsilon}\frac{d}{d\Delta}\sigma(\varepsilon,\Delta)f(\mathbf{V})f(\varepsilon) = \frac{4\pi n V'^2}{\mu\varepsilon'}\frac{d}{d\Delta'}\sigma(\varepsilon',\Delta')f(\mathbf{V}')f(\varepsilon'),$$

where V' and Δ' correspond to the inverse process. (Thus $\Delta' = (\varepsilon - \varepsilon')/\varepsilon'$, $= -\Delta/(1 + \Delta)$.) Using the result of Problem 19.3, and taking $f(\mathbf{V})$ to be Maxwellian, deduce that

$$V^2\frac{d}{d\Delta}\sigma(\varepsilon,\Delta)\varepsilon^{-7/2} = V'^2\frac{d}{d\Delta'}\sigma(\varepsilon',\Delta')\varepsilon'^{-7/2}.$$

(3)   Use Eq. (21.5) to show that the mean time to formation of a hard binary in a virialised system is of order Nt_r. (In a 'hot' system enclosed by a box it is even longer.)

(4)   Let t_1 be the time for a particle to move a distance equal to the $90°$-deflection distance, t_2 the crossing time, t_3 the relaxation time, t_4 the time to form the first binary (Problem 3), and t_5 the time to convert the entire cluster into binaries. Neglecting the Coulomb logarithm, show that the N-dependence of these times is in geometric progression.

(5)   Obtain the estimate $\delta h/h \sim (m_3/\sqrt{M_{12}M_{123}})(a/R_p)^{3/2}$ for the relative change in angular momentum of a hard binary of semi-major axis a which is subject to a nearly parabolic encounter with a third star at minimum distance $R_p \gg a$. (Because the change in a decreases exponentially with some power of R_p, as shown in Box 21.1, it follows that the change in eccentricity can be much larger than the relative change in energy. For numerical results on this, see Hut & Paczynski 1984.)

(6)   For rectilinear motion of the third body in an adiabatic encounter its distance from the centre of mass of the binary is $R = \sqrt{R_p^2 + V^2t^2}$. Deforming the contour in Eq. (21.12), to avoid $t = 0$, use integration by parts to show that

$$I = \frac{1}{V^2}\int\frac{\exp(int)}{\sqrt{R_p^2 + V^2t^2}}\left(\frac{in}{t} - \frac{1}{t^2}\right)dt.$$

For a very hard binary, for which $n \gg V/R_p$, the second term in the bracket may be neglected. By deforming the contour further round the branch point at $t = iR_p/V$ and making a cut from there to $t = i\infty$, show that

$$I \asymp \sqrt{\frac{2\pi n}{R_p^3 V^3}}\exp\left(-\frac{nR_p}{V}\right)$$

asymptotically.

(7)   An 'isosceles' triple system in the plane consists of three stars of equal mass
      $m = 1$ at the points (x, y), $(-x, y)$ and $(0, -2y)$. Write down equations of
      motion in units such that $G = 1$, and verify that these are satisfied by the
      collision solution $x = A(-t)^{2/3}$, $y = B(-t)^{2/3}$, where $A = 3/2^{4/3}$ and
      $B = \sqrt{3}/2^{4/3}$. Obtain the linearised equations for small relative perturbations
      δx, δy. Show that these have solutions in which δx, δy vary as $(-t)^{\lambda}$, for four
      possible values of λ (which you should find).

22

Laboratory Experiments

The analytical approaches sampled in the previous chapter have some good uses, but providing accurate useful numbers is not always one of them. They may provide suitable scaling laws, showing how the statistics of three-body scattering depends on the masses involved, but there is usually an overall coefficient that must be determined in some other way. We now turn to a technique which can fill in such gaps. Actually it is marvellous how complementary the two techniques are. Numerical methods are not good at determining cross sections of very rare events, e.g. very close triple approaches, but it is often precisely these little corners of parameter space where analytical methods are feasible.

Numerical methods offer astrophysicists a tool quite analogous to the kinds of particle colliders in use by high-energy physicists. There beams of particles are fired at targets (or other particles), and the relative frequencies of different kinds of collision debris can be observed. Using numerical methods we can see what happens when a binary (the target) is fired on by a single particle, and the experiment may be repeated as often as we care.

Numerical studies of three-body encounters go back almost one hundred years. In 1920 L. Becker published results on exchange encounters (see Chapter 19) which were carried out with the aid of 'mechanical quadrature'. But it was not until the era of electronic computing that the investigation of triple scattering orbits became a sizable industry. In 1970 Harrington reported results from a few hundred encounters (Harrington 1970). By 1975, however, it was possible for Jack Hills (Hills 1975a) to publish some very useful data on the basis of about 14 000 numerical experiments, albeit with some significant restrictions on the number of free parameters (see below). These restrictions were removed by Piet Hut and John Bahcall in research which began to be published in 1983, by which time they had accumulated data on over a million experiments (Hut & Bahcall 1983). Much of the organisation of these runs was still done by hand, and the process was eventually fully automated by Piet Hut and Steve McMillan by 1995 (McMillan & Hut 1996).

There are two issues to be confronted with numerical techniques for the scattering problem. One is the question of how to integrate the equations of motion, and the other is the matter of how to organise the results. These methodological issues are the themes of this chapter.

Numerical integration

At first sight it is a simple matter to integrate the three-body equations numerically. One can take an integration method off the shelf, like the fourth-order Runge–Kutta method. Indeed this usually works quite effectively, though a sensible method of choosing the time step is a big help. One effective method is to repeat each step with two steps of half the size, which allows an estimate of the truncation error (see Press *et al.* 1992). The time step is controlled by keeping this error within prescribed bounds. The error estimate also allows the error to be corrected approximately.

As an example we may consider the well-known case of the three-body problem posed by Burrau (1913). It takes particles of mass 3, 4 and 5 initially at rest at the vertices of a $3-4-5$ Pythagorean triangle. Burrau conjectured that this configuration would lead to a periodic orbit of the general three-body problem. During its evolution there is one very close approach of two of the bodies, and it was thought at one time (Szebehely & Peters 1967) that a special treatment of close approaches (Chapter 15) was needed to cope with this, but in fact the Runge–Kutta method with adaptive step size control works well.

Other standard numerical methods which are often adopted are higher-order Runge–Kutta methods, Hermite integration (Makino & Aarseth 1992, and Problem A.1), which requires computation of both the acceleration and its derivative,[1] and the method of Bulirsch and Stoer (see Press *et al.* 1992). The latter repeats the technique of interval-halving several times for each integration step.

None of these methods take account of several important features of the equations of motion of the three-body problem. In the first place the equations may be written in the form

$$\ddot{\mathbf{r}}_i = \mathbf{a}_i(\mathbf{r}_1, \mathbf{r}_2, \mathbf{r}_3), \qquad (i = 1, 2, 3), \tag{22.1}$$

---

[1] The derivative of the acceleration is sometimes called *jerk*. Perhaps the position and its successive derivatives should be referred to by the letters x, v, a, j, s, c and p (up to the sixth derivative), as we have been given to understand that some physicists refer to the last three as *snap, crackle* and *pop*. See www.weburbia.com/physics/jerk.html

where $\mathbf{a}_i$ is the acceleration of the ith star. The usual forms of the above methods force the user to rewrite the equations as first-order equations:

$$
\begin{aligned}
\dot{\mathbf{r}}_i &= \mathbf{v}_i, \\
\dot{\mathbf{v}}_i &= \mathbf{a}_i(\mathbf{r}_1, \mathbf{r}_2, \mathbf{r}_3),
\end{aligned}
\tag{22.2}
$$

where $\mathbf{v}_i$ is the velocity of the ith star. The methods can, however, be specially adapted to equations of the form (22.1). In the case of Runge–Kutta algorithms the corresponding methods are referred to as 'Runge–Kutta–Nystrom' methods.

Next, none of these methods pay attention to the fact that Eqs. (22.2) are *canonical*, i.e. they are Hamilton's equations for an appropriate Hamiltonian. Now one of the analytical properties of Hamilton's equations is that they are a 'symplectic' dynamical system. This is a generalisation to systems with n degrees of freedom of the idea of *area-preservation* for systems with one degree of freedom. For example, the one-dimensional simple harmonic oscillator has equations of motion which can be written in Hamiltonian form as

$$
\begin{aligned}
\dot{x} &= v, \\
\dot{v} &= -x.
\end{aligned}
\tag{22.3}
$$

Now a solution of these equations can be plotted as a moving point in phase space (with coordinates x and v), and the velocity of this point has components given by the right-hand sides of Eqs. (22.3). Thus phase space is equipped with a velocity field, and the important point to notice is that this field is *divergence-free*: $\partial \dot{x}/\partial x + \partial \dot{v}/\partial v = 0$. Thus if we follow the motion of a two-dimensional region in phase space as it is carried along by the flow, its volume does not change.

It might seem desirable to invent an integration algorithm which also exhibited this property of area preservation (or, in the case of interest to us, its generalisation to the concept of symplecticness). Runge–Kutta methods are not symplectic in general, but there is another class of algorithms which is. The simplest example is the 'leapfrog' algorithm, often called the Verlet algorithm in computational physics. For equations of the form (22.2) this may be written as

$$
\begin{aligned}
\mathbf{r}_i^{n+1} &= \mathbf{r}_i^n + h\mathbf{v}_i^{n+1/2}, \\
\mathbf{v}_i^{n+3/2} &= \mathbf{v}_i^{n+1/2} + h\mathbf{a}_i^{n+1}.
\end{aligned}
\tag{22.4}
$$

Here the superscript indicates the time, in multiples of the stepsize h after some initial time. In the case of the simple harmonic oscillator the equations reduce to

$$
\begin{aligned}
x^{n+1} &= x^n + hv^{n+1/2}, \\
v^{n+3/2} &= v^{n+1/2} - hx^{n+1}.
\end{aligned}
\tag{22.5}
$$

By computing the Jacobian determinant $\dfrac{\partial(x^{n+1}, v^{n+3/2})}{\partial(x^n, v^{n+1/2})}$ it is easy to show that this is area-preserving.[2]

---

[2] Strictly speaking, an integrator for a problem with one degree of freedom is symplectic if the map $(x^n, v^n) \to (x^{n+1}, v^{n+1})$ is area-preserving. The fact that v is discretised at intermediate times complicates the issue, but we shall not explore this detail here.

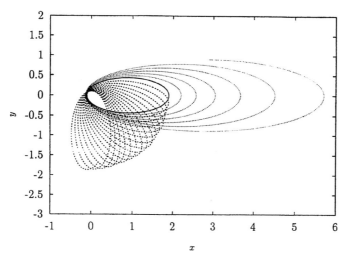

Fig. 22.1. Three numerical computations of a Kepler orbit with $e = 0.9$ and $a = 1$: (a, dotted) Euler (cf. Problem 1), with time step proportional to $r^{3/2}$, (b, short-dashed) leapfrog, with fixed time step, and (c, solid) using the method of Mikkola & Tanikawa (1999). (This is a leapfrog algorithm based on a slightly more complicated Hamiltonian than the usual Kepler Hamiltonian, cf. Chapter 15.) Runs (b) and (c) used similar numbers of steps, and run (a) used about three times as many. The purpose of the figure is not, however, to compare efficiency, but to illustrate the properties of different algorithms. Euler appears to preserve the direction of pericentre but not the energy, and with leapfrog the behaviour is reversed. The third, i.e. (c), preserves both, but not the phase (though this is not evident in this plot).

It might seem to have nothing to do with area preservation or symplecticness, but in practice an algorithm like the leapfrog method is very good for long-term energy conservation. The fact that the method is area-preserving essentially shows that the numerically generated phase path does not spiral out from (or in to) the origin, which is precisely the behaviour exhibited by the Euler method (Problem 1 and Fig. 22.1). In terms of *order*, i.e. the way in which the local truncation error depends on h, the leapfrog is very primitive. It is not hard, however, to invent symplectic methods of higher order (Kinoshita *et al.* 1991).

Our discussion of the symplectic property of the leapfrog integrator made the unstated assumption that h was constant. In practice, as we have already seen, h should be allowed to vary, and then it is not clear that the symplectic property of the algorithm is preserved. This stumbling block to the implementation of effective symplectic integrators was removed by Hut, Makino and McMillan (1995), who realised that the way out was to incorporate another exact property of Eqs. (22.1): their time-reversal invariance. Observe that, if $\mathbf{r}_i(t)$ is a solution of Eqs. (22.1) then so is $\mathbf{r}_i(-t)$. One consequence of this fact is that, if the velocities of three stars in a triple system are reversed at some instant, then the stars will exactly retrace their orbits. Here is another property which we should perhaps try to incorporate into a numerical integrator.

Let us look at Eqs. (22.4) from this point of view. By inverting these we immediately find

$$\mathbf{r}_i^n = \mathbf{r}_i^{n+1} + (-h)\mathbf{v}_i^{n+1/2},$$
$$\mathbf{v}_i^{n+1/2} = \mathbf{v}_i^{n+3/2} + (-h)\mathbf{a}_i^{n+1}.$$

Since these are true at each step, we can reduce the superscript on the second equation by one, and obtain

$$\mathbf{r}_i^n = \mathbf{r}_i^{n+1} + (-h)\mathbf{v}_i^{n+1/2},$$
$$\mathbf{v}_i^{n-1/2} = \mathbf{v}_i^{n+1/2} + (-h)\mathbf{a}_i^n.$$

Note that these equations are of exactly the same form as Eqs. (22.4), except that h is replaced by $-h$ and the superscript is running downwards. Thus if we stop a leapfrog integration at some point, reverse h, and continue using the leapfrog algorithm, we recover the previous results (except for rounding error). This is the analogue (for algorithms) of the time-reversal invariance of the equations of motion.

Now the above argument is ruined if we allow the time step to vary, for example as a function of the accelerations $\mathbf{a}_i$, unless we ensure that the time step respects the time-reversal invariance as well. One possibility is to ensure that h depends on a symmetric function of the values at the beginning and the end of the time step. It turns out that such a procedure works very well.

It is still possible to break a time-symmetric, symplectic integrator with sufficiently close approaches between two or three of the bodies. There is a formulation, based on a different Hamiltonian, which will accurately compute even a collision orbit (Mikkola & Tanikawa 1999, see Chapter 15). The traditional way out, however, is the technique of two-body *regularisation*, also discussed in Chapter 15. Its extension to three- and other few-body systems is best carried out with the 'chain regularisation' of Mikkola & Aarseth (1993), which is an extension of an old three-body method devised by Aarseth & Zare (1974).[3] We recommend the survey by Mikkola (1997) for a fascinating glimpse of the history behind these methods. In recent years, however, it has even become possible to combine these regularisation techniques with time-symmetrisation (Funato *et al.* 1996, Leimkuhler 1999).

The final point to be mentioned in connection with numerical methods is the accuracy which it is feasible or desirable to insist on. This is really the same question for $N = 3$ as we discussed in general in Chapter 13. There we already saw that there are situations in which the slightest change in the initial conditions of a three-body interaction can lead to drastic changes in the final result. Dejonghe & Hut (1986) have discovered examples in which an error in initial conditions may be magnified by over 100 orders of magnitude before the final fate of the triple system is settled.

---

[3] For $N = 3$ there is an alternative due to G. Lemaitre (1955), though he is better known nowadays for his work in cosmology.

These considerations show that one cannot hope to devise any method which will produce an accurate solution for *all* triple interactions.

Organisation of the calculations

Each triple encounter requires the specification of initial conditions. As the three-body problem has nine degrees of freedom, it might be thought that no fewer than 18 parameters are required. However, it is possible to work in a barycentric frame, reducing the tally to 12 parameters. One more may be removed by choice of a unit of length (e.g. the initial semi-major axis of the binary). We can rotate coordinates so that the binary initially sits in the x, y plane with its major axis along the x-axis, reducing the count by three (recall that the orientation of the orbit could be specified by the three Euler angles). Next, it is usual to start the integration with the third body at some fixed distance from the binary, sufficiently large that the effects on the binary of the incomer up to that point are negligible; this leaves only seven parameters to be specified. These may be the initial eccentricity and phase of the binary, and all but one of the initial coordinates and all velocity components of the incomer. Against these simplifications, however, we must not forget that the result depends on the masses, though scaling (e.g. so that the total mass is unity) removes one.

In a given problem it is usually clear how these parameters are to be distributed. In a simple situation, for instance, we are interested in encounters between stars of given masses, and the initial speed of the intruder and eccentricity of the binary are specified. Then the initial phase of the binary corresponds to a uniformly random choice of the initial time, while the remaining parameters of the intruder correspond to a uniform distribution in space and an isotropic distribution of velocities. It is customary to select such initial conditions using a well designed random number generator, but there is some advantage in these problems in using *quasi*-random numbers (Press *et al.* 1992), which aim to sample the domain of initial conditions more purposefully.

One of the parameters commonly used to specify the initial conditions of the intruder is the *impact parameter*, p. i.e. the distance of closest approach between it and the barycentre of the binary, if it was undeflected by the binary. This differs from all other parameters in that it is unbounded, and it is the most troublesome parameter of all to deal with. Some upper limit must be imposed, and it is legitimate to do so, for if p is too large the intruder flies wide of the binary and does not affect it significantly. But where should the limit be set? The problem with the choice of maximum impact parameter is this: allowing too large an impact parameter can imply a large waste of computer time on uninteresting orbits; while choosing too small an impact parameter (e.g. one much smaller than the initial semi-major axis of the binary) will yield a systematic underestimate of some cross sections, since some encounters of interest will be missed.

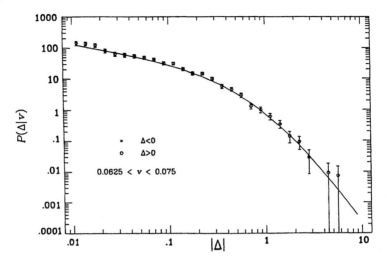

Fig. 22.2. Numerically determined cross section for hard binary scattering for equal masses, and a semi-analytic fit. The scaled differential cross section $P = \dfrac{v^2}{\pi a^2} \dfrac{d\sigma}{d\Delta}$ is plotted, with error bars, against the relative increase in binding energy, Δ, where v is the initial relative speed of the third body and the binary, in units of the minimum speed required for ionisation. From Heggie & Hut (1993).

One manual way of handling this situation is to carry out a pilot study to determine the range of impact parameters in which the encounters of interest are observed to take place. Only then does one embark on production runs, intended to compute cross sections with desired statistical accuracy (Fig. 22.2).

It is now possible to automate the entire process (McMillan & Hut 1996). Rather than relying on human inspection of pilot calculations, software has been written which includes an automatic feedback system ensuring near-optimal coverage of the range of impact parameters while guaranteeing completeness. The basic idea is to maintain a safety zone outside the 'bull's eye' region, defined as the area where interesting reactions take place (where 'interesting' is defined according to the particular purpose underlying a given set of experiments). Allowing for the possibility of *dynamically* enlarging this safety zone guarantees rapid convergence towards accurate cross sections.

The automated scattering software we are describing was designed in a multi-layer object-oriented approach and is implemented in C++. At the lowest layer, the orbit integration engine consists of a fourth-order, time-symmetric, variable-time-step Hermite integrator. This technique provides excellent long-term numerical stability, something that is essential for the treatment of long-lasting resonance scattering events.

On top of this lowest layer, there are several layers that contain: (1) checks to determine whether a given scattering experiment has reached its final outgoing state; (2) checks to allow optimisation features to be activated, such as analytical integration of inner and outer orbits of hierarchical triple systems in which the outer orbital period vastly exceeds the inner orbital period; (3) diagnostic functions to store information describing the build-up of energy errors; (4) various bookkeeping functions that chart the overall character of the orbits (e.g. democratic versus hierarchical resonance states); and (5) checks for overlap of stellar radii, in which case merging routines are invoked that can replace two colliding stars with a single merger product.

On top of these layers, the first user-accessible layer contains a single-scattering command, with a large number of options. The masses and radii of the stars can be specified, as well as the orbital parameters of the binary, the impact parameter of the encounter, and the relative velocity, asymptotically far before the encounter. The initial distance from which the integration starts is determined automatically (and will be much larger for, say, a $10M_\odot$ black hole approaching a given binary compared to a $0.5M_\odot$ dwarf). In addition, an overall accuracy parameter gives a handle on the cost/performance tradeoff. In practice, typical relative energy errors can be easily kept as small as 10^{-10}; production runs, however, usually aim at errors of order 10^{-6}, allowing a speed-up of a factor ten in computer time with respect to the most accurate integrations attainable.

The next layer contains all the management software to conduct a series of scattering experiments. Depending on the type of total or differential cross section requested, the user can choose an appropriate command to activate a 'beam' of single stars aimed at the 'target plate' of binary stars. After a few minutes, a preliminary report appears on the screen, with estimates of all relevant cross sections plus their corresponding error bars (Box 22.1). Thereafter, subsequent reports appear, each one after a four times longer interval. Because of the Monte Carlo nature of the orbit parameter sampling, each following report carries error bars that are half the size of the corresponding ones in the previous report. Thus reasonable estimates can often be obtained in ten or fifteen minutes, with more accurate results following in an hour (on a fast modern work station). This is in marked contrast with the duration of experiments in a particle collider in high-energy physics!

Box 22.1. Example of computational determination of scattering cross sections

The command `sigma3 -d 128 -v 1.2 -m 0.3` uses part of the Starlab package (www.manybody.org/starlab.html) to estimate the cross section for various outcomes of encounters between a single star and a binary. The binary component masses are $1 - m$ and m, and the mass of the third star may be set with a further option. The other two parameters are d, which is a measure of the density of experiments in a central trial zone (the fractional error in the largest cross section will be of order $1/\sqrt{d}$), and v, the speed of the third body when at a large distance, in units of the value at which the total barycentric energy of the triple system is zero. Here is a summary of the results. It may not look beautiful to the reader, but it does to anyone who has organised this kind of thing by hand.

```
scatter_profile:
m1=0.7 m2=0.3 m3=0.5 r1=0 r2=0 r3=0 v_inf=1.2
---------------------------------------------
total scatterings=25, CPU time=7.45
   exch1     exch2      ion     error      stop
   2.220     0.888     3.108    0.000     0.000
  +-1.83    +-.628    +-2.03    +-.000    +-.000
---------------------------------------------
total scatterings=60, CPU time=16.94
   exch1     exch2      ion     error      stop
   0.444     0.999     0.777    0.000     0.000
  +-.222    +-.333    +-.294    +-.000    +-.000
---------------------------------------------
total scatterings=312, CPU time=105.6
   exch1     exch2      ion     error      stop
   0.768     1.224     0.712    0.000     0.000
  +-.158    +-.199    +-.153    +-.000    +-.000
---------------------------------------------
total scatterings=1560, CPU time=487.97
   exch1     exch2      ion     error      stop
   1.025     1.181     0.861    0.000     0.000
  +-.101    +-.105    +-.089    +-.000    +-.000
```

The values are those of $\dfrac{v^2\sigma}{\pi a^2}$, where v is defined as above. Notice that the cross section for exchanging the heavier binary companion is smaller.

Problems

(1) Solve Eqs. (22.5) exactly, for initial conditions $x^0 = 1$, $v^{1/2} = 0$. Find out in what sense the solution is superior to that of the corresponding Euler algorithm

$$x^{n+1} = x^n + hv^n,$$
$$v^{n+1} = v^n - hx^n$$

with initial conditions $x^0 = 1$, $v^0 = 0$.

(2) Compare the total of the cross sections at the bottom of Box 22.1 with the estimate for the cross section of close encounters (Eq. (21.4)).

(3) For fixed masses, how many parameters are needed to specify a three-body scattering experiment (a) in two dimensions, (b) in one dimension?

23

Gravitational Burning and Transmutation

The arguments of the last few chapters apply to three-body stellar interactions any-where. Now we have to take into account the specific environment in which we are interested, i.e. to study the behaviour of binaries and triples inside much larger stellar systems. For these purposes it is very useful to have an idea of the *mean* rate of change in the energy of a binary, and also the effect of a single encounter on the motion of the binary.

Let us consider the mean rate of change first. Relative to the current energy of the binary, ε, this may be defined by

$$\langle \Delta R \rangle = \int \Delta R(\Delta) d\Delta, \tag{23.1}$$

where R is the rate (number per unit time) at which the binary experiences encounters in which the relative change of its binding energy is Δ. The function R becomes infinite as $\Delta \to 0$, but in practice we can impose a small positive cutoff on Δ, at the point where positive and negative energy changes nearly cancel.

As an example, consider the formula

$$R \propto |\varepsilon|^{-1} \Delta^{-\alpha} (1 + \Delta)^{-4.5+\alpha}, \tag{23.2}$$

which, with $\alpha \simeq 0.54$, has been shown to provide a satisfactory fit to numerical data for $\Delta > 0.01$ for equal masses in the very hard binary limit (Heggie & Hut 1993). Then the integral in Eq. (23.1) is quite insensitive to the choice of cutoff, and the result is that the absolute rate of change of ε is independent of ε. This result accounts for the values at the extreme right in Fig. 23.1. Equation (23.2) can also be used to show that about half the hardening rate of a hard binary is accounted for by encounters

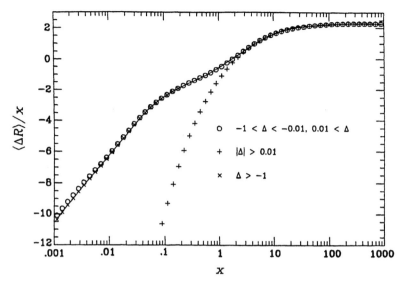

Fig. 23.1. The mean rate of change of energy of a binary subject to encounters, from Heggie & Hut (1993). The initial relative speed of the binary and the single star have a Maxwellian distribution corresponding to temperature T, and x is the ratio between the binding energy of the binary and kT, where k is Boltzmann's constant. Thus hard binaries are at the right, and soft binaries are at the left. The reaction rate R in the axis label differs (by scaling) from the quantity R defined in the text, and the ordinate is actually proportional to the mean rate of change of energy. This is constant in the hard binary limit, as shown (see Problem 1). The circles and crosses show the result if 'ionising' encounters, i.e. those leading to disruption of the binary, are excluded. While it may be useful to consider ionising encounters separately, it leads to a graph with a more complicated shape than in the case where such encounters are retained (symbol $+$).

in which

$$\Delta > \Delta^* \equiv 0.53. \tag{23.3}$$

Keeping this result for later, let us now turn to the effect of an encounter on the *motion* of the binary. Let the initial and final velocity of the binary relative to the intruder be $\mathbf{V}$ and $\mathbf{V}'$, respectively, and let ε be the initial energy of the binary. By energy conservation (in the rest-frame of the barycentre of the three bodies, assumed to have equal mass m) it follows that

$$\frac{1}{3}mV^2 + \varepsilon = \frac{1}{3}mV'^2 + \varepsilon(1 + \Delta).$$

(The factor $1/3$ arises because the reduced mass of a system consisting of a binary of mass $2m$ and a single star of mass m is $2m/3$.) For a sufficiently hard binary and an average encounter, $\Delta\varepsilon \gg mV^2$, and so this reduces approximately to $V'^2 \simeq -3\Delta\varepsilon/m$. In this same frame the velocities of the binary and single star are $\mathbf{V}'_b = \mathbf{V}'/3$

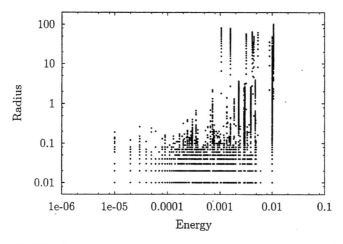

Fig. 23.2. Radius and binding energy of all hard binaries in an N-body simulation of a star cluster, with $N = 8192$. The horizontal banding at the bottom results from the fact that the radius was given to only two decimal places; similarly the vertical banding at the left. The vertical banding at the right results from a few persistent hard binaries escaping to large radii and (in most cases) drifting back to the core. The model was initially a Plummer model, with equal masses and no tide.

and $\mathbf{V}'_s = -2\mathbf{V}'/3$, respectively, since the relative velocity must be $\mathbf{V}'$ and the total momentum in the barycentric frame must vanish. Finally, to compute the new speed of the binary and single star in the rest frame of the whole cluster we must add the velocity of the triple system, but again this is negligible for an average encounter with a sufficiently hard binary. Thus the final speeds of the binary and single star in the cluster are approximately $\sqrt{-\Delta\varepsilon/(3m)}$ and $2\sqrt{-\Delta\varepsilon/(3m)}$, respectively. (Remember that $\varepsilon < 0$ and $\Delta > 0$ typically.)

Armed with this information, let us now follow the fate of a three-body binary as it experiences a succession of encounters, starting from the time of its formation in the core of a stellar system. Initially we do not expect the predicted speeds to be significantly in excess of the velocity dispersion of the core. The calculations of the preceding paragraph are very approximate in this case, but the reaction products continue to move around in the core, and their kinetic energy (in which, for a binary, we include only the kinetic energy of the motion of its barycentre) has increased by $|\Delta\varepsilon|$ per encounter.

As time goes on and further encounters take place, the binary steadily hardens, and typical values for the predicted speeds of the reaction products increase. The first thing to happen is that the single star can now escape from the core, and eventually the time comes when the binary achieves enough speed to do so also (Fig. 23.2; see also Giersz & Spurzem 2000). This is no dream. There is a known binary (with an eight-hour period) in the outskirts of the globular cluster M15 which probably got there by ejection from the core in a three-body encounter (Phinney & Sigurdsson 1991).

At least one of the components is a neutron star, and the binary is sufficiently close to display a relativistic perihelion advance, just as the planet Mercury does, but about 30 000 times faster.

When the binary is ejected from the core it finds itself in a region of lower density, and encounters are less frequent. If, as an example, we consider a binary moving on a radial orbit, the total number of encounters during one orbit (from the core to the maximum radius which it can reach and back to the core again) varies as

$$N_{\text{enc}} \sim \int n\sigma v dt, \tag{23.4}$$

where n is the number-density of single stars, σ is the cross section for close encounters, and v is the mean encounter speed. (Note the use of an 'n-sigma-v' argument; cf. Box 13.1.) Now for a hard binary we may estimate that $\sigma \sim Gma/v^2$ (cf. Eq. (21.4)), where a is the semi-major axis of the binary, and it turns out, for any reasonable cluster model, that the dominant variation in the integrand in Eq. (23.4) comes from the density profile $n(r)$. This causes the rate of encounters to diminish rapidly as the size of the orbit of the binary increases. Therefore the subsequent hardening of the binary depends vitally on its orbit shrinking back to the core as quickly as possible. In fact mass segregation will accomplish this in due course (on a time scale no longer than the two-body relaxation time).

At long last the binary will become so hard that the single star it interacts with will leave not only the core but the whole cluster. As it climbs out of the potential well of the cluster it loses speed, emerging with a kinetic energy $-\frac{2}{3}\Delta\varepsilon + m\phi_c$, where the first term is an estimate of the kinetic energy of the star after the encounter, and ϕ_c is the potential in the core (assuming $\phi \to 0$ at large radius). Since the encounter itself contributed an energy $-\Delta\varepsilon$, the change in energy of the bound system is the difference of these terms, i.e. $-\frac{1}{3}\Delta\varepsilon - m\phi_c$. One can think of this as consisting of two contributions; the first is the normal contribution of the binary, which we still assume to remain bound, and the second is the escape energy of the single star. The latter is the maximum contribution which the single star can make, no matter how energetically it escapes.

A fruitful way of looking at this is that the term $-m\phi_c$ is the work done on the cluster as the single star moves out of the potential well. This interpretation also suggests that the energy contributed to each part of the cluster equals the contribution which this part makes to the central potential. Thus for a shell of material of mass M and radius r, the heating is GMm/r. By this argument we also see that, if the star escapes from the core but not from the cluster, it does not heat the core alone, but instead all the material between its initial and final radii.

Though it is tangential to our present purpose, it is worth stating that similar considerations apply to other forms of heating through mass loss. S. Portegies Zwart has pointed out to us that mass loss from stellar evolution is more important than one might think, because the mass is lost by the most massive stars (which evolve fastest), and mass segregation concentrates these in the core. It is here that the maximum energy is generated per unit mass ejected.

Back to the binary. Eventually it is so hard that not only does it eject the single stars it interacts with, but it also ejects itself. Thus the influence of each binary is self-limiting. Though a binary consisting of two point masses can in principle release an infinite amount of energy, the process by which it releases energy is its own bottleneck.

How much energy does the binary contribute before retiring? The result is fairly easily estimated by adding up the contributions from the various phases in its evolution (Box 23.1), which evaluates to about $-24m\phi_c$. It is usual to express the central potential in terms of the central one-dimensional velocity dispersion, σ_c^2,[1] though strictly there is no upper bound on the ratio $-\phi_c/\sigma_c^2$. Usually it is assumed that energy generation by binaries is important in systems with small cores of very high density, especially in late core collapse. In this case, as we have seen (Chapter 18) the evolution of the core is self-similar, and in this case the above ratio is constant: $-\phi_c/\sigma^2 \simeq 13.5$ in anisotropic Fokker–Planck models with stars of equal mass (Takahashi 1995).

Box 23.1. Energy generation by a single hard binary

Until the single star is capable of escaping, binary heating is 100% efficient: the entire change in energy of the binary goes to heating the cluster. This phase comes to an end when the new energy of the single star is comparable with the escape energy, i.e. when $2\Delta\varepsilon/3 \simeq m\phi_c$. We may suppose that this loss becomes important when half of the hardening of the binary is caused by encounters that eject the single star, i.e. when $\varepsilon \simeq 3m\phi_c/(2\Delta^*)$ (see Eq. (23.3)). Since the initial energy of the binary is comparatively negligible, it follows that the energy contributed in this first phase is itself about $-3m\phi_c/(2\Delta^*)$.

By a similar argument the binary (treated as a particle of mass $2m$) escapes at the last encounter when its energy is of order $\varepsilon \simeq 6m\phi_c/\Delta^*$. During the second phase (i.e. up to this point) all energy given to the binary is held within the cluster; this is one third of the change in internal energy of the binary, i.e. $-3m\phi_c/(2\Delta^*)$. It takes approximately $\log(4)/\log(1+\Delta^*) \simeq \log(4)/\Delta^*$ encounters to increase the binding energy of the binary by the necessary factor of four, and each contributes an energy $-m\phi_c$ via the single star (see text). Adding all these contributions together we find that the energy given by a hard binary to the cluster before escaping is roughly $-m\phi_c(3 + \log(4))/\Delta^*$.

Putting our various estimates together, we find that each newly formed hard binary eventually contributes an energy of approximately $110m\sigma_c^2$. Combining with an estimate for the rate at which such pairs form (see Eq. (21.5)), we readily find that

---

[1] The subscript should avoid confusion with a cross section.

the rate of heating (per unit mass) is given by

$$\epsilon \simeq 85 \frac{G^5 m^5 n^2}{\sigma_c^7}. \tag{23.5}$$

The surprisingly large coefficient is a warning against reliance on dimensional analysis alone. Nevertheless, a considerable number of complexities are still swept under the carpet if we use such a formula (and indeed it is common to use a formula of this type when modelling the effect of binaries in the evolution of stellar systems, unless one uses an N-body method). Quite apart from the fact that it would be better to express the result in terms of the local potential rather than the local velocity dispersion, we are assuming (i) that all the energy yielded by a binary is injected at the location where it was born; (ii) that the energy is inserted instantaneously, the moment it is born; and (iii) that binaries form continuously (in time). All these assumptions are wrong, and something can be done to patch them up. For instance, it is possible, in the simulation of cluster evolution, to inject binaries at times determined stochastically, but in such a way that the mean rate of formation is as predicted. It is also possible to arrange for the binaries to emit energy in discrete events, just as in reality. And it is also possible to mimic the effect of each encounter on the spatial location of each binary, and, in particular, their ejection from the core following particularly energetic interactions (Giersz & Spurzem 2000). We stress again, however, that these difficulties are entirely circumvented in an N-body simulation.

One final aspect of this problem which is less well understood and less easy to model is how to handle the formation and evolution of binaries in systems with a distribution of stellar masses. One interesting contrast with the equal-mass case is the behaviour of the cross section for close encounters (Eq. (21.4)). For stars of equal mass, as a binary hardens its semi-major axis shrinks and so σ decreases. For stars of unequal mass, we have already seen that a binary with one component of relatively low mass is liable to have this component exchanged with a more massive incomer. Now the energy of the binary need not change much, and an increase in the mass of one component may cause a to increase. Thus the close encounter cross section increases, because of both the increase of a and the increase of the masses. Binary evolution has a runaway tendency in these circumstances, at least until the components are about as heavy as they can be.

Problems

(1)   For a binary on the hard/soft threshold, moving in a cluster in virial equilibrium, use Eq. (23.4) and reasonable estimates for n, σ and v to show that the number of close encounters in a relaxation time is of order $1/\log \Lambda$, i.e. the reciprocal of the Coulomb logarithm. How does the result depend on the energy of the binary if it is hard? Use your answer to show that the mean rate of hardening of a hard binary is independent of the semi-major axis a. Use

this result in turn to estimate the shape of the distribution of binding energies of hard binaries.

(2)   Assume that a binary is born with small energy, that its energy increases in increments which are distributed as in Eq. (23.2), that the resulting energy heats a cluster (with or without ejection of the single star, as appropriate), and that the binary remains in the core until it escapes. Perform Monte Carlo simulations to compute the mean energy yield per binary, and compare with the result of Box 23.1.

(3)   Problem 18.2 showed that stellar-mass black holes would form a dense core within the larger core of normal stars. In this core, black hole binaries form by three-body encounters. Using arguments of Box 23.1, estimate the number of single black holes which are ejected in the hardening of each black hole binary, and hence estimate the approximate fraction of black holes which are ejected as binaries.

Note: the black holes in some of those escaping binaries will later spiral into each other, under emission of gravitational radiation. These binaries form a promising category for detection by LIGO, a pair of laser-based gravitational wave detectors (Portegies Zwart & McMillan 2000).

Part VII

Primordial Binaries: $N = 4$

The following three chapters complicate the million-body problem for astronomically motivated reasons. Chapter 24 explains these by tracing the history of the discovery of *binary stars* in star clusters, in numbers which imply that they are *primordial*, i.e. they were born along with the cluster itself. They are associated with several of the remarkable phenomena which help to explain why globular star clusters are so important to astrophysicists, such as the sources of X-rays within them. We contrast their behaviour in star clusters with the much milder behaviour of binaries in less extreme environments.

In systems with many binaries, four-body encounters between two binaries are common. Chapter 25 discusses in detail one of the commoner outcomes: *hierarchical triple systems*. They are one class of three-body problem where the motion is both non-trivial and amenable to detailed calculation. Since these systems are stable and very long-lived, but may have tiny orbital time scales, such results are important for efficient computer simulation of N-body systems with many primordial binaries.

Chapter 26 discusses the effect of binary–binary encounters on the rest of the system. In important ways they can dominate the effect of the three-body encounters discussed in earlier chapters, though not forever, as binaries are also destroyed in these encounters. The outcomes of the interactions are also more complicated than in three-body encounters, and we show how to classify these. We also show that, in astrophysical applications, the fact that stars have finite diameter has an important role to play in encounters, when several stars are temporarily confined to a small volume of space.

24

Binaries in Star Clusters

In the original *Star Wars* movie there is a brief but memorable scene of two stars shining down from the afternoon sky. We find it striking because we are so used to seeing just a single star in the sky, but in fact it is the Sun that is unusual. In the neighbourhood of the Sun most stars are binaries, and some belong to triple systems or even little groups with still larger numbers of stars (Duquennoy & Mayor 1991). Our nearest neighbours, for example, are a binary (α Centauri) with a distant third companion (Proxima Centauri, see Matthews & Gilmore 1993). Such systems are unlikely to arise by chance encounters (Problem 1), and so the abundance of binaries and triples suggests that most stars are born that way. Indeed binaries and other multiples are most common among the youngest stars (see Kroupa 1995).

A brief observational history

With this background, astronomers were perplexed by how difficult it was to find any binaries in globular star clusters. Admittedly, wide visual pairs were not expected, as such binaries would be destroyed by encounters with other stars in the dense environment of a globular cluster (Problem 21.1). Indeed, there is observational evidence that this has happened (Côté *et al.* 1996). But it should have been possible to detect closer binaries by observing periodic variations in their radial velocity, or else through the discovery of eclipsing variable stars. Nothing was found by these methods, and it was even being argued that close binaries must have destroyed themselves. One possibility was that a close binary, even though it does not tend to be destroyed by encounters with other stars, may have its eccentricity significantly changed. If the eccentricity becomes high enough (and arguments of statistical mechanics show that this is not unlikely), the separation of the stars at pericentre may become so low that the stars will collide and merge (see Chapter 31). Note that, though this argument turned out not to be needed, it takes the dynamics of star clusters beyond the point-mass approximation for the first time in this book.

Firm evidence of binaries eventually arrived, but it came from a most unexpected direction: the discovery of variable sources of X-rays (see Verbunt & Hut 1987, Grindlay 1988). Such sources were already known elsewhere in the Galaxy, and those discovered in globular clusters were interpreted in the same way; i.e. as binaries containing neutron stars. The standard model is that the other star in the binary is a more-or-less normal star whose surface layers are just on the point of becoming gravitationally unbound by the attraction of the neutron star.[1] As this normal star evolves its material expands slowly, and so there is a small but steady stream of gas which leaves its surface and, by conservation of angular momentum, orbits the neutron star in a disk. Viscosity in this disk then slowly drags the material of the disk towards the neutron star, and the conversion of the potential energy of the gas, as it sinks in the gravitational field of the neutron star, into other forms of energy, gives rise to the X-ray emission that makes these objects observable.

The discovery of these sources was the first clear evidence for the presence of neutron stars as well as for binaries, and was a major factor in turning astrophysicists' attention and interest to the globular clusters. Though the inferred number of binaries was not yet large by Galactic standards, and though it was not yet clear whether they were born in the clusters or manufactured there, the numbers of X-ray sources were somehow far higher in globular clusters than in the rest of the Galaxy. Suddenly these old and placid members of the Milky Way community turned out to be fertile breeding grounds for some of the most interesting and energetic objects in the nearby Universe.

In due course a few more binaries were eventually discovered by the radial velocity technique, though only a few per cent of all observed stars were found to be binaries (see Hut *et al.* 1992a for early developments in this subject). Perhaps the first observers (see Gunn & Griffin 1979) had just been unlucky. At any rate it was realised that there were strong selection effects which prevented more than a fraction of the binaries from exhibiting observable changes in their radial velocity. One reason for this is that it was only possible to observe the brighter stars. These are giants, with radii considerably exceeding those of most normal stars. Therefore very close binaries could not be expected to survive this giant stage of stellar evolution, for the same reason that highly eccentric binaries would not be expected. There was thus a

---

[1] This process is referred to as *Roche lobe overflow*. The donor star is surrounded by an imaginary surface, outside which its gravitational attraction is overcome by that of the neutron star. The equipotential surfaces closely resemble those which control escape from a star cluster on a circular orbit about a galaxy (cf. Fig. 12.1, which can be thought of as depicting the 'Roche lobe' of a star cluster).

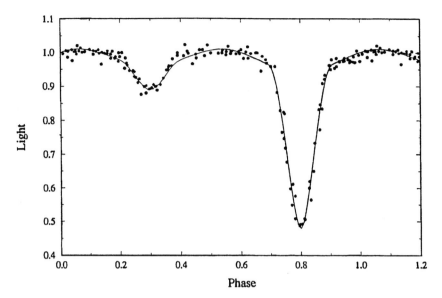

Fig. 24.1. The light curve of an eclipsing binary star in the globular cluster M71 (after McVean *et al.* 1997).

rather limited range of separations (between the two stars in the binary) where radial velocity binaries with giant companions could be expected: too small, and the stars collide; too large, and the binary would be destroyed by encounters. Difficult though it is to allow accurately for such selection effects, by now it was no longer at all certain that the proportion of binaries in globular clusters was very much smaller than in the neighbourhood of the Sun, except for the destruction of wide binaries by encounters.

The hurdle having once been cleared, the evidence for binaries in globular clusters mounted steadily. Pulsars were observed in binary systems (again by detecting a periodic Doppler shift, this time in the pulse frequency; see Phinney 1996). One of these is even a triple system (see Thorsett *et al.* 1999). X-ray sources of lower luminosity, thought to be binaries where a white dwarf takes the place of the neutron star in the brighter sources, were discovered (see Livio 1996). In some clusters it may be possible to detect binaries in colour-magnitude diagrams: unresolved binaries have similar colours to the other stars but are noticeably brighter (see Rubenstein & Bailyn 1997). Eclipsing binaries were discovered at last (e.g. Fig. 24.1), many as a by-product of searches for gravitational lenses (e.g. see Kaluzny *et al.* 1998)! And in recent years it has become possible to find evidence for radial velocity binaries among the normal stars which have not yet evolved into giants (see Côté & Fischer 1996).

In short, there is now abundant evidence that a substantial fraction of stars in globular clusters are binaries. Satisfactory as this may be to the astrophysics community as a whole, stellar dynamicists have been a little less wholehearted in their response.

It is still easier to understand and model clusters containing single stars, while the introduction of binaries brings in its wake a host of new issues. Simplest, but still irksome, is the need to decide, when constructing a dynamical model, what is the distribution of semi-major axes, eccentricities and mass ratios of the binaries. Much more complicated are the issues involving the finite radii of the stars, the changes in their internal structure caused by their gravitational interaction, and finally a whole host of questions concerning the internal evolution of stars in binaries.

For the time being we shall continue to treat the stars in a binary as point masses, and study the dynamical consequences of the existence of primordial binaries. The other complications we defer to the final part of the book. Of course, it is by no means clear that these complications really are so superficial that they can be treated as a minor perturbation of the picture provided by the point-mass approximation. But, as Donald Lynden-Bell once remarked, 'Only a fool tries to solve a complicated problem when he does not even understand the simplest idealisation.'

Where the action is

Why should we care, from the point of view of dynamics, that many stars are binary? After all, in studying the dynamics of a galaxy, the formation of spiral arms or bars, infall of a companion galaxy, and many other problems, the duplicity of the stellar populations is ignored. The fact is that, nearly everywhere in our Galaxy, the local dynamics of multiple star systems and the global dynamics of the Galaxy as a whole can be neatly separated. In the solar neighbourhood, a double star (or triple or other multiple star system) with a diameter of a few AU will have a negligible chance to interact with neighbouring stars, even on a time scale comparable to the current age of the Galaxy. A rough estimate suffices here. With a density n of one star per cubic parsec, a relative velocity v of 50 km/s, and a target area σ with a radius r of 5 AU, the rate of close encounters is $n\sigma v = \pi r^2 vn = 10^{-13}/\text{yr}$. This gives a probability of only 0.1% for such an event to occur during the next Hubble time.

More distant encounters do occur more frequently, of course, but a single encounter is relatively harmless, and it typically takes many encounters to unbind a wide binary (Fig. 24.2). As a result, only the widest binaries in the solar neighbourhood, with separations of order $10^3 - 10^4$ AU, are potentially subject to dissolution by encounters with passing stars and molecular clouds. However, even in this case the local–global interaction is a one-way street: the effect of the environment on the binary can be dramatic, but the feedback effects on the environment are negligible. A binary with a separation of 10^3 AU typically has a binding energy that is three orders of magnitude lower than the kinetic energy associated with the *relative* motion of single stars in the solar neighbourhood (and even less compared with the energy of motion around the Galactic Centre). Whether or not such a binary is dissolved hardly makes a mark

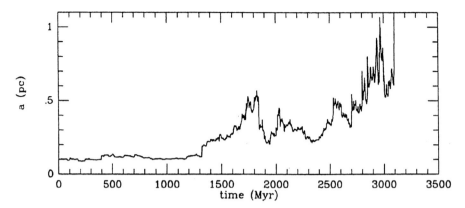

Fig. 24.2. Evolution of the semi-major axis of a soft binary in the Galaxy. From Bahcall *et al.* (1985).

on the motion of the field stars. Clearly, the dynamical interactions with binaries and multiple star systems form an utterly negligible component in the energy budget of the Galaxy as a whole.

Even in the densest parts of the Galaxy, in the inner parsec of the galactic nucleus, the energy locked up in the internal motions of binaries is likely to be at least an order of magnitude less than the energy available in the motions of the single stars and the centre-of-mass motions of the binaries and higher multiple star systems. With a velocity dispersion of more than 100 km/s, only the tightest binaries will have an orbital speed exceeding the typical centre-of-mass motion, and these binaries will only contain a small fraction of all the stars in the field.

The situation is just the opposite in the case of star clusters. Both open clusters and globular clusters have a much lower velocity dispersion than the Galactic Centre, and a much higher density than the solar neighbourhood. The combined effect gives a situation in which a typical binary can easily have an orbital velocity far exceeding the velocity dispersion of the cluster, and therefore have an energy $\gg 1$ kT. Combined with the fact that observations show us that a fair fraction of the stars in clusters have been formed as binaries, as we have seen, it is clear that binaries play an important role in the dynamics of star clusters.

As a result, the total energy locked up in binary binding energy is at least comparable to, and in some cases may well exceed, the total energy of the cluster as a whole (in the form of the kinetic and potential energy of the single stars and of the centres of mass of the binaries). The binding energy of a single binary may easily dominate that of the core. Given this situation, changes in binary properties that take place during the course of normal stellar evolution will have a repercussion on the dynamical evolution of a star cluster as a whole. While it is the macroscopic energy budget that drives this dynamical evolution, this budget can be significantly modified through the strong coupling with the comparable microscopic energy budget of internal degrees of freedom of binary stars. In addition, close encounters involving a combination of single stars and binaries can affect the parameters of the binaries in very complex ways. Either type of process, internal evolution in relatively isolated binaries, or three-body and four-body encounters, will modify the balance between the two energy budgets of a cluster, governing the external and internal degrees of freedom (bulk energy and total binary binding energy, respectively).

Problems

(1) Estimate the rate of formation of hard binaries in the Galaxy, using Eq. (21.5).

(2) Suppose that the giants in a globular star cluster have mass $0.8 M_\odot$ and radius 0.3 AU (cf. Harris & McClure 1983), and that the threshold between hard and soft binaries occurs when the mean square relative speed of the components is 10 km/s. For binaries with one giant component and one small main sequence component of comparable mass, and eccentricity e, compare the semi-major axis for contact of the components at pericentre with the semi-major axis of a binary on the soft/hard threshold.

(3) Binaries are distributed uniformly on the sky with surface density Σ_0 per unit area. Their semi-major axes a are distributed uniformly in $\log a$. If distance r is measured from the location of a particular star, the *mean surface density of companions*, $\Sigma(r)$, is defined to be the average value of $\delta N(r)/(2\pi r \delta r)$, where $\delta N(r)$ is the number of stars with distances between r and $r + \delta r$. Show that $\Sigma(r) \propto r^{-2}$, approximately, for small r, and $\Sigma(r) \simeq 2\Sigma_0$ for sufficiently large r.

25

Triple Formation and Evolution

Triple systems are very familiar. The motion of the Earth and Moon around the Sun is lightly perturbed by the other planets, and if such effects are neglected it is a nice example of a triple system. Furthermore, the distance between the Earth and Moon is much smaller than their distance from the Sun, and so it is an example of what is called a *hierarchical* triple system. The dynamics of such a system can be understood, to a satisfactory first approximation, as two Keplerian motions. In the case of the Earth–Moon–Sun system, one of these is the familiar motion of the Moon relative to the Earth, and the other is the motion of the barycentre of the Earth–Moon system around the Sun. The barycentre lies within the Earth, in fact, and we are more familiar with the picture that the Earth orbits around the Sun, but it is more accurate to say that it is the motion of the barycentre that is approximately Keplerian. This was realised by Newton (*Principia*, Book I, Prop. LXV), and it was he who really originated the study of hierarchical triples.

The mass ratios in the Earth–Moon–Sun system are rather extreme. Even though the Sun is so distant, its mass is so great that it exerts a much greater force on the Moon than the Earth does. Therefore it is not immediately clear why the Moon continues to orbit the Earth, and why the Keplerian approximations are successful (Problem 1). In the same way the gravitational pull of the Sun on the Earth greatly exceeds that of the Moon on the Earth, and so it is strange that the tides in the Earth's oceans generally rise and fall in step with the diurnal motion of the Moon and not that of the Sun.

Beyond the solar system there are triple systems in which the masses are more comparable. In the globular cluster Messier 4, for example, there is a triple consisting of a neutron star in a binary with what is presumed to be white dwarf, and a more distant substellar object (a massive planet) in orbit about the inner binary. The masses are thought to be about $1.35 M_\odot$, $0.35 M_\odot$ and $\sim 0.01 M_\odot$ (Thorsett *et al.* 1999).

Triple formation

How can such a system arise? Well, maybe they were born like that (see Clarke & Pringle 1991). Otherwise the obvious answer, at first sight, is *capture*: a third star approaches a binary, and its perturbing action slightly unbinds the binary and binds it to the binary. Unfortunately we have already seen a mathematical argument (Chapter 20) that three-body capture will not work. This is also readily understood physically (Chapter 19): if the perturbation of the third star is strong enough to bind it, on successive orbits the energy of the third star will perform a random walk and eventually become unbound again. But if binaries are sufficiently abundant in a stellar system, binary–binary interactions are as common as binary–single interactions. It is then natural to ask whether binary–binary encounters could lead to the formation of hierarchical triples with non-zero probability.

Consider first the case in which one of the binaries is far harder than the other. It is then very unlikely that either component of the wider pair will make a close approach with the close pair; close, that is, compared with the smaller semi-major axis. Therefore in most such encounters the close pair will act like a single object with mass equal to the total mass of its two components. Since we are assuming that the masses of all the single stars are comparable, this binary behaves like a relatively massive object. Now we have already seen, in our study of binary–single interactions (Chapter 19), that an exchange interaction occurs with substantial probability, and that a massive intruder is very likely to displace a less massive component of the binary. In the case of binary–binary encounters, then, we may expect that the close pair will eject one component of the wider pair, probably the less massive one. The end result is that the tight binary pairs up with the other component, which acts as a third body orbiting the close pair in a hierarchical triple.

It turns out, as numerical scattering calculations show, that the probability of formation of a hierarchical triple system in these circumstances is substantial. Even for equal masses it occurs in roughly 50% of all encounters in which the minimum distance of the two binaries is of order the sum of their original semi-major axes. And even when the original semi-major axes are equal, so that neither pair can be thought of as acting as a single object, the probability is reduced by a factor of only two or so (Mikkola 1984). Nor is hierarchical triple formation in binary–binary encounters of only theoretical interest: it is plausible that the triple system in the globular cluster M4 formed in this way (Rasio *et al.* 1995).

The formation of triples seems relatively straightforward and common. It is, therefore, worth investigating what happens to them subsequently. Some are not quite hierarchical enough to survive for long: the mutual perturbations of the three stars

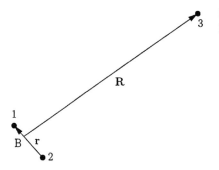

Fig. 25.1. Notation for hierarchical triple systems.

leads to a destruction of the hierarchy or the escape of one of the stars. Finding the boundary line between stable and unstable hierarchies is an important problem. Eggleton & Kiseleva (1995) have given a rather reliable formula[1] determined empirically from many numerical integrations. The analytical approach has proved harder, generally providing criteria that may be rigorous but are wide of the mark in practice. Recently, however, Mardling (2001) has developed a new approach (based essentially on Fig. 10.3) which gives a sharp criterion with a sound theoretical foundation.

Now we turn to the hierarchical triples which pass the stability test. Though there are many kinds of three-body problem, this is the one where the traditional methods of celestial mechanics are most useful. We shall see that the results justify the effort.

Triple evolution

As before, we describe a hierarchical triple as a combination of two Keplerian motions: the relative motion of the 'inner' binary, and the motion of the 'outer' star relative to the barycentre, B, of the inner pair (Fig. 25.1). Let the masses of the components of the inner pair be m_1 and m_2, and let $\mathbf{r}$ be the position vector of the first relative to the second. Let the third star have mass m_3 and position vector $\mathbf{R}$ relative to B. Then the energy of the triple, in the rest frame of its centre of mass, is

$$E = E_i + E_o + V, \tag{25.1}$$

where E_i and E_o are the energies of the inner and outer Keplerian motions, and V is a correction for the fact that the outer star is not actually in orbit about a star of mass $m_1 + m_2$ at B; thus

$$V = -\frac{Gm_1m_3}{|\mathbf{R} - \mu_2\mathbf{r}|} - \frac{Gm_2m_3}{|\mathbf{R} + \mu_1\mathbf{r}|} + \frac{G(m_1 + m_2)m_3}{R},$$

where $\mu_i = m_i/(m_1 + m_2)$ and $R = |\mathbf{R}|$. Now the most systematic way of handling a problem like this is some technique of canonical perturbation theory. We, however, shall proceed by more elementary methods which at least make the results plausible.

---

[1] Except for a misprint: in their Eq. (2), for 2.2 read -2.2.

First let us consider the evolution of E_i. Its rate of change is

$$\dot{E}_i = -\dot{\mathbf{r}} \cdot \frac{\partial}{\partial \mathbf{r}} V(\mathbf{r}, \mathbf{R}). \tag{25.2}$$

In a significantly hierarchical triple with comparable masses, the time scale on which $\mathbf{r}$ varies is much smaller than the time scale for the variation of $\mathbf{R}$. Therefore, when integrating Eq. (25.2), we may hold $\mathbf{R}$ constant to good approximation. At lowest order, then, the evolution of E_i is given by $E_i(t) = E_i(0) - V(\mathbf{r}(t), \mathbf{R}) + V(\mathbf{r}(0), \mathbf{R})$. The right-hand side here is periodic at lowest order, and so there is no secular (long-term) evolution in E_i and hence in the semi-major axis, a_i, of the inner binary.

When it comes to the perturbations in the motion of the third body, we may average V over the fast motion of the inner binary, as the difference between V and its averaged value has negligible effect when integrated over the period of the inner binary.[2] The averaged value depends on the elements of the inner binary, which vary very slowly, even by comparison with the outer period, as we shall see. It follows, then, from an argument similar to that applied to Eq. (25.2), that E_o also has no long-term variation, and so neither has the outer semi-major axis, a_o.

Now we know that the left-hand side of Eq. (25.1) is constant and that E_i and E_o have only periodic perturbations, but do not evolve secularly. It follows that the orbit-averaged value of V, i.e. averaged over the motion of both the inner and outer binaries, is constant. The averaging itself is carried out in Box 25.1, and it turns out that the result is independent of the orientation of the orbit of the third body within its plane. This is remarkable, because it implies that the magnitude of the orbit-averaged angular momentum of the outer binary, c_o, is constant. The argument leading to this conclusion is analogous to the familiar fact that the angular momentum of a system is conserved if the Lagrangian of the system is invariant under rotations. It may also be viewed as an application of Noether's Theorem (Box 7.1), or as resulting from the fact that the angular momentum of the outer orbit is conjugate to the orientation of the orbit, in the sense of Hamiltonian dynamics.

Now we are in a position to exploit the fact that the total angular momentum of the whole triple system, $\mathbf{c}$, is constant, for $\mathbf{c} = \mathbf{c}_o + \mathbf{c}_i$, and so

$$c^2 = c_o^2 + 2c_o c_i \cos i + c_i^2 \tag{25.3}$$

is constant, where i is the relative inclination of the inner and outer orbital planes. Now, as we have seen, c_o is constant on average, and so the sum of the last two terms on the right of Eq. (25.3) is also constant on average. However, we see from Eq. (25.5), and the corresponding expression for c_i, that the last term on the right of Eq. (25.3) is negligible, since $a_i \ll a_o$ in a hierarchical triple, and so the average value of $c_o c_i \cos i$ is constant. Finally, using the definition of c_i in terms of the elements

---

[2] Care is needed if the inner and outer binaries are in *resonance*. In practice this exception is important only when the ratio of the outer and inner semi-major axes is small enough that the hierarchy is close to being unstable (cf. Kiseleva *et al.* 1994).

Box 25.1. Averaging the perturbation potential

In order to evaluate the average of V, we first approximate it by the first non-zero term in a multipole expansion, which gives

$$V \simeq -\frac{Gm_1m_2m_3}{2(m_1+m_2)R^5}(3(\mathbf{r}\cdot\mathbf{R})^2 - r^2R^2). \tag{25.4}$$

In order to perform the orbit average over the motion of the outer binary, we shall choose a coordinate system aligned with the major and minor axes of the outer orbit. In this frame we write $\mathbf{R}$ in components as $\mathbf{R} = R(\cos\theta, \sin\theta, 0)$. Then it is easily seen that, except for factors depending on G, $\mathbf{r}$ and the masses, a typical term to be averaged is of the form $V_{ij} = X_iX_j/R^5$, where X_i and X_j are two of the components of $\mathbf{R}$. For example, if $i = j = 1$ its average is $\langle V_{11}\rangle = P_o^{-1}\int R^{-3}\cos^2\theta dt$, where P_o is the outer period. From standard formulae of Keplerian motion we have

$$R^2\dot\theta = \sqrt{G(m_1+m_2+m_3)a_o\left(1-e_o^2\right)} \tag{25.5}$$

(specific angular momentum) and $R = a_o(1-e_o^2)/(1+e_o\cos\theta)$ (the equation of an ellipse). It follows that

$$\langle V_{11}\rangle = \frac{1}{2a_o^3\left(1-e_o^2\right)^{3/2}}, \tag{25.6}$$

where we have used a standard Keplerian result for the orbital period in the form $P_o = 2\pi a_o^{3/2}/\sqrt{G(m_1+m_2+m_3)}$.

　Similar calculations show that the only other non-zero average is $\langle V_{22}\rangle$ and that

$$\langle V_{22}\rangle = \langle V_{11}\rangle. \tag{25.7}$$

This demonstrates that the averaged perturbation is independent of the orientation of the orbit of the third body within its plane of motion.

---

of the inner orbit, as well as the fact that a_i and c_o are constant on average, we see that $c_\perp = \sqrt{1-e_i^2}\cos i$ is constant on average. The notation reminds us that $c_\perp$ is proportional to the component of the inner angular momentum perpendicular to the outer orbit, but we have omitted a factor involving several quantities which are either constant or constant on average.

　This is our first interesting non-trivial result. To investigate its consequences, let us denote initial values by (0). Then, neglecting periodic variations, what we have found is that

$$\sqrt{1-e_i^2}\cos i = \sqrt{1-e_i^2(0)}\cos i(0). \tag{25.8}$$

Suppose, for example, that the inner and outer orbital planes are nearly orthogonal, so that the right side is nearly zero. It follows that, as i and e_i evolve, e_i may reach values close to $e_i = 1$ (depending on the maximum value of $\cos i$). Then the distance of closest approach of the components of the inner binary may be much smaller than their semi-major axis, and may even be smaller than the sum of their physical

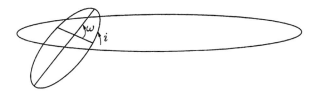

Fig. 25.2. Geometry of a
hierarchical triple system.

radii. (We are remembering now that these are not point masses, but stars.) This has obvious astrophysical consequences, but of course it does not follow, just because high eccentricities are possible, that they are inevitable. We still have some work to do.

Before we go on it is interesting to mention another context in which similar phenomena arise. It concerns a fascinating group of comets named the Kreutz group, after the nineteenth century astronomer who first drew attention to them. They are more usually called the 'sungrazers', because they approach the Sun's surface to within a distance of about a solar radius. It has been proposed (Bailey *et al.* 1992) that they originated in orbits of very high inclination and much larger (i.e. more normal) perihelion distance, and that the perturbations by the planets have evolved their orbits so that they now have relatively low inclination and very high eccentricity.

We return to the averaging of V. In Box 25.1 we substitute Eqs. (25.6) and (25.7) into Eq. (25.4), and soon find that V, averaged over the outer motion, takes the form

$$\langle V \rangle = -\frac{Gm_1 m_2 m_3}{4(m_1 + m_2)a_o^3 \left(1 - e_o^2\right)^{3/2}}(r^2 - 3z^2),$$

where z is the component of $\mathbf{r}$ orthogonal to the outer orbital plane. Now it is straightforward, but a shade tedious, to perform the orbit average over the inner motion. As variable of integration it is best to use the *eccentric anomaly* E,[3] and all that is needed are some formulae of Keplerian motion (Box 7.2), and a little three-dimensional geometry (Fig. 25.2). We simply state the result, which is that the doubly orbit-averaged expression for V is

$$\langle\langle V \rangle\rangle = -\frac{Gm_1 m_2 m_3 a_i^2}{8(m_1 + m_2)a_o^3 \left(1 - e_o^2\right)^{3/2}}$$
$$\times \left(2 + 3e_i^2 - 3\sin^2 i \left(5e_i^2 \sin^2 \omega + 1 - e_i^2\right)\right), \tag{25.9}$$

where ω (the 'argument of pericentre') is the angle between the major axis of the inner orbit and the line in which the two orbital planes intersect (the 'line of nodes').

Now we can visualise the secular evolution of the triple system. In Eq. (25.9) there are really only three variables: e_i, i and ω. Also, $\langle\langle V \rangle\rangle$ is a constant, which can be evaluated from the initial conditions, and so Eq. (25.9) states that the evolution takes place on a certain surface in the space defined by these three variables (Fig. 25.3). Equation (25.8) defines another surface in the same space, and so the evolution must take place on their intersection, which is a curve. It can be seen from the intersections

---

[3] Not to be confused with energy!

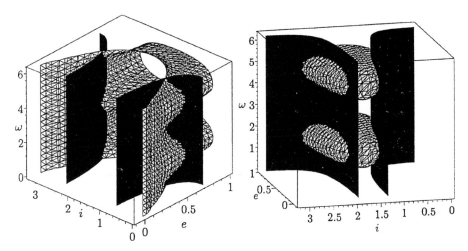

Fig. 25.3. Evolution of a hierarchical triple system according to the method of averaging. The evolution takes place along a curve of intersection of the two surfaces, which are given by Eqs. (25.8), solid, and (25.9), wire frame. The argument of pericentre, ω, may travel through $360°$ (left) or oscillate (right).

that, though the evolution takes two rather different forms, in both forms the evolution of the eccentricity and the inclination is oscillatory. What distinguishes the two forms is the evolution of the pericentre ω, which may either oscillate or 'circulate' (Fig. 25.3), rather as a simple pendulum may either oscillate or swing through $360°$. In either event the periodic oscillations in e_i and i are often called *Kozai cycles* (Kozai 1962).

We also notice from Fig. 25.3 that the maximum value reached by the eccentricity is less, in general, than would be predicted naïvely from Eq. (25.8), because the values are also constrained by Eq. (25.9). In our previous discussion it was assumed that the maximum value occurs when $i = 0°$ or $180°$, but in fact the values of i are constrained within narrower limits, in general, by Eq. (25.9). Nevertheless, it is still true that large eccentricities may occur if the initial inclination is high, i.e. near $90°$ (Fig. 25.4).

It is worth making an observation which is, at first sight, very surprising, even to experts, and that is that these variations in i, e_i and ω do not depend on the masses or the dimensions of the orbits. One might have thought that if the third body were sufficiently far away or of sufficiently low mass its influence would remain very small. In fact, these affect only the time scale on which the evolution takes place, and not the size of its effects.

We have not quite finished, for Eq. (25.9) is the potential which the inner orbit experiences, on average. It depends on two angles, and so the inner orbit experiences a torque. Variations of ω correspond to rotation about an axis normal to the inner orbital plane, and so the corresponding torque causes variations in the magnitude of

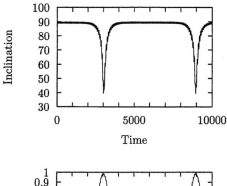

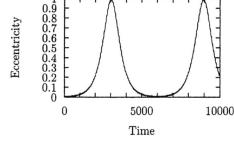

Fig. 25.4. Evolution of the inclination (degrees) and inner eccentricity in a hierarchical triple system in which the initial inclination was 90°. Both inner and outer orbits were initially circular.

the angular momentum. It is clear that this torque vanishes if $e_i = 0$, as is easily verified by computation of $\partial\langle\langle V\rangle\rangle/\partial\omega$. Variations of i correspond to rotation about the 'line of nodes', which is the imaginary line in which the two orbital planes intersect. It is easy to see that this torque causes the angular momentum of the inner orbit, which is orthogonal to the inner orbital plane, to precess about the normal to the outer orbital plane (at least, at the present level of approximation).

Though we have obtained this result by simplified perturbation theory, Newton gave a very simple qualitative explanation of this precession[4]. In Fig. 25.2, suppose that m_3 is at the extreme left-hand side of its orbit. After m_1 moves up through the plane of motion of m_3, it is attracted to m_3 less than m_2 is. Hence m_1 descends through the plane of m_3 at a point displaced (in a sense opposite to the motion of m_1) from the original line of nodes. The same thing happens when m_1 passes below the orbital plane of m_3, and when m_3 itself is in other positions.

We can easily estimate the angular velocity of precession, although in Box 25.2 we do so only in the case in which $e_i \simeq 0$, by way of illustration. The result is that the period of precession is

$$\frac{2\pi}{\Omega} = \frac{4}{3}\frac{m_1 + m_2 + m_3}{m_3}\frac{P_o^2}{P_i}\left(1 - e_o^2\right)^{3/2}/\langle\cos i\rangle,\qquad(25.10)$$

where P_i and P_o are the inner and outer periods, respectively, and the average of $\cos i$ is taken over one Kozai cycle. This result implies that, for given masses, e_o and mean

---

[4] *Principia*, Book I, Prop. LXVI, Cor. XI. See also Airy (1884), Moulton (1914).

Box 25.2. Precession of an inner circular orbit in a hierarchical triple system

When $e_i = 0$, the torque is simply

$$\frac{\partial \langle\langle V \rangle\rangle}{\partial i} = \frac{3Gm_1m_2m_3a_i^2}{4(m_1 + m_2)a_o^3 \left(1 - e_o^2\right)^{3/2}} \sin i \cos i,$$

acting parallel to the outer orbital plane. Now the component of the inner orbital angular momentum parallel to this plane is

$$c_{\parallel} = \frac{m_1m_2}{m_1 + m_2} \sqrt{G(m_1 + m_2)a_i} \sin i$$

if e_i is negligible. The angular speed of precession, Ω, is obtained by dividing this into the torque, and so

$$\Omega = \frac{3m_3}{4a_o^3} \sqrt{\frac{G}{m_1 + m_2}} \left(\frac{a_i}{1 - e_o^2}\right)^{3/2} \cos i.$$

This is rather unmemorable, and it is better to express the result in terms of the periods (see text, Eq. (25.10)).

---

inclination, the periods are in geometric progression: the outer period is of order the geometric mean of the inner and precession periods. This fact, which was known to Newton,[5] is a general rule of thumb for the time scale of long-period perturbations in hierarchical triples, and for some observed triple systems it leads to a time scale short enough that the effects are detectable within a few years (Jha *et al.* 2000). It also explains the sense in which the perturbations by a distant or low-mass companion are weak: the evolutionary effects are just as large as for a close or massive companion, but take much longer to develop.

Anyone who has taken more than a casual interest in eclipses of the Sun and Moon will be aware of our last result on time scales. Eclipses occur at two times of the year, called 'eclipse seasons', when the Sun lies close to the line of nodes, i.e. the line of intersection of the orbital planes of the Moon (about the Earth) and the Earth (about the Sun). Because of the perturbation of the Sun on the Earth–Moon system, the orbital plane of the Moon precesses, the line of nodes rotates in the plane of motion of the Earth, and so the time of year at which eclipse seasons occur slowly changes. The period of the precession in this case is 18.6 yr. Since the ratio of the year to the month is about 13.4, Eq. (25.10) predicts 17.8 yr.

It is appropriate to end where we began, with the most familiar of all hierarchical triple systems. It was here, moreover, that the peculiar property of high-inclination orbits found one of its earliest applications (Lidov 1963). It is little wonder that the Moon is not on a high-inclination orbit!

---

[5] *Principia*, Book III, Prop. XXIII (Newton 1729).

Problems

(1) Explain why the Moon continues to orbit the Earth even though the force
     exerted by the Sun on the Moon is much larger than that exerted by the Earth
     on the Moon. Explain why the lunar tide is higher than the solar tide, even
     though the gravitational attraction of the Moon on the Earth is much weaker
     than that of the Sun.

(2) Use Eqs. (25.8) and (25.9) to compute the maximum eccentricity of the inner
     orbit if it is initially nearly circular.

(3) Complete the derivation of Eq. (25.9).

26

A Non-Renewable Energy Source

As we have seen, primordial binaries exist in some abundance in globular clusters, and so there is good reason to study the behaviour of a million-body problem in which there are many binaries. How do such clusters evolve?

To some extent the binaries may be treated as heavy point masses, provided we are interested in interactions with other stars at distances much greater than the typical semi-major axis of a binary. Thus the binaries behave like a heavy species, and the process of mass segregation (Chapter 16) ensures that they become heavily concentrated towards the core of the cluster. It follows that, even if we start with a cluster in which only ten per cent of the stars are binaries, we quickly find ourselves looking at part of the system (the core) where the abundance of binaries is more like 90%. Incidentally, it has even been argued that young star clusters may consist entirely of binaries. For one reason or another, then, the study of stellar systems in which essentially all stars are binaries is an important one.

Something was already said about the interaction between two binaries in the previous chapter. There we quickly focused on the formation of hierarchical triple systems in such encounters. In the present chapter we focus on the energetics of these interactions. We shall see in the next chapter that this is the main way in which these interactions feed back into the overall dynamics of a stellar system.

It is very instructive to compare the energetics of binary–binary interactions with that of three-body interactions (Chapter 23). Conservation of energy immediately tells us that the formation of binaries in three-body interactions is necessarily exothermic, because a bound system is formed from three unbound particles. Primordial binaries, on the other hand, are initially endowed with negative energy, and it is only in their subsequent interactions that this can influence the evolution of the N-body system. Again, once a hard three-body binary has been formed it is very difficult to destroy: it keeps hardening and heating the stellar system until it is ejected. By contrast, we have seen in the previous chapter that binary–binary encounters can easily destroy one of the participants, perhaps forming a hierarchical triple system, and this suggests that the useful energy inside a population of primordial binaries is limited. On the other hand, it is clear that the cross section for an encounter between one binary and another exceeds that for interaction between the same binary and a single star: gravitational focusing is somewhat stronger, because the participating masses are larger, and the physical size of the participants is somewhat greater. When we weigh these considerations together, it is not very clear whether or not binary–binary interactions will be as effective a source of heat as three-body interactions.

Let us look at these interactions in a little more detail. We saw in Chapter 22 that, even for three-body interactions, the number of parameters required to specify an interesting set of encounters is already rather substantial, and this problem is much worse for binary–binary encounters. Therefore we shall make matters as simple as possible by considering stars of equal mass. It might be thought also simplest to consider binaries with equal initial energy or semi-major axis, but for our present purposes it is helpful to return to the rather extreme form of binary–binary encounter considered in the previous chapter, in which one binary is much harder than the other.

Broadly speaking, the outcome of a binary–binary encounter may be classified in the same way as three-body interactions: fly-by, resonance and exchange (Box 26.1). (With our choice of parameters the closest analogy in three-body scattering is encounters between a binary and a single star in which the mass ratios are 1 : 1 : 2, the last of these being the harder binary.) Indeed, numerical scattering experiments show that the three outcomes occur in roughly comparable amounts, for impact parameters in the range in which exchange and resonance encounters can occur.

A fly-by is the easiest to understand by three-body analogies. The less hard binary is hardened (by an amount of order its initial energy, or $\langle\Delta\rangle\varepsilon$ on average, cf. Chapter 23) and both reaction products retire from the encounter at increased speed. Thus the encounter is exothermic (on average), and both binaries survive. The result is the same if the harder binary is captured in a 'hierarchical resonance', which will eventually decay again.

Prompt exchanges are also easy to understand. The massive incomer displaces one component of the initially wider binary, without necessarily altering its semi-major axis much, and increases the binding energy of the binary by roughly a factor of two. This reaction is also exothermic, therefore. The difference between this kind of encounter and a fly-by is that the reaction products are now a single star and a hierarchical triple system. Though this may be perfectly stable in isolation, it now presents a large cross section to interactions with other single stars. The effects of these will tend to harden the motion of the outer star in the triple, and eventually destroy the temporary stability of the hierarchy. The outcome is the same if the harder incoming binary is insufficiently hard, for then the triple is not hierarchical

Box 26.1. Interactions between two binaries

Fig. 26.1 shows the various outcomes of binary–binary scattering when one binary is so hard that its energy is unaffected by the encounter. In effect, this is three-body scattering, and the state referred to as 'democratic quadruple' is not *truly* democratic.

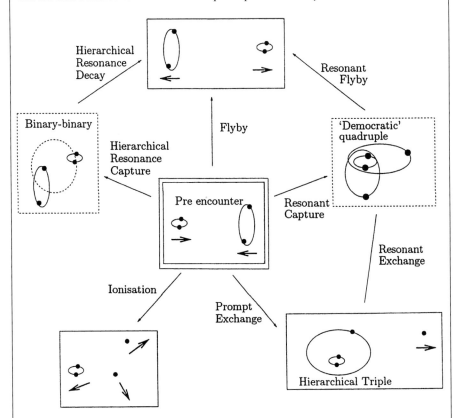

Fig. 26.1. Some of the reaction routes for interactions between one binary and another much harder one. Initial state: double box; intermediate state: dashed box; final state: solid box.

If we allow the harder binary to be slightly affected by the encounter, but not to be disrupted, the number of possibilities increases. Now it is possible for a binary–binary configuration to evolve into a 'democratic quadruple', which can now be truly democratic. Another possible temporary outcome of the encounter is a four-body hierarchy, consisting of an inner hierarchical triple surrounded by a distant fourth body.

In general (when there is no restriction on the hardness of either binary) the outcomes may be classified as follows. Let x, x denote the components of the first binary, and y, y those of the second binary. We use notation in which stars which are bound into a binary are enclosed in square brackets, whereas braces (i.e. {}) are used to group objects which are not bound to each other. Thus the initial configuration of a binary–binary encounter is represented by $\{[x, x], [y, y]\}$. With this notation, Table 26.1 lists the various possible final outcomes. For those with an asterisk, a new configuration is obtained by interchanging x and y. Some listed outcomes would be very rare, e.g. those involving a bound triple system such as $[x, x, y]$ (Fig. 26.2).

Table 26.1. *Stable outcomes of a binary–binary encounter*

$\{x, x, y, y\}$	Complete ionisation
$^*\{[x, x], y, y\}$	Ionisation of one pair
$\{[x, y], x, y\}$	Ionisation and exchange
$\{[x, x], [y, y]\}$	Fly-by
$\{[x, y], [x, y]\}$	Complete exchange
$^*\{[[x, x], y], y\}$	Hierarchical triple with escape
$^*\{[[x, y], x], y\}$	Hierarchical triple with exchange and escape
$^*[[[x, x], y], y]$	Hierarchical quadruple
$^*[[[x, y], x], y]$	Hierarchical quadruple with exchange
$^*\{[x, x, y], y\}$	Bound triple with escape
$^*[[x, x, y], y]$	Hierarchy with bound triple

The list can be extended to include temporary intermediate states (as in Fig. 26.1) by replacing braces with square brackets. Thus the first item becomes $[x, x, y, y]$ (democratic quadruple). The number of outcomes is also greater if we wish to distinguish the two components of each binary, in which case we must introduce different letters for each component. (Table 26.1 only distinguishes the two binaries.) With some effort, the notation can even be extended to provide a systematic classification of the outcome of encounters between arbitrary systems.

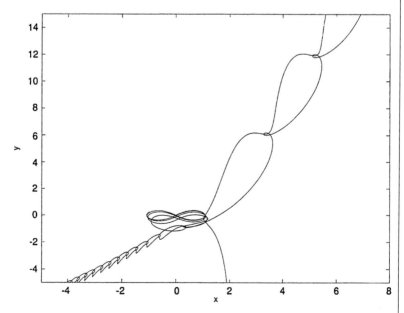

Fig. 26.2. A rare outcome of binary–binary scattering; formation of a stable triple system with one escaper (from Heggie 2000).

enough to be stable. (These refinements are not shown in Fig. 26.1 in Box 26.1, which restricts itself entirely to an isolated binary–binary interaction in which one binary is arbitrarily hard.)

Finally we turn to democratic resonant encounters, i.e. the formation of a temporary bound quadruple system. Eventually, just like a temporarily bound triple, this is expected to resolve with the ejection of one star (in which we continue to think of the harder binary as a single massive object). The ejected star may be the binary, in which case the outcome is just like a fly-by. In fact, however, the escaper is likely to be one of the stars of lower mass, i.e. one of the components of the initially wider binary. In that case the result is a hierarchical triple and a single escaper. But we must pause before concluding that the result is like that of an exchange encounter. During the evolution of the triple system, the three stars (one of which is a close binary) have many opportunities for a close encounter. Indeed, numerical studies show that it is quite likely that, at some stage before the evolution of a triple system is complete, the smallest separation of some pair of stars is no more than about 1% of the initial separation of the binary (Hut & Inagaki 1985), which here means the wider binary. Now we did assume that the other binary was much harder than the wider binary, but we now see that even a factor of 100 in hardness may not be enough to ensure that the binary–binary interaction behaves like a triple interaction, if a democratic resonance is formed. If indeed there is a close interaction between the hard binary and one component of the wider pair, this is likely to lead to the ejection of that component, and possibly also of the binary. If the latter happens, the result of the encounter is a single binary and two single stars; the energy released will be comparable with that of the initially harder binary. If, however, the binary remains bound to the other component of the originally wider binary, then the resulting triple system may be either hierarchical or unstable.

This has been a somewhat elaborate discussion, which, incidentally, illustrates that the problem of classifying these events automatically is non-trivial. The bottom line, however, is that binary–binary interactions release energy comparable with that of the less hard component (and sometimes that of the harder component), but that the probability of destroying one of the binaries is high, certainly above 50%. It follows that the maximum energy that may be extracted from a binary, up to the point of its destruction, is of order its binding energy. This is very different from the behaviour of three-body binaries, whose energy yield is determined by the central potential (Chapter 23). In other respects, however, much the same considerations apply as with three-body binaries. In particular, reaction products of binary–binary interactions are often found well outside the core, from which they drizzle back into the core by dynamical friction (Hut *et al.* 1992b).

With these conclusions in mind, let us return to the problem of a stellar system containing a modest proportion of binaries initially, say f by number.[1] The simplest

---

[1] This means that, if the cluster contains N objects, where an 'object' may be either a binary or a single star, there are fN binaries.

way of estimating their effect is by comparing the energy that they may release with the energy of the whole cluster. The latter consists of two contributions: the internal energy of the binaries, E_{int}, and all the rest, which by contrast we refer to as 'external' energy, E_{ext} (cf. Chapter 19). Approximately, E_{ext} is the energy we would compute by thinking of each binary as a single object, i.e. by replacing it by its total mass located at its barycentre. It is this energy which determines the virial radius of the cluster. Now a binary which is on the border between being soft and hard has an internal energy of approximately E_{ext}/N, where N is the total number of objects in the system (Chapter 19). Denoting by x the 'hardness' of each binary, i.e. its energy in units of this threshold energy, we see that $E_{int} \simeq f x E_{ext}$, since the number of binaries is $f N$. Now we have already seen that the energy which these binaries may yield is of order E_{int} itself, and therefore this may have a significant effect on the cluster as a whole only if $f x \gtrsim 1$.

From this calculation we see that, if $f \ll 1/x$, the primordial binaries can have only a modest influence on the overall evolution of the cluster. They may well burn up rapidly and yield all the energy they can, but they are a non-renewable energy source, and if more energy is needed to sustain the evolution of the system (Chapters 27, 29) it must come from elsewhere.

From what has been said, it seems that primordial binaries may have a dominant influence on the energetics of a stellar system if either they are of modest hardness and substantial abundance (e.g. $x \sim 5$ and $f \sim 20\%$), or there are a few very energetic binaries. In the latter case, however, it is not enough simply to consider the energetics of the situation; the time scale on which the energy is released is also relevant. Roughly speaking, the time scale for close interactions of a binary of hardness x varies as $x t_r$, where t_r is the relaxation time (Problem 23.1). Therefore it is clear that, for extremely hard binaries, the time scale greatly exceeds that on which other dynamical processes, and the evolution of the system as a whole, are taking place. Indeed, there are some non-trivial dimensionless factors in the expression for the time scale of binary hardening, and even binaries with $x \gtrsim 100$ are relatively inefficient at powering the evolution of a stellar system. The most important binaries are those with a hardness of order 10 (plus or minus about half a dex), and it is clear that their abundance must be not much less than about 10% by number if they are to provide enough energy to feed the evolution of a stellar system. It is rather curious that the observations of binaries in globular clusters imply a population of roughly this size in the appropriate range of semi-major axes.

All of our discussion has been based on the point-mass approximation. Close binary–binary interactions with stars of finite radius can be studied with suitable simulations, and reveal a variety of important differences (Goodman & Hernquist 1991).

Problems

(1) Following the guidance in Box 26.1, construct a list of temporary (intermediate) outcomes of a binary–binary encounter. Hence try to construct

pathways, as in Box 26.1 (Fig. 26.1), leading to each permanent outcome on the list in Table 26.1. (For some practical examples, see Aarseth & Mardling 2001.)

(2) Estimate the cross section for close interactions between two hard binaries. If one is much harder than the other, how is the result altered if one seeks the cross section for significant perturbation of *both* binaries?

(3) Obtain the Starlab package (www.manybody.org/starlab.html and Box 22.1). The command

```
scatter -i "-M 1 -r 0 -v 1 -t -a 1 -e 0 -p -a 1 -e 0"\
-t 100 -D 0.01|xstarplot
```

will generate a random head-on, equal-mass, binary–binary encounter. By viewing the plot and inspecting the analysis of the final configuration, try to follow the interaction via Box 26.1, Fig. 26.1, or its extension to binaries with comparable energies (Problem 1).

Part VIII

Post-Collapse Evolution: $N = 10^6$

The following three chapters complete the story of the evolution of a million-body system, in its purely stellar dynamical form. Chapter 27 begins by estimating the rate at which the formation and evolution of (non-primordial) binaries effectively generates energy within the system. The first application is to show that this is sufficient to halt core collapse. Then we consider other ways of generating the energy: binaries formed in dissipative two-body encounters between single stars, and primordial binaries; we quantify the extent to which the effectiveness of primordial binaries depends on their abundance and their energy.

In Chapter 28 we consider how a balance can be struck between the creation of energy (by binary interactions) deep in the core and the large-scale structure of the rest of the cluster. We first describe a standard argument which implies that conditions in the core, where the energy is generated, are governed by the overall structure. We outline the core parameters and overall evolution which this argument implies. Next we give arguments to show that this balance can be unstable, and describe the phenomenon of *temperature inversion* which is associated with the generation of *gravothermal oscillations*. The manner in which they depend on N (in idealised models) is an example of a *Feigenbaum sequence* of 'period-doubling bifurcations' in this context.

Chapter 29 rounds off the evolution of a million-body system by focusing on the evolution of gross structural parameters: total mass, and a measure of the overall radius. The result depends on the boundary conditions, and is discussed separately for an isolated system and one limited by a tidal field.

27

Surviving Core Collapse

The last eight chapters, dealing as they have done with interactions between only three or four stars, might seem a long digression away from the subject implied by the title of this book. Yet we shall see, as we take up the thread of the million-body problem where we broke off at the end of Chapter 18, that an understanding of the behaviour of few-body systems is crucial in following the evolution of the system through core collapse and beyond.

We left the system rushing towards core collapse, its central density rising inexorably, so that it would reach infinite values in finite time. How is this catastrophe averted? In fact there is no shortage of choices, for at least five different mechanisms have been proposed over the years. Admittedly, two are rather out of favour at present: a central black hole (e.g. Marchant & Shapiro 1980), or runaway coalescence and evolution of massive stars (Lee 1987a, and Problem 1). The other three involve binary stars in one guise or another, and it is not hard to see why this is attractive. After all, the mechanism responsible for core collapse is a two-body one (Chapter 14). Therefore it is clear that higher-order interactions, which we have neglected so far, might in principle eventually compete with two-body relaxation when the density becomes high enough. And three-body interactions can create binaries (Chapter 21 and Fig. 27.1).

While the creation of binaries in three-body interactions at least brings binaries into the picture, it is not the principal mechanism by which core collapse is halted. Nor is this the only way in which binaries may enter the scene. In this chapter we examine three possibilities: besides 'three-body binaries', which is the name given to binaries formed in the way we have just mentioned, we consider 'tidal-capture binaries', formed in dissipative encounters between two single stars (Chapter 31) and 'primordial binaries', formed along with the single stars at the birth of the stellar system (Chapter 24).

From the astrophysical point of view the second of these is an 'also-ran'. In fact, since the discovery of primordial binaries, the third option is the hot favourite. But there is plenty of room for a side bet on three-body binaries, and the reason for this is that there are some observed clusters whose structure may be hard to understand if we rely on primordial binaries to halt core collapse. As we shall see, in the presence of primordial binaries, the core does not shrink that much before the collapse is halted. But there is a sizable fraction of the globular clusters in our Galaxy whose cores are unobservably small (Trager *et al.* 1995). The classic example is M15 (Fig. 1.2; Guhathakurta *et al.* 1996; for a more extended history of ideas on this cluster, see Meylan & Heggie 1997). At one time the dense core in this object was explained by supposing that it surrounded a black hole (Newell *et al.* 1976). Then it was explained by a static model containing a dense core of neutron stars (Illingworth & King 1977). More recently still it has been modelled successfully by an evolving Fokker–Planck simulation (e.g. Dull *et al.* 1997), in which the core is dominated by neutron stars, and the post-collapse evolution of the cluster is powered by three-body binaries (Lee 1987b). The number of neutron stars lies well within observationally based limits (Wijers & van Paradijs 1991). None of these models have a need for primordial binaries, and it may be that primordial binaries are somehow ineffective in clusters like M15.

Three-body binaries

The formation of three-body binaries is an exothermic reaction, in the sense described in Chapter 23. Two single stars are replaced by a binary, a third single star acting as a catalyst (Fig. 27.1). Conservation of energy in this reaction leads to the relation

$$\frac{1}{2}m_1 v_1^2 + \frac{1}{2}m_2 v_2^2 + \frac{1}{2}m_3 v_3^2 = \frac{1}{2}M_{12} v_{12}^2 + \frac{1}{2}m_3 v_3'^2 - \frac{Gm_1 m_2}{2a},$$

where the subscript 12 refers to the barycentre of the binary (i.e. $M_{12} = m_1 + m_2$), a is its semi-major axis, and v_3' is the new speed of the catalyst. The negative energy locked up in the relative motion of the components of the binary (the last term) is compensated by increased kinetic energy of the catalyst and the barycentre of the binary.

Formation of the binary is only a small part of the story. We have already seen (Chapters 19, 23) that a hard binary, once formed, continues to harden, and this further increases the kinetic energy of the single stars and the barycentre of the

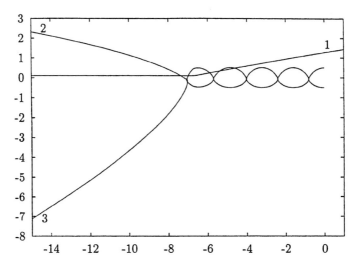

Fig. 27.1. Formation of a binary in a triple encounter. The stars are labelled at the start of the motion. After a close encounter between 2 and 3, particle 1 absorbs enough energy to bind them.

binary. Intuitively it seems plausible that the resulting enhancement of the kinetic energy of the particles in the core should tend to resist continued collapse.

The rate at which energy is generated by binaries is not at all the same thing as the rate of hardening. Though the mean rate of hardening of an individual binary is nearly independent of its energy, provided that it is hard (Chapter 23), a given binary is an effective source of energy only for a limited time. The increasing kinetic energy imparted to its barycentre gradually expels it to larger and larger distances from the core (Fig. 23.2), and then its further interactions become less and less frequent. In fact each binary yields an energy of order a multiple of the individual stellar kinetic energy in the core before it is effectively switched off by its increasingly far-ranging orbit (if it is not expelled from the cluster altogether).

The rate at which energy is created by the formation and evolution of 'three-body binaries' was estimated in Eq. (23.5). It is important to note, however, that, although binaries are predominantly formed in the core, relatively little of the energy they release is deposited there. As soon as the energy of a binary significantly exceeds the mean kinetic energy of single stars in the core, the products of energetic three-body interactions are expelled from the core, and deposit most of their energy elsewhere. In fact the energy added to the core (per binary) may be only of order a few times the mean kinetic energy of one core member, i.e. of order a few times $m\sigma_c^2$ (cf. Chapter 23 and Spitzer 1987, Sec. 7.1), where σ_c is the rms one-dimensional velocity of stars in the core. Therefore we shall simply neglect numerical factors, and, denoting by r_c the radius of the core (Eq. (8.9)), we see that

$$\dot{E} \sim G^5 m^6 n^3 r_c^3 \sigma_c^{-7}. \tag{27.1}$$

In order to establish that the role of binaries can be important we must compare this with the energetics of core collapse. One approach is to think of the star cluster as a conducting stellar gas (Eq. (9.6)[1]), and apply simple thermodynamics to the core. In the first law ('$TdS = dQ$'), we recall that the velocity dispersion is proportional to the temperature, T, while the change in entropy, dS, is the change in specific entropy times the mass. The change in thermal energy, dQ, results from conduction (denoted by a conductive luminosity in the gas model) and from energy generation (caused by formation and evolution of binaries). Applying the first law to the matter inside radius r_c, we see that

$$4\pi \int_0^{r_c} mn\sigma^2 \frac{D}{Dt} \ln \frac{\sigma^3}{n} r^2 dr = \dot{E} - L, \qquad (27.2)$$

where the logarithm is essentially the specific entropy of the gas, the symbol D/Dt denotes a Lagrangian time-derivative, and L is the conductive flux of energy across the Lagrangian shell of instantaneous radius r_c.

Until energy generation by binaries becomes important, the balance of this equation is struck between the left side and the flux L. Estimating the Lagrangian derivative factor by the reciprocal of the current time scale of core collapse, t_{cc}, which shrinks as the core collapses, we deduce the estimate

$$L \sim mn_c r_c^3 \sigma_c^2 / t_{cc}, \qquad (27.3)$$

where n_c is the number density of stars in the core. (This estimate may be equally well obtained by equating L to the rate of change of kinetic energy of the core.) During core collapse t_{cc} is related to the two-body relaxation time in the core, and in late collapse the relation is close to proportionality (Chapter 18). Thus we estimate

$$L \sim G^2 m^3 n_c^2 \ln \Lambda r_c^3 / \sigma_c, \qquad (27.4)$$

where Λ is the argument of the 'Coulomb logarithm' (Chapter 14).

Now it is clear that generation of energy by binary formation and hardening becomes important if the two terms on the right of Eq. (27.2) become comparable, i.e. if $L \sim \dot{E}$. It follows from Eqs. (27.1) and (27.4) that this leads to a relation of the form $(Gm)^3 n_c \sim \sigma_c^6 \ln \Lambda$. To make this more meaningful, we recall (Eq. (8.9)) that the core radius is often defined by $8\pi Gmn_c r_c^2 = 9\sigma_c^2$. It follows, then, that the energy supplied by binaries is important when $n_c r_c^3 \sim 1/\sqrt{\ln \Lambda}$. The dependence on the Coulomb logarithm here is unimportant. Also, in our derivation we have ignored all numerical factors. Therefore our conclusion should be that *binary heating becomes important when the number of stars in the core has shrunk to a certain value.* More careful estimates summarised by Hut (1996) suggest that the 'certain value' is actually in the range 50 to 80, whereas numerical models, such as those of Heggie & Ramamani (1989), yield smaller values. For N-body models with unequal masses

---

[1] A term $4\pi r^2 \rho \epsilon$ would be added to the right-hand side, where ϵ is the rate of generation of energy per unit mass, much as in Eq. (9.4).

(unpublished) the value is around 25, with no noticeable N-dependence in the range in which it has been tested, i.e. $4096 \leq N \leq 32\,768$.

Less important than the precise value is the fact that, within the limitations of our estimates, the value is independent of the initial number of stars in the cluster. Thus a small star cluster, with 100 stars, say, will not collapse very far before three-body binaries become important. In our million-body problem, however, binaries can influence the collapse of the core only when the core has reached very high densities. This dependence on N can be observed in the models displayed in the lower right frame of Fig. 9.3.

Two-body binaries

In the problem we have just been discussing, the stars are effectively point masses, and all are single stars initially. This is the cleanest scenario for the investigation of 'core bounce', but it is essentially unrealistic. In fact stars have finite radii (though the point-mass approximation is adequate for the degenerate stars, such as neutron stars), and the proportion of primordial binaries is non-negligible (Part VII).

Consider first a system consisting of single stars with finite radii, i.e. we continue to ignore primordial binaries. Binaries can now form by tidal interactions, and no third body is needed to carry off the energy required to bind them (Chapter 31). The important point is that formation of binaries by this process is a *two*-body process. It is not immediately clear, therefore, that it becomes increasingly important as the core collapses, because collisional relaxation (which is driving the collapse) is also a two-body process. Nevertheless, it turns out that tidal capture does eventually dominate. Indeed, studies show that, in almost all star clusters, this process would become important well before binary formation in three-body interactions had a chance to halt core collapse (Inagaki 1984). We shall not pursue these questions further, however, partly because the subsequent fate of a binary formed by tidal capture is a controversial area (see Chapter 31 again), but also because this process is itself dominated by interactions involving primordial binaries.

Primordial binaries

Now we return to the assumption that the radii of the stars can be neglected, and consider what happens if a certain proportion of the stars present initially are binaries. As in the previous case, interaction between binaries, or between binaries and single stars, is really a two-body process, in the sense that its rate is proportional to the square of the space density. Just as with tidal-capture binaries, therefore, it is not quite obvious that the energetics of these reactions can compete with that of relaxation, but in fact it does, at least if the proportion of binaries is high enough.

To understand why this is so we have to consider two effects which enter the picture but work in opposite directions. The first effect is that primordial binaries act in many respects like a species of heavy star. By the process of mass segregation, therefore, primordial binaries accumulate in the core, where their abundance steadily rises above its initial value as the core collapses. This greatly enhances the efficiency of binary–binary encounters in particular. The second effect is the destructive nature of binary–binary encounters, which often convert the two binaries into one binary and two single stars (Chapter 26). It turns out, however, that this is insufficient to prevent the binaries from halting core collapse.

It seems very likely that it is primordial binaries which would bring core collapse to a halt in a real star cluster, rather than either three-body or tidal-capture binaries. Therefore it is worth re-estimating the conditions under which core collapse will cease if we now consider binary–binary heating rather than that due to the formation and hardening of three-body binaries. The result (Box 27.1) is a condition of the form

$$f^2/(\eta \log \Lambda) \sim 1, \tag{27.5}$$

Box 27.1. Core bounce and primordial binaries

In order to decide whether primordial binaries can arrest the collapse of the core we need to estimate the rate of generation of energy in binary–binary encounters. We assume that all individual stellar masses are comparable, of order m, and that the binaries are hard, with semi-major axes of order a. As an estimate of the cross section for close encounters (within a distance of order a) we may adopt an expression of the form of Eq. (21.4). Without numerical factors this is simply Gma/v^2, where v is a typical stellar speed in the core. (Strictly, we mean the typical speed of a binary.) Let n be the number density of all objects, each binary being regarded as a single object, and let f be the proportion of objects which are binary. Each close encounter between two binaries will release an energy of order Gm^2/a, and so the rate at which energy is created is of order $v(fn)^2(Gma/v^2)(Gm^2/a)$ per unit volume. Estimating the volume of the core in terms of its radius r_c, as before, we easily obtain an estimate of the total rate, $\dot{E}$, at which energy is generated. It is, if anything, an underestimate, as we have ignored binary–single interactions (see Goodman & Hut 1989).

As in the case of three-body binaries, the condition for core bounce is obtained by equating $\dot{E}$ with the luminosity of the core, Eq. (27.3). For heating by binary–binary encounters this leads to the condition $t_{cc} \sim v^3/(f^2 G^2 m^2 n)$. Now t_{cc}, which was defined to be the time scale of core collapse, may be expressed as a multiple, $(1/\eta$, say,) of the central relaxation time (cf. Chapter 18). (The quantity η is the one numerical coefficient we keep track of, because it varies with time.) Thus $t_{cc} \sim v^3/(\eta G^2 m^2 n \log \Lambda)$, where the argument of the Coulomb logarithm is $\Lambda \propto N$ (Chapter 14). Putting the estimates from energy generation and the relaxation of the core together, we obtain the relation (27.5).

where η is a constant (the central relaxation time expressed in units of the time until collapse t_{cc}) and f is the proportion of objects which are binaries; if the quantity on the left is less than unity then core collapse wins (and must be arrested by some other process, such as formation of binaries in triple encounters), whereas heating wins in the opposite case.

Before we discuss this formula, we should consider whether sufficiently many of the binaries can survive the carnage. Doing this precisely is bedevilled by the difficulty of estimating how much of their energy is deposited in the core. If the binaries are not too hard, however, we may estimate that this is a large proportion of all the energy they may emit before destruction. Even if the softer binary is destroyed in every interaction, the energy each binary emits will be of order $Gm^2/(2a)$ per binary. Finally, we err on the safe side by ignoring the substantial contribution made in interactions between binaries and single stars in the core.

In view of these considerations we may safely estimate that the total energy which the binaries in the core can release is of order $f N_c Gm^2/a$, where N_c is the number of objects in the core. This is sufficient (in amount) to halt core collapse if it exceeds the energy released by the core on its collapse time scale, which in turn is of the order of the energy of the core itself, $\sim N_c m \sigma_c^2$. Thus enough binaries survive if

$$f \frac{Gm^2/a}{m\sigma_c^2} \gtrsim 1. \tag{27.6}$$

The coefficient of f here is a measure of the hardness of the binaries. The most effective binaries are not too hard, otherwise the time scale on which they interact is too long compared with that of core collapse, and they deposit little of their energy in the core, where it is needed. For somewhat hard binaries, then, it is not implausible that enough will survive to halt core collapse provided that f is not too low.

Though this discussion gives the impression that consumption of binaries is a bad thing, it has interesting observable consequences. The single stars which are created in binary–binary interactions may be expelled from the system with high speeds, and this has often been proposed as an explanation of 'high-velocity stars' (e.g. Leonard & Duncan 1988). It has also been argued that the only reason we see single stars at all is that they were created (or released) in binary–binary interactions in star clusters (Kroupa 1995).

Now we return to the discussion of Eq. (27.5), which is the condition for the rate of heating to be adequate. Notice that n, the number density of stars, is not involved, because both collisional relaxation and binary–binary heating are two-body processes, in the sense already explained. It follows that the arrest of core collapse by binary–binary heating is a slightly delicate matter. Of the three quantities in Eq. (27.5), all work in the required direction. As core collapse proceeds, f increases because of mass segregation, as already explained. Also, η decreases as core collapse advances (as mentioned briefly in Chapter 18), as the thermal coupling between the core and the cooler, outer parts of the cluster becomes weaker. Finally, if we assume

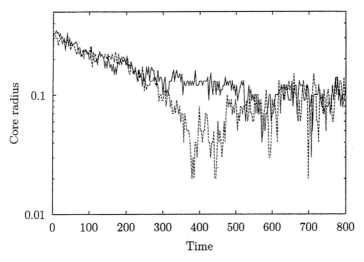

Fig. 27.2. Core collapse with primordial binaries. Two *N*-body models contain 2425 objects. In one model (dashed line) 75 of these are single stars with twice the mass of the other stars, in the other (solid line) they are binaries. For further details see Heggie & Aarseth 1992, where they are referred to as models I and S, respectively.

that the effective value of Λ is proportional to the number of stars *in the core*, as has been argued (Spitzer 1987), then this factor decreases also. However, the variation of this term is the weakest.

A crucial point to notice is that all these factors vary on a time scale which scales with the relaxation time. Suppose, then, that we have a system with a given N in which core collapse is terminated at some point, measured by a certain increase in the central density. Now consider a system with quite different N, but the same initial structure and the same initial proportion of binaries. (The binaries must also be equally hard in both systems initially.) In the second system the same increase in central density occurs after the same number of initial relaxation times. And at this point the increase in f and the decrease in η will be the same in both systems. Even the fractional decrease in the number of stars in the core will be the same. It follows that the two systems will reach core bounce at about the same point, i.e. after the same increase in the central *density* of stars. This is in marked contrast with the arrest of core collapse by the production of three-body binaries, where the *number* of stars in the core at core bounce is independent of N. In that case, the depth at which core collapse is arrested (as measured by the increase in the central density) increases with N. When primordial binaries are present, however, the depth of collapse at core bounce is almost independent of N, and deep collapses are prevented (Fig. 27.2). One may say that the essential physical reason for the difference in the two cases is that binary–binary heating and mass segregation are two-body processes, while binary–single heating is a three-body process.

There are, it is true, considerable approximations and simplifications behind this assertion. In particular, primordial binaries do not last forever. Some are consumed in arresting core collapse, and we must consider whether enough survive to prevent a later recollapse before the cluster dies (Chapter 29). For the meantime, however, core collapse has been brought to a halt, and in the next chapter we consider what options the system has for its subsequent behaviour.

Problems

(1)   Loss of mass m from the core leads to an effective heating of the cluster by an amount $|m\phi_c|$, where ϕ_c is the central potential (Chapter 23). Suppose that some kind of stellar explosion leads to mass loss at a rate $\dot{M} < 0$ from the core. Neglecting numerical factors, show that this is sufficient to arrest core collapse if $-\dot{M} \sim M_c/t_{rc}$, where M_c is the mass of stars in the core and t_{rc} is the central relaxation time. If the mass ejection is associated with stellar evolution, we may suppose that the associated luminosity of the core is

$L_c \sim \dfrac{-\dot{M}}{M_\odot/10^{10}\,\text{yr}} L_\odot$, since the lifetime of the Sun is of order 10^{10} yr. If the cluster is almost isothermal (Chapter 8) out to the half-mass radius, deduce

that $L_c \sim L \left(\dfrac{10^{10}\text{yr}}{t_{rh}} r_h/r_c \right)$, where L is the luminosity of the cluster (in the usual astrophysical sense, but *not* as in Eq. (27.3)). (This would imply a huge core luminosity; cf Goodman (1989), who considers the same issue *after* core collapse.)

(2)   Write down the *equilibrium* equations in the gaseous model (Box 9.2). Show that no reasonable solution is possible if the rate of energy generation, ϵ, takes the form in Eq. (23.5). Try to formulate a physical understanding of this result.

28

Gravothermal Oscillations

In the last chapter we saw that core collapse ends in what is (by stellar dynamics standards) a blaze of energy, which is emitted in interactions involving primordial binaries. Dramatic though that sounds, the real climax of the whole book is reached in the present chapter, where for the first time we catch a glimpse of the entire lifespan of a cluster. Admittedly we concentrate here on a highly idealised cluster, isolated from the rest of the Universe, consisting of stars of equal mass, and totally devoid of primordial binaries. In the next chapter we shall relax some of these idealisations.

Steady post-collapse expansion

What happens to the energy generated at the close of core collapse? To answer this question it is easiest to think of the cluster as a conducting, self-gravitating mass of gas, with its temperature decreasing from the core to the tenuous boundary. Because of the assumed temperature gradient, the heat flux is outwards, and the time scale for the transport of thermal energy is of order the local relaxation time, t_r. Therefore, after an interval of order t_{rh}, i.e. the value of t_r at the radius r_h which contains half the mass, the thermal energy generated at core bounce has diffused throughout the bulk of the cluster. In the same time interval M, the mass of the cluster, has hardly altered (cf. Problem 16.1). Therefore the effect of the generation of energy follows from the estimate

$$r_h \propto GM^2/|E|, \tag{28.1}$$

where E is the energy of the cluster.[1] Because of energy released in binary formation and evolution this increases, i.e. it becomes less negative, and so the cluster
expands.

There is an interesting analogy here with what happens to stars as they settle on
to the main sequence. They have contracted sufficiently that the core is hot enough
to sustain equilibrium through burning fuel in nuclear reactions. In much the same
way, the core of a cluster has contracted sufficiently to supply enough energy by
the consumption of primordial binary stars.[2] The main difference is that a star is able
to radiate away the energy it generates. Star clusters, on the other hand, can only
conduct heat (relaxation) or convect it (escape), but conduction is poor where the
density is low, and convection is poor because the escape rate is so slow, and so the
heat cannot get out. The only option open to the cluster is expansion.

Though this conclusion is correct, the argument gives the impression that the
expansion of the cluster is a consequence of the generation of energy. It is easy to
imagine that the binaries are the engine which drives the subsequent evolution of the
cluster, rather like a rocket engine under the launch vehicle of a payload, and that
the cluster simply has to respond to whatever energy the binaries choose to provide.
But more careful consideration of what happens during core collapse will show us
that it is the behaviour of the core that is governed by the flow of energy across the
half-mass radius, and not the other way around.

The idea that the core responds to the energy needs of the rest of the cluster was
a remarkable insight gained by Michel Hénon (1975). It was a vital breakthrough
which allowed theorists to overcome the impasse of core collapse.

The mechanism by which the core responds to the cluster as a whole can be
summarised as follows. As we have seen, the temperature gradient and relaxation
time at r_h determine the energy flux there. This flux must be supplied from within –
somehow. If there are no binaries in the core then the core will tend to collapse,
just as in the collapse phase we studied in Chapter 18. This drives the stellar density
sufficiently high that the stars are *forced* to produce enough power, by any of the
mechanisms outlined in Chapter 27, e.g. by making and hardening binaries. If, on
the other hand, the core is *over*-productive, then the energy it produces will heat the
core faster than the heat can be transported towards r_h; the core expands, reducing
the stellar density, and the rate at which the stars are interacting. Therefore there are
self-regulating mechanisms which prevent either too much or too little energy being
released by binaries in the core.

Actually, the argument we have just given was already well known in the context
of a problem we have often used as an analogy for star cluster evolution: the theory of
stellar structure. Stars are 'secularly stable' because the core of a star automatically
adjusts itself to produce as much energy as is required by conditions of energy
transport at larger radii (e.g. Schwarzschild 1958, Sec. II.6; Prialnik 2000).

---

[1] More precisely, it is what is sometimes referred to (e.g. Giersz & Spurzem 2000) as the *external*
energy, i.e. the part not locked up inside binary stars (which is referred to as the *internal* energy).
[2] Perhaps a better analogy is the burning of deuterium.

Box 28.1. Core radius in steady expansion with three-body binaries

The flux of energy from the core is given by Eq. (27.1). In steady expansion this must be comparable with the flux at r_h, which is of order $|E|/t_{rh}$. Using Eq. (28.1) and Eq. (14.12), and assuming that the velocity dispersion v^2 does not vary much inside r_h, we deduce that $G^2 m^2 \rho_c^3 r_c^3 r_h \sim M^2 v^4 \rho_h \log \Lambda$, where ρ_c and ρ_h are the density in the core and at r_h, respectively, m is the individual stellar mass and Λ is the argument of the Coulomb logarithm.

This estimate is not very illuminating, but we note that the core radius is defined so that $G\rho_c r_c^2 \sim v^2$ (Eq. (8.9)), and so our condition may be written as

$$\left(\frac{m}{M}\right)^2 \frac{\rho_c}{\rho_h} \frac{r_h}{r_c} \frac{1}{\log \Lambda} \sim 1.$$

Finally, from the fact that the conditions inside r_h are nearly isothermal, it follows that $\rho \propto r^{-2}$ between r_c and r_h (Chapter 8). Therefore, if we neglect the slow N-dependence of Λ, we conclude that $r_h/r_c \propto N^{2/3}$. Since the mass inside a given radius is nearly proportional to r in an isothermal model, it follows also that the number of stars inside the core in this post-collapse expansion varies as $N_c \propto N^{1/3}$.

We return to stellar dynamics. Before going on, however, we should point out where the energy comes from during the core collapse that we studied in Chapter 18, when there may be no binaries. In this phase of evolution the temperature gradient at r_h, and the value of t_{rh}, are quite comparable to the values immediately after core bounce, and so the energy requirements of the cluster are very similar. The difference is that, during core collapse, this energy is supplied by the collapse itself. This form of energy supply also differs from binary heating in that it does not alter the external[3] energy of the cluster, and so there is no overall expansion of the cluster; the half-mass radius is almost static, in fact.

Now we return to what happens after core bounce. Hénon's argument suggests that the core now reaches a kind of equilibrium, settling into a state where it provides just the energy required by the rest of the cluster. On this basis it is not hard to estimate the conditions in the core. The first point to observe is that the rms speed of stars cannot be very different from that at r_h: we have asserted that the flux from the core is comparable with that at r_h, where the time scale of heat transport is t_{rh}, but the value of t_r is very much shorter in the core. Therefore the vicinity of the core must be nearly isothermal, otherwise the flux from the core would be far too high. Besides the rms speed, only one other parameter is needed to specify conditions in the core, such as its mass or radius, and its value can be determined if we balance the heat generated by binaries in the core with the flux at the half-mass radius. It turns out (Box 28.1) that the number of stars in the core in steady post-collapse expansion is of order $N^{1/3}$. The average number of binaries present in the core varies as $N^{-1/3}$ (Goodman 1984).

---

[3] See footnote 1.

On the basis of Hénon's picture the entire evolution of the cluster, and not just of the core, now falls into place. First, the time scale of the expansion is of order t_{rh}, and so it follows that $t_{rh} \sim t - t_{\mathrm{coll}}$, i.e. the time since the end of core collapse. Using Eq. (14.13), we may deduce that

$$t - t_{\mathrm{coll}} \sim \frac{M^{1/2} r_h^{3/2}}{G^{1/2} m \log \Lambda}, \tag{28.2}$$

where m is the individual stellar mass and Λ is the argument of the Coulomb logarithm. Actually this estimate should really be read backwards: M varies on a rather longer time scale, and Eq. (28.2) shows that $r_h \propto (t - t_{\mathrm{coll}})^{2/3}$, roughly.

The slow escape of stars in post-collapsing isolated models is an interesting and little-explored problem. Some time ago N-body models showed that $M \propto (t - t_{\mathrm{coll}})^{-\nu}$, where $\nu \simeq 0.085$, and this alters the time-dependence of the expansion to nearly $r_h \propto (t - t_{\mathrm{coll}})^{(2+\nu)/3}$ (Giersz & Heggie 1994b). By today's standards the values of N in these simulations were modest, and this research did not address the N-dependence of the escape rate. If mass is lost on the relaxation time scale then the time for a fixed fraction of mass to escape should vary nearly in proportion to N, like the relaxation time. Recent N-body models by H. Baumgardt, however, show that the time taken to lose about 70% of the mass is almost independent of N. The tentative reason for this appears to be the higher concentration of models with larger N (Box 28.1), which leads to a larger escape rate per relaxation time in larger models.

It is interesting that the dependence of r_h on t follows from dimensional analysis, given only G, M and r_h. The same expansion law is obeyed in two-body collisions (Box 21.2), cold collapse (Fig. 2.1) and even the Einstein–de Sitter cosmological model (e.g. Peebles 1993). There is, of course, a large dimensionless number (N) in our problem, and this can make a big difference in the coefficient.

At last we can glimpse the eventual fate of the million-body problem. It expands on an increasing time scale, slowly losing mass as it does so.[4] After much time has elapsed it looks like a small N-body system with just a few tens of stars, but of immense size, and evolving almost imperceptibly. Moving away, and dominating at great distances, are the escapers, not only single stars but also binaries and sometimes even triples and higher multiples. Eventually the last escaper is ejected, leaving behind one last binary or triple.

Unsteady post-collapse evolution

Apart from the title of this section, perhaps the reader has spotted one or two other veiled clues that our picture of post-collapse evolution has some gaps. One of these is the mismatch between the number of stars in the core after core collapse (Box 28.1) and the number at core bounce, where we estimated (Chapter 27) that N_c is independent of N. It follows that, in a sufficiently rich system, the core must re-expand in

---

[4] Continuing an earlier remark, it is tempting to draw an analogy with the evolution of red giants. That is too controversial a topic, however, for the analogy to be enlightening.

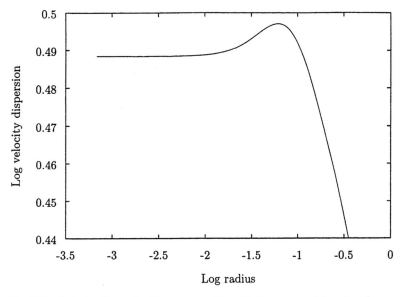

Fig. 28.1. Velocity dispersion ('temperature') profile in a gas model undergoing a gravothermal expansion (after Heggie *et al.* 1994).

the early phase of post-collapse evolution. Core bounce really is a bounce, and there must be a period of readjustment if the core is to settle down to the equilibrium size required after core bounce.

The second clue is that Hénon's argument does not, despite appearances, imply that the required equilibrium is ever reached. Within the terms of this argument it is equally possible that the balance is reached only in a time-averaged sense, and that the core really does alternate between high- and low-density phases.

The third clue is, in physical terms, the most telling. We mentioned that the inner parts of the cluster must be nearly isothermal if the energy produced in the core is to balance that required at r_h. Now we saw in Chapter 17 that such a system is gravothermally unstable, and so we should expect a time-dependent behaviour different from the steady expansion we have assumed. It was Daiichiro Sugimoto and Erich Bettwieser who first understood the consequences of this (Sugimoto & Bettwieser 1983, Bettwieser & Sugimoto 1984).

The first and third clues are related. At the end of core bounce the core is producing enough energy to halt the collapse, and this is too much for the subsequent expansion of the cluster as a whole, which is determined by the heat flux at r_h. Therefore the thermal energy generated builds up in and around the core faster than it can be conducted away. This causes an expansion and cooling of the core and its immediate surroundings, because of their negative specific heat. If core collapse is sufficiently deep, which simply requires sufficiently large N, the expanding core can actually cool to temperatures below that of its surroundings. At this point there is a *temperature inversion*, i.e. there is a range of radii in which the temperature increases outwards (Fig. 28.1). In this region the thermal flux is actually inwards. This flux enhances the expansion of the core and its cooling, which simply reinforces the driving force behind the expansion. It is the gravothermal instability again, this time working in reverse. Eventually, however, the expanding core comes in thermal contact with the cooler parts of the cluster around r_h. Now the heat generated in the core can flow out as required, and the temperature inversion disappears. But by this point the core is so distended that it is producing insufficient energy. By Hénon's argument it is forced to recollapse so that its density is sufficiently high to produce the required energy. But this collapse is more-or-less like the first core collapse: the core overshoots, creates too much energy, and a temperature inversion, and the cycle recurs.

What we have outlined is a picture of what are now called *gravothermal oscillations* (Box 28.2). The detailed nature of these oscillations can be studied in

---

Box 28.2. A brief history of gravothermal oscillations

These were discovered by Bettwieser and Sugimoto using a gaseous model (Chapter 9). At that time, however, other gaseous models (computed by one of us (Heggie 1984)) and Fokker–Planck models (Cohn 1985) were showing no indication of this behaviour, and so the discovery was greeted with much scepticism when it was announced and discussed at a conference in 1984 (which was organised by the other of us (Goodman & Hut 1985)). Behind-the-scenes discussions at that meeting drew attention to the importance of the choice of time steps in the numerical integrations. In particular, M. Schwarzschild mentioned to one of the authors the resemblance to the numerical problems which had alerted him to the He core flash in a Population II giant (Schwarzschild & Härm 1962). Not long after the symposium, further work by the main disputants confirmed the oscillations. Jeremy Goodman (Goodman 1987) then showed that the steady expansion, though possible, was unstable if N was large enough.

All of this work was conducted with simplified models, and there was considerable doubt whether N-body models could exhibit such behaviour. It was argued that the fluctuating temperature in a discrete model could mask the subtle temperature inversion that plays such a crucial role in the expansion phase. Limited N-body modelling (of a specially tailored system in a reflecting sphere (Heggie *et al.* 1994)) at least removed this objection.

It was still not possible to check that the full cycle of oscillations would actually occur; computers were not powerful enough. A detailed analysis, however (Hut *et al.* 1988), showed that the computer power necessary to exhibit gravothermal oscillations, if they existed, was of order 1 Teraflop. Eventually the GRAPE-4 computer was finished at the University of Tokyo (Chapter 3). It had been specifically constructed to settle this point, and was the first computer in the world to reach the required speed. Its first results were presented at another conference (Hut & Makino 1996), this time in Tokyo and organised by Daiichiro Sugimoto, just over ten years after the first announcement of gravothermal oscillations. The existence of gravothermal oscillations in sufficiently large N-body systems was now established (Makino 1996; and Fig. 9.3, lower right panel).

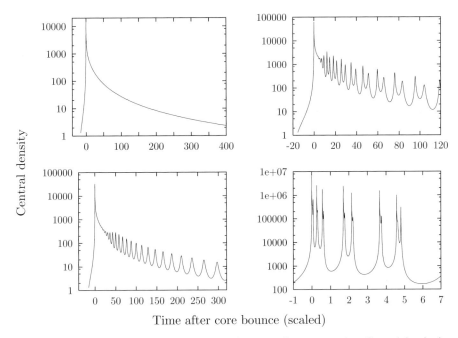

Fig. 28.2. Period-doubling and chaos in post-collapse expansion. Central density is plotted against time (in units of the initial half-mass relaxation time, with origin shifted to the time of core bounce) for gas models with $N = 6000$ (top left), 8000 (bottom left), 10 000 (top right) and 50 000 (bottom right) stars (recomputed and redrawn, after Heggie & Ramamani 1989).

various ways, but analogies with other problems suggest what we might expect to find. We have argued that it is steady expansion which is unstable, but it is easy to transform the problem into a problem on the stability of equilibrium (Goodman 1987). In steady post-collapse expansion of an isolated model of constant mass we expect that radii scale as $(t - t_{coll})^{2/3}$, by Eq. (28.2). Therefore, if we write $\tilde{r} = r(t - t_{coll})^{-2/3}$, and treat the other variables such as the space density similarly, we shall turn what is really a steady expansion into what looks like an equilibrium, just as the expansion of the Universe can be frozen out in comoving coordinates.

Now the arguments of the present section of this chapter imply that this equilibrium is unstable, for N sufficiently large. Thus we are studying a system in which an equilibrium becomes unstable as some parameter (N in this case) changes. One of the commonest scenarios in such problems is that the instability takes place by a sequence of *period-doubling bifurcations*, which accumulate at a parameter value at which chaotic oscillations set in (e.g. Cvitanović 1989). This is roughly what happens in detailed models of the post-collapse behaviour of star clusters, both gas models (Fig. 28.2) and Fokker–Planck models (Breeden *et al.* 1994). When N is small enough (less than a few thousand, in fact, if all stars have the same mass), the post-collapse expansion is stable and steady. Just above some value of N the expansion exhibits a simple periodic behaviour. At somewhat larger N, alternate cycles have different amplitudes: the period has doubled. This period-doubling presumably cascades rapidly as N increases, until at a somewhat higher critical value of N one sees for the first time non-periodic oscillations. Tests show that these have the hallmarks of chaotic variations.

The analogies with other systems exhibiting this behaviour can be pushed further. In particular, although the core, being a continuum, is a system with infinitely many degrees of freedom, it is found that its evolution is nearly restricted to a low-dimensional (actually two-dimensional) subspace (Breeden & Cohn 1995). Further, inspection of the flow on this subspace shows that it can be represented by a one-dimensional first return map with the same qualitative properties as the famous logistic map, in which so much of the basic theory of period-doubling was developed (see Cvitanović 1989).

Admittedly, there are some important respects in which this picture must be qualified. For one thing, it is complicated by the fact that the generation of energy in N-body systems is stochastic in nature, as it depends on the accidental formation and evolution of individual binaries. It is not clear that period-doubling could be observed in N-body simulations in these circumstances, though it is certain that there is a change in character in the oscillations of the core at around the values predicted by the simple models (Fig. 9.3, lower right panel). Below those values the fluctuations look simply random, and some are clearly associated with energetic few-body interactions. Above the predicted values, the oscillations, while chaotic, have a systematic character, independent of specific identifiable binary events. The second qualification is that, when we consider slightly more realistic models of stellar clusters, e.g. with a spectrum of masses or primordial binaries, it turns out that conditions for the onset of gravothermal oscillations are less favourable. The next chapter explores some of these less idealised models, and little more will be said of unsteady post-collapse evolution.

Problems

(1)   According to an isolated model of Hénon (1965) the time scale of evolution in post-collapse expansion is given by $r_h/\dot{r}_h \simeq 10.8 t_{rh}$, where t_{rh} is the *current* half-mass relaxation time. N-body models of this phase show that the number of particles diminishes (by escape) roughly according to $d \ln N / d \ln r_h \simeq -0.12$ (Giersz & Heggie 1994b), while the loss of mass in the pre-collapse phase is a few per cent and may be neglected here. Fokker–Planck models show that the duration of core collapse for a Plummer model is about $18 t_{rh}(0)$ (Spitzer & Thuan 1972, Takahashi 1995), where $t_{rh}(0)$ is the *initial* value. Estimate the entire lifetime of the model (i.e. until $N \simeq 2$) in units of $t_{rh}(0)$.

(2) Several toy models illustrating gravothermal behaviour were investigated by
     Allen & Heggie (1992). Among them was the simple system

$$\frac{dT_1}{dt} = \frac{\Gamma}{\Gamma - 1}\xi(T_1 - T_2) - cT_1^{(4-\alpha)/(\alpha-2)}$$

$$\frac{dT_2}{dt} = \frac{1}{\Gamma - 1}\xi(T_1 - T_2) - (T_2 - 1)T_1^{-(6-\alpha)/[2(\alpha-2)]},$$

where T_1 is the temperature of the core, T_2 is the temperature of the region
where a temperature inversion may develop, and $\Gamma \simeq 1.28$, $\alpha \simeq 2.23$ and
$\xi \simeq 0.00037$ are constants. The parameter $c \propto N^{-2}$, and this term corresponds to binary heating. The forms of this and the other terms were devised
by simple phenomenological considerations of gravothermal behaviour.
    Using appropriate numerical or analytical tools, demonstrate the following:
(a) for sufficiently large c there are no equilibria; (b) for small enough c there
are two equilibria; (c) as c decreases, the hotter equilibrium first changes
from a stable node to a stable focus; (d) as c decreases further, the hotter
equilibrium loses stability and throws off a stable limit cycle in a Hopf
bifurcation.
    These changes correspond physically to the following: (a) when N is very
small, the presence of a binary leads to the rapid expansion of the core; (b) for
larger N a stable balance is possible between binary heating and gravothermal
cooling; (c) for still larger N, the approach to this stable balance has the
nature of a damped oscillation (as is observed in gas models); (d) for large N
the balance between heating and cooling is a dynamic (oscillatory) one.
    A further variable is needed before a simple model like this can
demonstrate chaotic behaviour similar to well developed gravothermal
oscillations.

(3) Detailed models show that gravothermal oscillations mainly affect stars
     inside the radius $r_{0.01}$ containing the innermost 1% of the mass (Cohn et al.
     1989). If the region from there out to the half-mass radius is nearly iso-
     thermal, show that the relaxation time there is $t_{r,0.01} \simeq 10^{-4}t_{rh}$. If $t_{r,0.01}$
     governs the time scale of gravothermal oscillations, and if each binary
     releases an energy amounting to about $110m\sigma_c^2$ (Chapter 23), show that
     about one binary per oscillation is needed for $N = 10^6$.

29

Dissolution

The previous chapter dealt mostly with a highly idealised model. All stars were single and had the same mass, and the system was isolated. As we saw, even the presence of a spectrum of stellar masses changes the picture, as it is found that gravothermal oscillations set in only for considerably larger values of N. In the present chapter we shall also see that the presence of primordial binaries further weakens their probable relevance. Even when gravothermal oscillations do occur, they seriously affect the structure of only the innermost 1% or so of a cluster. Therefore, in this chapter we concentrate once more on the steady post-collapse evolution of a stellar system. Also, we mainly have in mind a system with a significant population of primordial binaries, and boundary conditions set by the tidal field of the surrounding galaxy. First, however, we consider the simpler case of an isolated cluster.

Isolated clusters

The first thing that is changed in post-collapse evolution when we add primordial binaries is the radius of the core. A similar argument to that of Box 28.1 shows that the ratio r_h/r_c is now almost independent of N (cf. Problem 1). In fact the ratio depends more on the proportion of binaries (which decreases as the binaries are consumed).

These statements greatly weaken the clues to the occurrence of gravothermal oscillations which we discussed in the case of post-collapse evolution powered by three-body binaries (Chapter 28). In the first place, the ratio r_h/r_c at core bounce is also independent of N, if core collapse is halted by primordial binaries (Chapter 27).

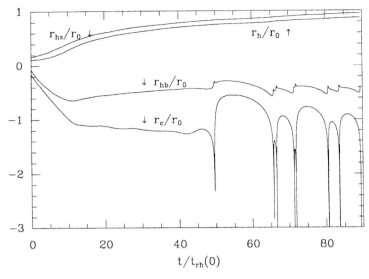

Fig. 29.1. Evolution of a cluster with primordial binaries, after Gao *et al.* (1991). From top to bottom the curves are the half-mass radius of single stars, all stars, binaries, and the core radius. All are plotted against time. Mass segregation during core collapse is followed by a relatively steady post-collapse phase until the binaries are exhausted, at which point gravothermal oscillations begin.

Therefore, if there has to be a period of adjustment to post-collapse conditions, at least it does not get more serious as N increases. Second, the post-collapse core does not get smaller and smaller as we consider systems with larger and larger N. Therefore there is no need to posit a large (and unstable) isothermal region within the half-mass radius. Finally, it is quite possible that this post-collapse core is as large as that in a modest N-body system powered by three-body binaries, and we know that these do not exhibit gravothermal oscillations.

The only worry is that a system may not have enough primordial binaries to sustain a sufficiently long period of post-collapse expansion. A substantial fraction were already burned simply in halting core collapse (Chapter 27), and the residual population is gradually used up in powering the post-collapse evolution. After that, the only effective heating mechanism left would be the formation and hardening of three-body binaries. The point at which this first happens clearly depends on the initial proportion of binaries. In early simulations (Spitzer & Mathieu 1980) it was found that core collapse was delayed by less than the time of core collapse, even for large binary fractions. Nowadays, however, there is a consensus (with various methods and authors) that, if the initial proportion of binaries is of order 10% (depending on the distribution of binary parameters), then the binaries can keep gravothermal oscillations at bay for a time about three times the length of the phase of core collapse (Fig. 29.1; see also McMillan & Hut 1994). Even this may, however, be less than the lifetime of a number of globular clusters.

With this caveat we assume that the evolution of the core need not distract us, and turn to the evolution of the cluster as a whole. We already saw in Chapter 28 what happens if the system is isolated and has stars of equal mass. When a spectrum of masses is present we might expect our simple picture to be spoiled by the continued action of mass segregation. In fact, however, N-body simulations suggest that our simple picture holds up remarkably well. They show that mass segregation slows dramatically towards the end of core collapse, and that thereafter the profile of mass segregation does not alter very much (Fig. 16.1). Simple relations like

$$t - t_{\text{coll}} \sim \frac{M^{1/2} r_h^{3/2}}{G^{1/2} m \log \Lambda} \tag{29.1}$$

(cf. Eq. (28.2)) apply with unexpected accuracy even when there is a broad spectrum of stellar masses, though the reasons for our good fortune in this respect are not understood.

Tidally bound clusters

Next, let us replace our idealised picture of an isolated system by one with a somewhat more realistic boundary condition. Tidal truncation (Chapter 12) has the characteristic that the radius of the system, r_t, must vary with its mass as

$$r_t \propto M^{1/3}, \tag{29.2}$$

which implies that the mean density of the system is constant. We have seen, however, that an isolated cluster expands in post-collapse evolution, and thus its density reduces. What happens with a tidal boundary condition is that the expansion pushes the tenuous outer layer over the tidal boundary, leaving behind the denser material at smaller radii. A balance is struck between the truncation by the tide, which leaves behind denser material, and the expansion, which reduces the density, and the net effect is that the mean density remains constant.

In isolated systems the mass is nearly constant, and so $E \propto -1/r_h$. In tidally bound systems, however, the half-mass radius will be comparable with (but smaller than) the tidal radius, and so it follows from Eq. (29.2) that $E \propto -Gr_h^5$. Thus the increase in E caused by binary interactions in the core is accompanied by a *decrease* in r_h: the cluster shrinks as it is heated. This is quite unintuitive for someone brought up on the evolution of isolated clusters.

Now we consider the time scale of this evolution. The right-hand side of Eq. (29.1) is an estimate of t_{rh}, and so Eq. (29.2) shows that this is proportional to M. Arguing as before that t_{rh} determines the time scale of evolution, we see that the evolution speeds up as the system loses mass, and in fact should come to an end at a finite time in the future, t_{end}, say. The time dependence is therefore obtained by equating the right-hand side of Eq. (29.1) to the time *remaining*. Using Eq. (29.2)

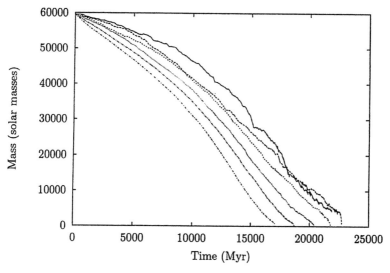

Fig. 29.2. Loss of mass in tidally limited N-body models. From right to left the initial number of stars is $N = 1024, 2048, 4096, 8192, 16\,384$ and $32\,768$. The models have been scaled to a real cluster with initial mass $6 \times 10^4 M_\odot$ in a tidal field of a certain strength, and with the same mass function and initial structure (see Heggie *et al.* 1998). The scaling is imperfect, for reasons which are discussed in Baumgardt (2001). The modest increase in the escape rate after about half the mass has escaped is associated with the change from pre-collapse to post-collapse evolution.

again, it follows that $M \propto t_{\mathrm{end}} - t$ (see Fig. 29.2), and that the remaining lifetime of the cluster is proportional to the current half-mass relaxation time. In the model of Hénon (1961) the constant of proportionality is approximately $22.4 t_{rh}$.

The tidal truncation progressively removes the stars of highest energy. By mass segregation these tend to be the stars of lowest mass. The result is that a dying cluster contains an increasing concentration of the more massive objects. As in panning for gold (or neutron stars), the tide washes out the lighter particles. Incidentally, this helps to compensate for the destruction of binaries in the binary–binary reactions that we have relied on to sustain the post-collapse evolution.

Stars which escape from the cluster, or which are washed across the tidal boundary, do not necessarily leave the vicinity of the cluster quickly. Unless of high energy, e.g. escapers resulting from three-body interactions, they cross the tidal boundary at low speed in the vicinity of the Lagrangian points (Chapter 12). Thereafter they move off under the action of tidal and inertial forces to form 'tidal tails' of escaping matter (Fig. 12.3). When the cluster as a whole dissolves, in the sense that its tidal radius shrinks to zero, there is still a residual concentration of stars – the smile on the face of the Cheshire cat.

It is remarkable that post-collapse evolution in a tidal field was the first evolutionary problem in the subject to be understood in detail (Hénon 1961). At that stage the theory was developed for the case of a system in which all stars have the same mass, the distribution of velocities is isotropic, and the tidal field is modelled as a cutoff at the tidal radius. Empirically, however, it has been found that much the same conclusions apply to much more realistic systems: unequal masses, anisotropic velocity distribution, anisotropic tidal field, primordial binaries, etc. The nearly linear dependence of the mass on time is one of the most robust and memorable results in the whole of this subject, and a nice example by J. Stodólkiewicz appears prominently on the front cover of the proceedings of IAU Symposium No. 113 (Stodólkiewicz 1985).

Problems

(1) Use estimates in Box 27.1 to show that the total heating rate of primordial binaries in the core is of order $f^2 m \sigma_c^3 / r_c$, where f is the binary fraction, m is the individual stellar mass, σ_c is the central velocity dispersion, and r_c is the core radius. Deduce that, in steady post-collapse expansion powered by primordial binaries, the ratio r_c / r_h is almost independent of N.

(2) (Spitzer 1987, p. 59) In homologous evolution in a tidal field show that the total energy varies with mass according to

$$\frac{d \ln |E|}{d \ln M} = \frac{5}{3}.$$

If stars escape across the tidal boundary, modelled as a cutoff at the tidal radius r_t, with negligible excess energy, show that $\dot{E} = -(GM/r_t)\dot{M}$, and deduce that $E = -0.6GM^2/r_t$. Infer that $r_t = 2.4$ in N-body units (Box 1.1). (In the limit $W_0 \to 0$ King models have this limiting radius (cf. Problem 8.6). Most models have larger tidal radii, however (cf. Fig. 8.3), and require an additional source of energy for homologous evolution.)

(3) Suppose the estimate in Problem 1, with r_c replaced by r_h and σ_c the rms one-dimensional speed, can be used to estimate the rate of heating in an isolated uncollapsed cluster with half-mass radius r_h, and that mass is lost by evaporation (Eq. (9.3), where t_r is estimated by the half-mass relaxation time t_{rh}). Using standard estimates for a virialised cluster, deduce that

$$\frac{d \ln r_h}{d \ln M} - 2 \simeq -\frac{f^2}{\mu \ln(\lambda M/m)},$$

where μ and λ are constants. Sketch solutions of this differential equation, assuming that f is also constant. (Cf. Hills 1975b, where heating between single stars and primordial binaries was considered.)

Part IX

Star Cluster Ecology

The remaining chapters of the book go beyond the N-body problem as it is understood outside astrophysics. Here the fact that the bodies are *stars* is essential.

Chapter 30 sets the scene by summarising the various dynamical processes that have been introduced so far, the various kinds of stars and other relevant topics which are of interest in astrophysics (such as colour–magnitude diagrams) and, most importantly, the relations between these two sides of the problem. Special attention is paid to those kinds of stars which are readily observed and where dynamical processes are most immediately relevant: *blue stragglers*, *millisecond pulsars* and *X-ray sources*.

Chapter 31 analyses in some detail the simplest process where the stellar nature of the bodies is vital: collisions and other encounters between two individual stars, where the gravitational interaction is not the whole story. We estimate the rate of collisions, and how it depends on the stellar density and the kinds of stars present. Non-gravitational interactions are also vital in understanding the role and evolution of binary stars, especially when interactions with other (single) stars occur frequently enough. The effects of collisions on the participating stars are outlined, and we consider the dynamics of *near-collisions*, where non-gravitational effects are important ('tidal capture').

The dynamics and evolution of binary stars are taken up in detail in Chapter 32. Special attention is paid to the ways in which *blue stragglers* can arise from interactions involving binaries. We give examples of the effects of encounters between a binary and a single star, and those with another binary. This is a growth area of the subject, and our outline of its current status is likely to be out of date by the time this book is in print.

The final chapter, Chapter 33, summarises the astrophysical million-body problem. We recap the various dynamical processes we have discussed and their interrelations, and present the overall evolution of a star cluster, from birth to death, in a simplified diagram which encapsulates the gross aspects of its spatial structure. We consider again the time scales on which the evolution takes place. Finally, we consider how it is that the clusters we see have managed to survive for most of the lifetime of our Galaxy, and the rate at which these last survivors are perishing.

30

Stellar and Dynamical Evolution

Globular star clusters have an important place in modern astrophysics for several reasons, but let us mention just two here. Firstly, they are a laboratory for the study of gravitational interactions and dynamical evolution, and this is the motivation for much of the research that we have written about in this book. Secondly, however, each cluster is also a sample of stars of very similar age and composition, and are an ideal test-bed for theories of stellar evolution.

Over the years these two aspects of cluster studies built up their own communities of theorists and observers, and their own suites of problems (Fig. 30.1). Now what is remarkable is that, for many years, research in these two areas proceeded in almost total isolation from each other. It was possible to pursue an active and successful research career on one side of this diagram (Fig. 30.1) without even being aware of the existence of the people working on the other side. Whenever one did need something from the other side, the most primitive tool for the job was used. Dynamicists would use mass functions which to any observational astronomer would seem distinctly bizarre, while observers, if they ever needed a theoretical model, would dust off an old one which ignored decades of subsequent theoretical development. Dynamicists were fascinated by the problems of stellar systems with stars of only two different possible masses, while fitting a Fokker–Planck model was something that no non-theorist ever attempted.

Within about the last decade all that has changed. The volume of proceedings of a meeting held in 1990 (Janes 1991) gives a snapshot of this shift in outlook. It was increasingly being realised that dynamics has an influence on the kinds of stars that are to be found in globular clusters, and that these in turn influence the dynamics that can occur. The feedback mechanisms between stellar dynamics

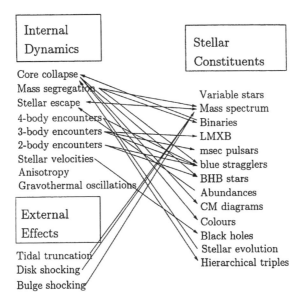

Fig. 30.1. Some of the problems of globular star clusters, and the interrelations between them. Arrows between topics on the same side of the diagram are not shown. Not all arrows imply cause and effect; the 'stellar velocities' to 'black holes' means that the former could be an observational indicator of the latter; in some cases the link is indirect. The diagram is considerably simplified: in fact the mass spectrum affects almost everything on the left, and any of the first six categories on the left could affect the CM diagram. Acronyms include: low mass X-ray binary (LMXB), blue horizontal branch stars (BHB stars, also known as 'faint' or 'hot' horizontal branch stars) and colour–magnitude diagrams (CM diagrams).

and stellar evolution therefore play a vital role in the evolution of star clusters. The term 'ecology', introduced by Heggie (1992), captures the essence of this interplay.

We have already seen one way in which the evolution of the stars affects the dynamics of a cluster. Towards the end of a star's life on the main sequence it loses much of its mass, and, as many stars complete their evolution, the resulting unbinding of the cluster may even cause its complete dissolution (Chapter 11). Other influences of stellar evolution on stellar dynamics may be more subtle. According to point-mass dynamics, for example, a hard binary tends remorselessly to get harder (Chapter 19f). When we bear in mind that stars have finite radii, however, and that these radii increase dramatically during certain phases of stellar evolution, the resulting shrinking of the binary orbit cannot continue indefinitely. This may limit the energy which each binary may release, and its potential importance for dynamical evolution.

Astronomers were also slow to realise the existence of links in the opposite direction, i.e. the role of dynamics in shaping stellar evolution. But the evidence that stellar evolution in star clusters was different from stellar evolution elsewhere gradually built up, and forced a paradigm shift that is still not complete.

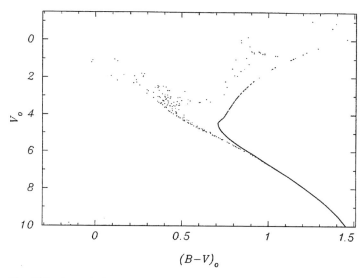

Fig. 30.2. A synthetic colour–magnitude diagram for an old stellar population (after Portegies Zwart *et al.* 1997). The simulation starts at 10 Gyr with a population of degenerate stars (neutron stars and white dwarfs), and main sequence stars with a relatively flat mass function, but no binaries. The results of stellar evolution and collisions are shown at about 12 Gyr. Besides the main and giant sequences, there is a substantial population of blue stragglers (to the left of the turnoff point) and yellow stragglers (to the left of the giant branch). For further examples of synthetic 'observational' data, see Portegies Zwart *et al.* (2001b).

The earliest evidence of environmental factors in stellar evolution was the discovery almost half a century ago by Sandage (1953) of 'blue stragglers', stars that *seem* to be substantially younger, and hence bluer, than the bulk of the stars in a star cluster (Fig. 30.2). The implications of this discovery were not absorbed by the community at large, and for the next twenty years, globular clusters still kept most of their secrets hidden. Both authors of this book remember how, when we took our first course in astronomy, globular clusters were portrayed as some of the oldest and dullest objects in astrophysics. Even as we began our research careers, the dynamics of star clusters was regarded as a quiet backwater in astrophysics, while all the excitement was being generated elsewhere; in high-energy astrophysics, for instance. Paradoxically, it was precisely from this direction that attention was once more directed to these 'old and boring' objects. In the mid seventies star-cluster research received an enormous boost from the unexpected discovery of globular cluster X-ray sources (cf. Fig. 1.3).

It soon became clear that these sources had a much higher relative abundance in globular clusters than in the Galaxy as a whole (see Katz 1975, Hut & Verbunt 1983, Bailyn 1996). While the family of globular clusters contain less than 0.1% of the stars in the Galaxy, they harbour more than 10% of the low-mass X-ray sources throughout the Galaxy. The search was on for an explanation, and an early suspect was what

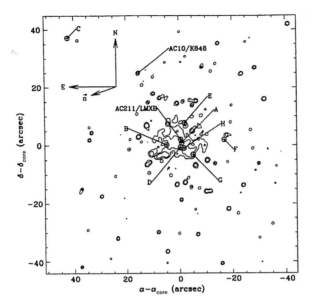

Fig. 30.3. A high-energy astrophysicist's view of the core of M15 (from Kulkarni & Anderson 1996). Eight pulsars are indicated with letter designations, and one low mass X-ray binary (AC211). Compare with Fig. 1.2.

makes globulars so special, dynamically: their high density. Various scenarios were proposed, including binaries formed in stellar collisions (Sutantyo 1975), but quickly the dominant explanation arose that neutron stars had been *tidally captured* by normal stars (Fabian *et al.* 1975), to form new binary stars. In these binaries, mass overflow from the normal star could then provide the fuel to turn the neutron star into an X-ray source.

More than ten years later, from 1987 onwards, another exciting type of object was discovered in globular clusters, in rapidly growing numbers: pulsars, rapidly spinning neutron stars that are discovered through the pulsed radiation they emit (Fig. 30.3; Kulkarni & Anderson 1996). Many of these pulsars were found to be a member of a binary system, and tidal capture was once again proposed as the main mechanism for the formation of this other class of unusual binaries.

Soon afterwards, it is true, two important aspects challenged the use of tidal capture as a panacea to explain any type of weird star system in globular clusters. On theoretical grounds, closer investigation of the mechanism of tidal capture cast doubt on its efficiency (mainly Kochanek 1992, but see also Mardling 1995, Kumar & Goodman 1996). On observational grounds, however, a new mechanism presented itself naturally, based on the discovery of large numbers of primordial binaries (Chapter 24).

Still, the shift in outlook was permanent. It was now clear that one could not understand the kinds of stars that were found in globular clusters without taking the stellar environment into account, and, equally, that the study of dynamics in isolation from stellar evolution led, at best, to an incomplete picture. Nor was the evidence for this confined to exotic kinds of star: even the relative abundance of the most ordinary stars appeared to depend on the stellar density, causing observational variations in the colour of a star cluster from centre to halo (e.g. Djorgovski *et al.* 1991). What we shall do in the next two chapters is to describe some aspects of this interplay between dynamics and the stellar nature of the bodies involved.

Despite the weakening of the original motivation for considering tidal capture, we begin there in the next chapter. This process is one of the simplest, best studied and most robust aspects of cluster dynamics in which the finite radii of the stars must be taken into account. Tidal captures and stellar collisions between single stars do take place at significant rates in globular star clusters, and are also an important aspect of interactions involving binaries, though by no means the whole story. Those interactions are the subject of Chapter 32.

Problems

(1)   In a system in which all stars have solar mass and radius, and the (three-dimensional) velocity dispersion is 20 km/s, compute the energy of the most energetic binary, in units of 1 kT. (The solar radius is about 7×10^5 km.) Compare the energy with that of a whole globular star cluster.

(2)   Consider the arrows in Fig. 30.1, and the mechanisms by which these influences occur.

31

Collisions and Capture

Once upon a time it was thought that novae arose from stellar collisions. In a normal galactic environment, however, where most novae are found, collisions are rare, in the sense that a given star is very unlikely to experience a collision during the age of the Universe. The stars are simply too far apart. In very dense environments, however, such as galactic nuclei, collisions may even be sufficiently frequent to dominate dynamical evolution. Globular clusters lie in between these extremes. In them, collisions do affect large numbers of stars, but are not normally thought to predominate over purely gravitational interactions (i.e. two-body relaxation). And the restriction to single stars is an unrealistic idealisation. Globular clusters possess large populations of binaries, which automatically place stars in close proximity. All in all, it is estimated that there is a collision between two stars somewhere in the Universe every ten seconds.

The rate of collisions

For a stellar system consisting of stars of equal mass m, radius R and number-density n, the collision time scale (whose reciprocal gives the rate at which a given star collides with any other) is given approximately by

$$t_{\text{coll}}^{-1} = 8\pi G m n R \langle V^{-1} \rangle, \tag{31.1}$$

where V is the relative speed of the two stars, and the average is taken over all stars that may be encountered. (This formula is based on Keplerian relative motion of the two stars, and assumes that gravitational focusing is large, for which the condition is that $V^2 \ll 2Gm/R$; this is satisfied in practice in globular star clusters.) For stars of solar type, the formula may be re-expressed more intelligibly as

$$t_{\text{coll}}^{-1} \simeq 2 \times 10^{-15} \left(\frac{n}{\text{pc}^{-3}} \right) \left\langle \frac{\text{km/s}}{V} \right\rangle \text{yr}^{-1}.$$

That very small coefficient shows why collisions are so rare in the neighbourhood of the Sun, where $n \sim 1 \text{ pc}^{-3}$ and $V \sim 20 \text{ km/s}$. In globular clusters V is not so different, but the density is much higher, typically by a factor of about 10^4 at the cluster centre. Hence the collision time scale is reduced to something of order 10^{12} yr. Since a cluster lives for roughly 10^{10} yr, this means that of order 1% of stars in the core of the cluster will have experienced a direct collision within its lifetime.

No one number can cover the complexity of this issue. In the first place the collision time scale varies with the stellar radius and mass. Almost all stars in globular clusters have smaller radii and masses than that of the Sun, and so the collision rate is smaller than we have estimated. The main exception is the evolved giant stars, whose radii may exceed that of the Sun by a factor of 100. Collisions are less effective than one might think, because of their low-density envelopes. Even so, in a collision at 10 km/s between a star of mass $1M_\odot$ and a giant of mass $2M_\odot$, the maximum distance of closest approach for capture is almost as large as the radius of the giant (Bailey & Davies 1999). More importantly, this phase of a star's life is relatively short, roughly 15% of its entire lifetime (Eggleton *et al.* 1989), and so the contribution of these stars is not dominant. For example, in models of the star clusters 47 Tuc and ω Centauri, Davies & Benz (1995) have estimated that the rate of collisions between two main-sequence stars exceeds the rate of collisions between a main-sequence star and a red giant by at least a factor of two. Such results depend on the adopted mass function, but the required computations are easily performed using a simple extension of Eq. (31.1). Furthermore, there are now convenient simple formulae for the evolution of single stars, of any relevant mass, metallicity and age, including formulae for the radius (Hurley *et al.* 2000).

The second complicating factor is that the collision time scale depends on the stellar density. This is smaller outside the core, but on the other hand there are more stars there. Still, collisions are a two-body process, and so the number of collisions is roughly proportional to $\int \rho^2 r^2 dr$, where $\rho(r)$ is the density at radius r. For any reasonable density profile this integral is dominated by the vicinity of the core (see Problem 3).

A third point for consideration is the fact that the stellar density and mass function, and therefore the collision rate, vary enormously because of the dynamical evolution of the cluster, especially during collapse of the core (e.g. Statler *et al.* 1987). The total number of collisions is given by the time integral of the rate, and an argument along the lines of our discussion of the spatial variation of the collision rate shows that late core collapse does not dominate the total number of collisions (Sugimoto 1996); though the collision rate is very high then, this phase is over too quickly for this to matter. On the other hand, mass segregation is a more-or-less permanent consequence of dynamical evolution, and in young, compact clusters (at least), it has a major effect on the collision rate (Portegies Zwart *et al.* 1999).

Fourth, the collision rate varies very much from one star cluster to another. It has been found that the largest collision rate among the Galactic globular clusters is roughly 10^4 times larger than the smallest (Hills & Day 1976). Put another way, half the collisions in the entire cluster system of the Galaxy (about 150 clusters) occur in just about six clusters (Verbunt & Hut 1987, Johnston & Verbunt 1996). Such facts are important in trying to understand why the abundance of possible products of collisions (see Chapter 30) varies so much from one cluster to another.

The discussion so far is typical of much discussion of the dynamics of globular clusters in assuming that all stars are single, forgetting that it has now been established that a sizable proportion are binaries (Chapter 24). The existence of binaries immediately puts stars in close proximity and it is obvious that this may enhance the rate of collisions. There are at least two mechanisms by which the components of a binary may be brought into contact.

The first mechanism is purely internal. We have already remarked that stars expand greatly when they become red giants, and when this happens to one component of a binary it is very likely that the two components will touch. More precisely, what happens is that the material at the surface of the large component exceeds the 'Roche radius', at which point it is more strongly attracted to its companion. The dynamics is very similar to escape of stars in the tidal field of the galaxy. Figure 12.1 shows the Roche surface of the cluster in this case. For stars, however, the process is referred to as *Roche lobe overflow*. It is really one aspect of the incredibly complicated problem of stellar evolution in the context of binary stars, which will be treated in the next chapter.

The second mechanism is external. When a binary is disturbed by a passing star, its eccentricity may increase so much that the new pericentric distance of the components is less than the sum of their radii. An ordinary fly-by gives the intruder one chance to work on the eccentricity of the binary, and the probability of a resulting collision need not be high. If, however, the intruder is temporarily captured by the binary in a resonance (Chapter 19), it is as if the eccentricity is constantly changing; for this reason, during such an encounter the minimum distance between two stars (and it may be attained by one binary component and the intruder as easily as by the two components) is about 1% of the initial semi-major axis of the binary (Hut & Inagaki 1985). Thus stars in binaries with semi-major axes less than about 100 stellar radii are seriously at risk. If a hard binary with a large red giant forms a resonance, coalescence is almost certain, since the initial semi-major axis cannot be large enough.

Now it is obvious qualitatively that this is a major source of collisions. The time scale for such encounters between a binary and a third star is much shorter than that for collisions between two single stars, because the radius (semi-major axis) of a binary so much exceeds the radius of a single star (see Eq. (31.1)). This more than compensates for the fact that the proportion of stars in binaries of the required size is not very large. Besides, mass segregation increases their concentration in the core, and therefore their interaction rate.

The effect of collisions

What happens to stars when they collide? It is obvious from everyday experience that speed has something to do with it. Place an egg on the kitchen table and you have a useful egg; drop it on the table and you have a mess.

If two stars, again of mass m and radius R, are dropped onto each other, i.e. we neglect their relative speed when far apart, then at the moment when they touch the kinetic energy of their relative motion is $mV^2/4 = Gm^2/(2R)$, where V is their relative speed at the instant when they come in contact.[1] Now the gravitational binding energy of a single star is $G \int (M/r)dM$, where M is the mass within radius r. For a star of uniform density we have $M \propto r^3$, and so the integrand varies as $M^{2/3}$. Therefore the binding energy is $\dfrac{3}{5}\dfrac{Gm^2}{R}$. Comparison with the energy of relative motion shows that, in a hypothetical star cluster with solar-type stars, there is not enough kinetic energy to unbind even one star, let alone two. Furthermore, in a normal star the density increases inwards, and so the binding energy required would be even larger.

Thus direct collisions in globular clusters cannot destroy stars. Instead, the two stars fuse, after a period of violent internal motions (Fig. 31.1). Beyond this broad result, the details of the merged remnant depend on the nature of the stars involved, and are somewhat controversial. If one star is a white dwarf, for example, by its high density it sinks more-or-less unscathed to the centre of its fellow-victim (see, for example, Ruffert 1992). Another possible kind of collision product, considered in the next chapter, is a blue straggler. Quite general recipes for the collisions of a variety of common types of star are given, quite independently, by Tout *et al.* (1997) and Portegies Zwart *et al.* (1997).

In the previous discussion we were concerned with a head-on collision, as the stars were dropped together from a great separation. In general, however, the stars will not be aimed directly at each other. If they still physically touch, and if we assume that they still merge, it is clear that one consequence will be that the remnant is rapidly spinning, rather like slicing a golf ball, or putting spin on a snooker ball.

What, though, if the two stars do not even touch when at their closest approach? They can still influence each other gravitationally. To understand what happens it is instructive to consider a famous old problem of dynamical astronomy which was first understood by Newton, that of the tides raised on the oceans of the Earth by the Sun and Moon.

---

[1] The factor of 4 on the left-hand side results from half the reduced mass of two stars of equal mass; cf. Chapter 19.

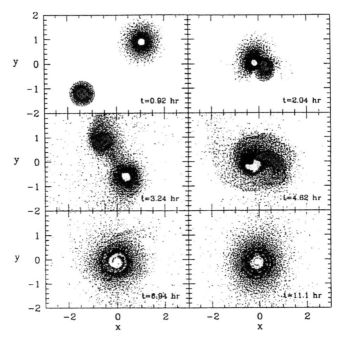

Fig. 31.1. Collision between two main sequence stars (after Sills *et al.* 2001). Material in different layers of the stars is coarsely distinguished by different shades of grey.

Let us consider just the lunar tide, for simplicity. There are two tides each day at most locations. To a first approximation, there is one high tide when the Moon is high in the sky, and another about 12 hours later. In fact it is unusual for one of the high tides to coincide with the time when the Moon is highest in the sky. In order for the tide to rise under the Moon, the water of the oceans must move across the bed of the ocean. The resistance to this motion is especially strong near narrow channels in shallow seas, as in the English Channel separating France from the British Isles. This motion also takes time, and so the maximum height of the water takes place some time after the Moon is highest in the sky. There is an analogy here with a mildly damped harmonic oscillator which is forced at a frequency much smaller than its natural frequency. The forced oscillation lags slightly behind the forcing.

Now return to the encounter between two passing stars, and let us think of one star as taking the place of the Earth with its oceans, while the other plays the role of the Moon. Again the second star raises tides on the first, and again the high tide at some location on the first star occurs some time after the second star has passed above that spot. Looking down at the first star, the second star sees a nearly spherical object, except for the tides he has raised. The nearer of these is behind him, in the sense that he has to look in the opposite sense from his orbital motion to see it (Fig. 31.2). Now this tidal material exerts a weak gravitational pull on the second star, in the sense

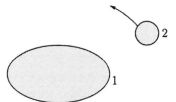

Fig. 31.2. The tidal bulge created by a passing star. Distortions of star 2 are not shown. See also the upper right panel of Fig. 31.1.

opposite to his orbital motion around the first star. Therefore it slows down his orbital motion.

What this argument shows is that the tidal interaction between the two stars removes energy from their relative motion. It is clear that most of it is removed around the time of closest approach. We are also now imagining that the two stars have approached from a large distance where they had a non-zero relative speed, of order 20 km/sec, say, and therefore a non-zero kinetic energy of relative motion. If this kinetic energy exceeds the energy removed tidally at closest approach, then the two stars will eventually separate again. If the reverse is true, however, the two stars are subsequently bound in a binary. This happens if the loss of energy is large enough, i.e. if the encounter is sufficiently close. *Tidal capture* has occurred.

What happens next is somewhat controversial. Certainly, because the two stars are bound to each other, further close encounters will occur. What effect these have depends on whether the tide raised by the first encounter has had time to subside (Kochanek 1992, Mardling 1995). If not, at the time of the next close approach the old tide could be in advance of the passing star, and then its effect is to unbind the pair; at least that would be the sense in which the interaction worked, though it might not be strong enough to succeed. In this scenario, the newly formed binary enters a phase of chaotic evolution, sometimes becoming less bound and sometimes more bound. The rotation of the stars presumably further complicates the interaction, however (see, for example, Witte & Savonije 1999).

At the same time the internal structure of the stars is being altered. The tidal interaction (except in some chaotic phases) is extracting energy from the relative motion of the two stars and transferring it to the tide. The dissipation of each tide further contributes this energy to the internal energy of the stars, which should expand somewhat in consequence (see McMillan *et al.* 1987, Tout & Kembhavi 1993). This clearly has the effect of enhancing the strength of the tidal interaction, and of leading to a situation where the two stars can be brought into contact. The upshot is that a single tidal encounter between two stars, even if it did not lead *directly* to physical contact between the two stars, can eventually lead to their merger. In other words the time scale for encounters leading to mergers is somewhat shorter than the collision time scale given by Eq. (31.1). The numerical size of the enhancement is, however, not large, as the effects of tidal dissipation decrease as a rather high power of the distance of closest approach (Problem 4).

Problems

(1)   Derive Eq. (31.1).

(2)   Two stars of mass m and radius R have relative speed V initially, when they
      are far apart. Estimate the mass loss in an encounter, by computing how much
      mass can be ejected (from the surface of one star to infinity) by the initial
      energy of relative motion of the two stars. (The result is of course zero if
      $V = 0$, in contradiction with results of numerical simulations; cf. Table 2 in
      Lombardi *et al.* (1996).)

(3)   In an isothermal model (Chapter 8) the density at large radius varies
      approximately as $\rho \propto r^{-2}$. Show that the rate of collisions is dominated by
      the contribution from the core.

      More precisely, recall that the scaled potential ϕ satisfies the equation

      $$\phi'' + \frac{2}{r}\phi' = 9\exp(-\phi),$$

      where r is the radius in units of the core radius and ϕ is assumed to vanish at
      $r = 0$ (Eq. (8.6)). In units of the central density, the density at a point where
      the potential is ϕ is $\exp(-\phi)$. In these units show (by numerical integration)
      that the integral

      $$\int_0^\infty 4\pi r^2 \rho^2 dr \simeq \frac{4\pi}{3} \times 0.55.$$

      (In general the result is $\rho_c^2 r_c^3$ times this factor, where r_c is the core radius
      and ρ_c is the central density; Spitzer 1987, p. 149.)

(4)   A star of radius r is disturbed by another star of equal mass passing on a
      parabolic orbit at a pericentric distance R. Assuming that the surface of
      the distorted star is an equipotential of the tidal field (cf. Eq. (12.6)), show
      that the relative difference between its largest and smallest radii in the plane
      of motion is $\delta r/r \simeq 6(r/R)^3$. By treating the distortions as two small equal
      masses added to the star, deduce that the work W done on the passing star
      varies as $W/E \propto (r/R)^8$, where E is the binding energy of one star.

32

Binary Star Evolution and Blue Stragglers

Once thought to be virtually devoid of binaries, globular clusters are now known to contain binaries in abundances not very much less than that of the galactic disk. Binaries in such large numbers, containing at least ten per cent of the stars in a typical globular cluster, cannot have resulted from dynamical interactions, and therefore must have formed at the same time that the bulk of the stars were formed. With so many binary stars around, all kinds of interesting reaction channels are possible, in three-body as well as four-body interactions. Binaries containing pulsars are just one example of the unusual objects that can result.

Even without dynamical interactions with other objects, binary star evolution in isolation is quite complicated enough. Compared to the evolution of single stars, a wide variety of new kinds of binaries and single stars can be created, through mass overflow from one star to the other, or through mass loss from the system, at various stages in their combined evolution. The stars can form a common envelope for a while, or one of the stars can explode as a supernova. Even if the explosion is symmetric the binary might not survive, as in the impulsive loss of mass in any stellar system (Chapter 11). Disruption is even more likely if the remnant receives a 'kick' (see Hills 1983), and if you find a neutron star in a binary in a star cluster it is likely to have got there by dynamical interactions (e.g. Kalogera 1996).

Complicated as such processes are, they have been studied for many years, and much is understood. Suitable introductions can be found, for example, in Iben (1985) and Iben & Tutukov (1997, 1998). For our purposes, what is even more useful is that those aspects of binary evolution that are reasonably well understood can be codified as relatively simple recipes (Portegies Zwart & Verbunt 1996, Tout et al. 1997).

What would be even nicer is some confirmation that these recipes produce the same answers.

Even without internal evolution of the components, tidal interactions between them can affect the dynamics of a binary. In the same way as tidal interactions occurring in close encounters of single stars (Chapter 31), these extract orbital energy without changing the angular momentum of the binary, and the result is circularisation of the binary orbit (Problem 1). The spins of the stars are also affected. Despite the low density in the envelope of a giant, these processes are especially important when the stellar radii are large.

The interaction between dynamical and stellar evolution of binaries is surely destined to be one of the growth areas in this whole subject in coming years. As yet few definitive results have emerged, and so our treatment serves only to give the flavour of what is possible. We concentrate on blue stragglers, as these occur in relatively great abundance, and even in small star clusters which are within our computational grasp at the present time.

Recycling stars: an example

In the dense cores of globular clusters, a typical star has a significant chance to undergo a collision during a Hubble time (Chapter 31). Since the velocity dispersion in globular clusters is one or two orders of magnitude less than the escape velocity at the surface of a typical main sequence star, almost all of the mass of two colliding stars is retained to form a merger product (Problem 31.2).

In a collision between two stars of relatively low mass, the merger remnant may have a mass below that of the main sequence turnoff, $0.8M_\odot$. In that case, it will be rather difficult to distinguish the merger product from an ordinary star of that mass. For a given candidate, detailed investigation might show unusual abundances, but spotting such stars against the background of normal stars would be akin to seeking a needle in a haystack.

In a collision where the sum of the masses of the original stars significantly exceeds that of the turnoff, by being larger than $1M_\odot$ say, the merger remnant will stand out in a Hertzsprung–Russell diagram, since it will be positioned on or near the main sequence, but bluewards of the turnoff. Such a star, if born at the same time as the rest of the stars of the cluster, should have evolved away from the main sequence. Its presence thus gives the impression of having straggled, and with its relatively bluer colour such a star is called a *blue straggler* (see Fig. 30.2).

The presence of such stars in old star clusters has been known for a long time. Now they are studied so intensively that a whole conference has been devoted to them (Saffer 1993; see also the review by Leonard (1996)). Even on observational grounds alone there is evidence that there is more than one way to make a blue straggler (Ferraro *et al.* 1993), and theorists are fortunately not short of ways of doing so too.

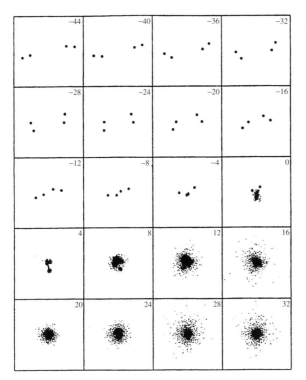

Fig. 32.1. Collision between two binaries (after Goodman & Hernquist 1991). Eventually all four stars merge into a single massive and extended stellar object.

Many, if not most, of the collisions between single main sequence stars in star clusters are likely to produce blue stragglers. One reason is that mass segregation will produce an excess of heavier stars in the central regions of a cluster. Other reasons are the larger radii of heavier stars, together with their enhanced gravitational focusing, both of which favour collisions for heavier, rather than lighter stars.

An even more efficient mechanism for inducing collisions is provided by binary–single-star and binary–binary encounters, especially those involving hard binaries, which have a significant chance to lead to resonant encounters where three or four stars are temporarily captured in a small area of space. Similar arguments again predict a larger collision probability for heavier stars. A special feature of three-body or four-body channels for blue straggler formation is that the mass of the final blue straggler may well be significantly larger than twice the turnoff mass, if more than two stars are involved in producing a single merger (Fig. 32.1).

A third way of using binaries to produce blue stragglers is through mass (Roche lobe) overflow from one component of a suitable binary onto the other member. This process comes in several flavours: coalescence, and mass transfer of various kinds (Leonard 1996). None of these mechanisms, however, require a high density of neighbouring stars, and so they are not special to the million-body problem.

As a simultaneous illustration of the last two types of blue straggler formation events, a calculation by Simon Portegies Zwart is reproduced in Fig. 32.2. The

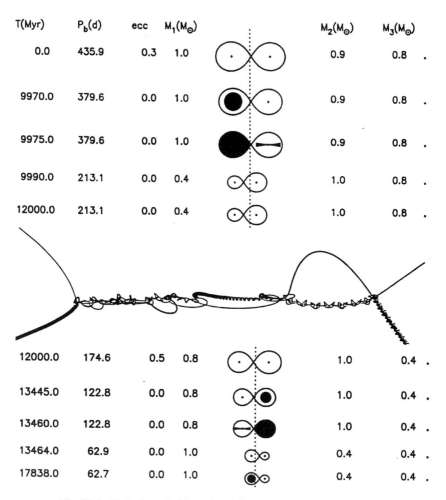

T(Myr)	P_b(d)	ecc	$M_1(M_\odot)$		$M_2(M_\odot)$	$M_3(M_\odot)$
0.0	435.9	0.3	1.0		0.9	0.8
9970.0	379.6	0.0	1.0		0.9	0.8
9975.0	379.6	0.0	1.0		0.9	0.8
9990.0	213.1	0.0	0.4		1.0	0.8
12000.0	213.1	0.0	0.4		1.0	0.8
12000.0	174.6	0.5	0.8		1.0	0.4
13445.0	122.8	0.0	0.8		1.0	0.4
13460.0	122.8	0.0	0.8		1.0	0.4
13464.0	62.9	0.0	1.0		0.4	0.4
17838.0	62.7	0.0	1.0		0.4	0.4

Fig. 32.2. Evolution of a binary involving one triple interaction (from Portegies Zwart 1996). For an explanation, see the text.

calculations were performed in the Starlab environment, using a version of the automated three-body scattering package described in Chapter 22, integrated with a simplified recipe for binary-star evolution.

There is nothing contrived about this particular set of initial conditions. Both the single star and the binary were drawn at random from a prescribed cluster core population of single stars and binaries. At each time step of a thousand years, the probabilities for close encounters between different objects were calculated, and actual collisions were carried out in Monte Carlo fashion by spinning a random number generator to see which stars would actually undergo an encounter. At each time step, those stars and binaries that did not undergo an interaction (the vast majority) were simply evolved with stellar evolution prescriptions: fitting formulae for stellar evolution tracks in the case of single stars, and simplified recipes in the case of binaries (for more details, see Portegies Zwart 1996, Portegies Zwart & Verbunt 1996).

Fig. 32.2 shows an example of an interplay between three stars which eventually leads to the formation of two blue stragglers. The first blue straggler is formed in isolation, as the result of binary star evolution (Roche lobe overflow), nearly 10 Gyr after its formation. At this time, the $1 M_\odot$ star attempts to climb the giant track. The first result is that its increased size leads to rapid tidal circularisation, erasing the initial eccentricity of 0.3 around $T = 9.97$ Gyr. Soon thereafter, the primary fills its Roche lobe, and around $T = 9.98$ Gyr dumps most of its mass on the secondary.

However, only $0.1 M_\odot$ of this supply of matter can be accreted onto the companion, and $0.5 M_\odot$ leaves the system. The reason for the inefficiency of mass transfer is that the donor star provides mass on its own thermal time scale, while the other star in turn can only accept mass on its thermal time scale, which is longer. By $T = 9.99$ Gyr, the primary turns into a low-mass helium star which then cools to become a white dwarf. From this time on, the secondary shows up as a blue straggler: it resembles a main sequence star, but has a mass and luminosity greater than that of a turnoff star. The secondary now evolves rather more quickly, but before it can reach the turnoff, a third star happens to pass through the system, 2 Gyr later, and is captured into a three-body resonance scattering event, displayed with time moving horizontally from left to right, for a duration of about a hundred years.

The outcome of the event, not surprisingly, leads to the lightest star being ejected, in this case the white dwarf, the end product of the original primary. The incoming $0.8 M_\odot$ star now takes the place of the primary. The orbit has tightened significantly, as could also be expected (hard binaries tend to get harder; cf. Chapter 19). After another 1.4 Gyr, the original secondary, saddled with the extra mass from the original primary, at last begins to climb the giant track. This leads to a second phase of tidal circularisation, in which the eccentricity induced in the three-body encounter is erased. After another 15 Myr, mass overflow takes place, again $0.6 M_\odot$ is lost, of which this time $0.2 M_\odot$ is accepted by the star which entered as a result of the exchange. This increases the mass of the latter from $0.8 M_\odot$ to $1.0 M_\odot$, thereby turning this star into a second blue straggler. The original secondary soon turns into a dwarf, orbiting the $1 M_\odot$ star. This second straggler phase finally ends around $T = 17.8$ Gyr, during the last phase of giant evolution.

The final state is not displayed in this figure as it takes place in the future: 17.8 Gyr exceeds almost all estimates of the age of the Universe. But we can guess what will happen. After the second blue straggler fills its Roche lobe, the binary will shrink significantly, and most likely a common envelope system will be formed. This is likely to lead to the merger of the two cores, resulting in a late type giant of $1.3 M_\odot$. A bit later, a white dwarf will then be left behind, with a mass of around $0.7 M_\odot$.

Observationally, the first blue straggler formed by the transfer of mass in a close binary will not be distinguishable from those formed by the 'classical' route, i.e. in collisions between single stars (Fig. 30.2). Only after the dynamical exchange interaction does the system become clearly different from 'ordinary' blue stragglers: the observation of a blue straggler in an eccentric orbit or in a detached binary with a main-sequence companion is a direct indication for formation by an exchange reaction.

The binary zoo

What we have seen is just one particular case study of the formation of blue stragglers. While it includes a stellar environment (which provided the interloper for the three-body interaction), there is no surrounding star cluster undergoing the many dynamical processes described in other chapters of this book. Such studies are now under way, and confirm that interactions involving both single stars and binaries are routes which greatly enhance the proportion of blue stragglers.

An example is the old open cluster M67, for which Hurley *et al.* (2001) have constructed an N-body model which includes the full panoply of stellar and dynamical evolution. Though open clusters are often regarded as rather low-grade fare in the world of N-body simulations, a 15 000-body simulation was needed to simulate the last half of its evolution. Twice that number might have been needed to cover the entire evolution, which would not have been feasible at that time for a stellar system with a rich complement of active binaries. Blue stragglers are present in M67 at an abundance roughly ten times higher than would occur in a population of isolated binaries and single stars. As this simulation showed, it is the dense environment in M67 which accounts for its wealth of blue stragglers.

Blue stragglers are just one of the unusual types of object that cannot form through single star evolution, but require either the presence of primordial binaries, or the interaction between two, three, or more stars in a dense stellar system. In addition, X-ray sources, pulsars, spectroscopic binaries and eclipsing variables have been detected in significant numbers, some of them right in the middle of the densest central regions of globular clusters. This diverse collection of objects forms a gold mine for dynamicists modelling the evolution of globular clusters. Another metaphor, coined by Melvyn Davies (Davies 1995) is also apt: the zoo of possible objects that can be formed in few-body interactions is bewildering, and rather little has been studied yet in this field – for those daring to brave this jungle, many species still remain to be discovered.

While all these binaries and binary products provide useful diagnostics of binaries and their interactions, some of them are even directly involved in the physical processes of energy generation which drive the expansion of the core after core collapse (Chapters 27–29). In this sense we are beginning to get a direct look at the central engines which power the later phases in the evolution of globular clusters.

Here is a final conceit. One may say that star cluster dynamics is like a ballet with about a million performers, but it is the binary stars which take centre stage. It is in their *pas de deux* that most of the action develops, and it is through their interactions with other players, in the form of occasional *pas de trois* and *pas de quatre*, that fascinating new patterns (and even new characters) arise.

Problems

(1)   A binary evolves at constant angular momentum but decreasing orbital energy. Deduce that the eccentricity decreases.

(2)   A stellar system has n single stars per unit volume, all with radius R. Assuming that two stars merge if and only if their closest distance is less than $2R$, show that the rate of formation of mergers in two-body encounters is $4\pi Gm Rn^2 \langle 1/V \rangle$, where V is the relative speed of the two stars.

   The system also has fn hard binaries per unit volume, all with semi-major axis a, and all components have radius R. The cross section for binary–single encounters in which at least two participants come within a distance d is known to be approximately

$$\sigma \simeq 8.5 \frac{\pi a^2}{V^2} \left(\frac{3}{2} \frac{Gm}{a} \right) \left(\frac{d}{a} \right)^{0.4}$$

(Hut & Inagaki 1985). Compute the rate of mergers in binary–single encounters, and compare with the rate of single–single mergers.

33

Star Cluster Evolution

It often happens in science that progress is made, not so much by the discovery of new facts, but by organising known facts in a new way. An illustration from chemistry is the periodic table, and in astronomy the HR (Hertzsprung–Russell) diagram is a perfect example. In its modern form this diagram, now called a CM (colour–magnitude) diagram, is a scatter plot of the luminosity or absolute magnitude of a sample of stars plotted against their colour (see Fig. 30.2). It is an immensely powerful tool, so familiar that its power is taken for granted, and it is invaluable for studying the evolution of stars. In this chapter our aim is to provide something comparable for star clusters, though our goal is the much more modest one of helping the reader to grasp in a few pictures the essentials of what has been described in greater detail in earlier chapters of this book.

The links between many of the dynamical processes we have discussed, and one or two others, are summarised in Fig. 33.1. Centre stage is mass loss, which occurs through the agency of several processes. Meanwhile, as the system is losing mass and heading towards oblivion, various processes are also causing its internal structure to evolve. Chief among these is two-body relaxation, which causes more massive stars to segregate inwards, and the core to collapse. Both mechanisms enhance the importance of interactions between primordial binaries, which eventually bring the collapse to a halt. Interactions between pairs of stars, and even triples and quadruples, also affect the kinds of stars we find in clusters, and the way in which they evolve. For astronomers, one of the interesting consequences is that the mass function changes. Often it is said that one simply needs sufficient destruction mechanisms (such as disk shocking) to achieve this. But disk shocking does not care about the mass of

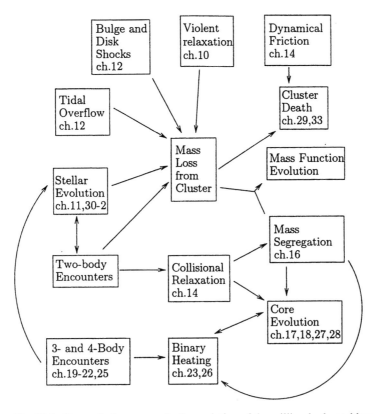

Fig. 33.1. Dynamical processes in the evolution of the million-body problem.

a star, and can only influence the mass function if two-body relaxation has already segregated stars of different mass.

Structural evolution

Next we shall try to picture how the system evolves under these processes, though we shall assume that certain of them are dominant. In particular, we imagine the evolution of a star cluster bathed in the tidal field of a parent galaxy and dominated by the following dynamical processes. First is the internal evolution of the stars, which causes them to lose mass; this weakens the potential well of the cluster, which may lead to much further loss of mass. Next, there is two-body relaxation, which leads to mass segregation and collapse of the core. Finally, collapse is arrested by the intervention of primordial binaries; the cluster enters the final phase of its life, continually shedding mass across the shrinking tidal boundary. Throughout, it is shaken intermittently by tidal shocks.

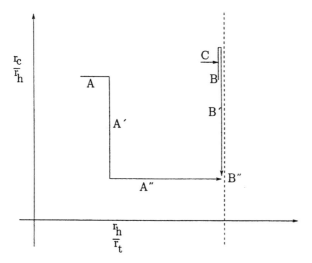

Fig. 33.2. Schematic evolutionary tracks of star clusters. See text.

We shall summarise these results in a diagram which reduces the structure of the cluster to its characteristic radii: the core radius, r_c, the half-mass radius, r_h, and the tidal radius, r_t. To reduce the problem to one that can be plotted on paper, we take the ratios r_h/r_t and r_c/r_h. The first of these tells us how liable the cluster is to tidal disruption, and the second states how evolved it is dynamically.

Phase I: stellar evolution

We begin with the early evolution, and assume that the time scale of relaxation is much longer than the time scale for major mass loss by the internal evolution of the stars. Provided that the latter is longer than the typical orbital periods of stars in the cluster, the cluster expands nearly homologously, unless it is already tidally limited. If it is not tidally limited, it behaves like an isolated cluster: r_c/r_h remains nearly constant, but r_t decreases a little with the loss of mass, and so r_h/r_t increases. If the cluster is tidally limited then the core may expand while the stars are losing mass, but r_h and r_t will contract because of the more copious loss of mass across r_t; thus r_c/r_h increases while r_h/r_t changes little. These two possibilities are illustrated in Fig. 33.2 as cases A and B. There is also an intermediate case (C), where the cluster is not quite tidally limited initially, but becomes so because of the expansion (Engle 1999).

We can place some numbers on this diagram. At the point where a cluster is tidally limited, the value of r_h/r_t is roughly 0.4 (Giersz & Heggie 1997). Also, how much evolution there is depends on the proportions of stars of different mass: if there are few stars of high mass, there is little evolution, and the cluster expands by as little as 10%. If there are many, the cluster may dissipate altogether. Roughly speaking, the fraction of mass lost also gives the factor by which a cluster would expand, though if it starts life nearly tidally limited then the mass loss may well result in the death of the cluster by tidal dissipation. What this fraction is may be seen in Fig. 11.1, though this completely ignores loss of mass by expulsion of gas, which is likely to dominate the first few million years.

Phase II: dynamical evolution

Eventually (and in quite a short time compared with the ages of globular clusters) the rate of mass loss caused directly by the evolution of individual stars becomes negligible. Those clusters that survive now enter a phase of evolution dominated by tidal stripping, relaxation, mass segregation, and core collapse, which means that r_c/r_h decreases. If the cluster already fills its tidal radius, it evolves with little change in the ratio r_h/r_t (which can hardly get bigger without the total dissolution of the cluster) but loses mass relatively copiously. Otherwise, little mass is lost in these phases, which mainly affect the core, and again r_h/r_t remains virtually unchanged. The two clusters shown in Fig. 33.2 now move along the evolutionary paths denoted by A' and B'. The time scale taken for this stage of the evolution depends on a number of factors. For a system which starts this stage of the evolution at a point where it is tidally limited, the relaxation time scale is proportional to the number of stars times the galactic orbital period at the location of the cluster. For other systems, the relevant time scale for the evolution of the core is smaller, i.e. the evolution of the core is faster.

As we have mentioned, stars escape as r_c/r_h drops. If the cluster lasts long enough, however, and assuming that it was born with a sufficient proportion of primordial binary stars, these begin to release energy in three- and four-body reactions when the mass inside the core has shrunk to a certain fraction of the whole, i.e. when r_c/r_h is small enough. What happens next depends on whether the cluster is living comfortably inside its tidal boundary (case AA'), or already tidally limited (case BB'). In the second case the cluster sheds mass, evolving in a self-similar manner first described many years ago by Hénon (B''). Otherwise, the release of energy drives an expansion of the cluster until it joins its tidally bound neighbour. This expansion is homologous out to the radius r_h, and so the evolution in our diagram is horizontal (A'') until the evolutionary track arrives at B''.

Though these tracks are schematic, more realistic treatments confirm the gross features. For example, Fig. 33.3 illustrates an N-body simulation on a diagram with the same axes. One can readily discern stages corresponding to the phases A, A' and A'' in Fig. 33.2, even though this cluster is powered by three-body binaries rather than primordial binaries. Stage A is quite short.

Destruction mechanisms

Finally, we would like to give a feel for the time scales on which all this action takes place, with special attention to the lifetime of a stellar system. A cluster is under

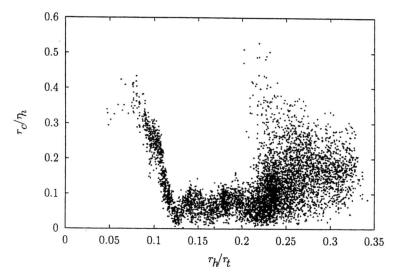

Fig. 33.3. Evolutionary track of an N-body simulation ($N = 4096$). It starts from a King model with $W_0 = 5$ whose limiting radius is 0.25 times the tidal radius. The initial mass function is a power law $n(m)dm \propto m^{-2}$ in the range $0.1 M_\odot < m < 15 M_\odot$. Stellar evolution uses the simple prescription introduced by Chernoff & Weinberg (1990). Plotting stops when the mass inside the tidal radius is less than 5% of its initial value.

attack both from the outside (tidal disruption, disk shocking, etc.) and from its own internal diseases (stellar evolution, two-body relaxation). To make matters simpler (without losing relevance to the old star clusters around us, which must have survived this phase) we shall ignore destruction on the time scale of stellar evolution, which must have become ineffective long ago. It is a disease of infancy.

Essentially the task is to estimate the time scales of the main processes which cause a cluster to lose mass. We ignore tidal overflow in the sense that this is not a primary cause of loss of mass; it would seem quite possible for a cluster to survive indefinitely in a tidal field if there were no other mechanism causing loss of mass (Muzzio *et al.* 2001). It is true that we have no reason to suppose that clusters are born in equilibrium with their tidal field, but again any outlying matter will be stripped on a short time scale.

The first process to consider, and the one on which much emphasis has been placed in this book, is two-body relaxation. For its time scale we take the half-mass relaxation time (Eq. (14.13)), i.e.

$$t_{rh} \simeq \frac{0.138 M_c^{1/2} r_h^{3/2}}{G^{1/2} m \ln \Lambda},$$

where m is the individual stellar mass, and M_c, r_h are the mass and half-mass radius of the cluster, respectively. If we set this equal to the ages of the typical old Galactic globular star clusters, we obtain a feel for the gross parameters of clusters which

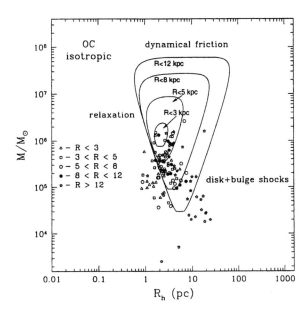

Fig. 33.4. The survival of Galactic globular clusters (from Gnedin & Ostriker 1997). The masses of 119 Galactic globular clusters are plotted against their half-mass radius. Different symbols are used for clusters within different ranges of galactocentric distance. The three stated destruction mechanisms are thought to control their distribution. The key on the top left refers to the assumed models for the Galactic potential and the distribution of the globular clusters.

can survive the destructive effects of this process to the present day. The result (with some refinements) is a line at the position of the lower left boundary in Fig. 33.4, which is a plot of cluster mass against half-mass radius.

The next process to consider is disk shocking (Chapter 12; see also Gnedin *et al.* 1999). This does different things on different time scales. The overall energy of the cluster is altered on the time scale given in Eq. (12.9). But the energies of individual stars are altered on the time scale of shock relaxation, which may, like ordinary two-body relaxation, lead to escape. Therefore we adopt a time scale of the form (see Problem 3)

$$t_{sr} \sim \frac{M_c R V}{G \Sigma^2 r_h^3},$$

where R is the galactocentric distance of the cluster, V is its orbital speed round the galaxy, and Σ is the surface density of the galactic disk. Setting this (with a suitable estimate of the numerical coefficient) equal to the lifetime of the globular clusters gives lines at the positions of the lower right boundaries in Fig. 33.4. Note that the time scale depends on location within the galaxy, especially on R: it decreases with decreasing galactocentric radius, not only because of the explicit factor of R, but also because Σ increases as R decreases.

Now we see that the observed clusters in our Galaxy fit rather comfortably into the pocket between these two constraints. They do not, it is true, explain the absence of clusters of very high mass. This could well be primordial, i.e. it may be that no such clusters ever existed. There is, however, another mechanism that might have depopulated this part of the diagram. We introduced dynamical friction (Chapter 14) as a process which slows down stars moving through clusters, especially massive

Fig. 33.5. A dying globular cluster E3. Image
obtained by use of
NASA's SkyView facility
(skyview.gsfc.nasa.gov)
located at NASA Goddard
Space Flight Centre.

stars. In the same way it acts on clusters moving through the Galaxy, and acts more
quickly on more massive objects. Estimates show that it would have dragged most
clusters of high mass (above the upper boundary on Fig. 33.4) into the Galactic
Centre. Together with the two processes of disk shocking and two-body relaxation,
this shows that there is a rather well constrained triangle of survival in this diagram,
where almost all of the present-day clusters are found.

The present distribution of cluster masses is well explained by a combination of
these dynamical processes and a suitably chosen initial condition (e.g. a power law).
It resembles a log-normal distribution (i.e. normal in the logarithm of the mass),
and can equally well be approximated by a flat distribution at low masses and a
power law at high masses (see Baumgardt 1998, Vesperini 1998 and references
therein, Fall & Zhang 2001). The radial distribution of clusters is also affected (e.g.
Capuzzo-Dolcetta & Tesseri 1999).

Estimates have been made of the rate at which clusters are dying (Aguilar *et al.*
1988, Hut & Djorgovski 1992, Murali & Weinberg 1997, Johnston *et al.* 1999). It is
likely that about half of the present population of the globular clusters in our Galaxy
will perish within the next 10^{10} years, which is about how long the present population
has lasted. There is nothing surprising in this, any more than in the fact that the age
of most individuals in the human population of the world is comparable with the
span of a human life. The clusters which will expire within the next billion years
are sparse objects. Likely candidates are the clusters E3 (Fig. 33.5) and Palomar 13
(Siegel *et al.* 2001). Even the names suggest that these are miserable objects, and not
in the same class as the brilliant clusters that made themselves known to Messier.
Almost all of the low-mass stars have left such an object, leaving behind the heavier
populations which escape least easily: binary stars, white dwarfs. Soon, by galactic
standards, nothing will be left.

Problems

(1) Add the line in Fig. 33.4 corresponding to the tidal boundary condition, assuming that the rotation curve of the Galaxy is flat, with rotation speed 220 km s^{-1}, and that $r_h \lesssim 0.4 r_t$. (There is a different line for each Galactic radius R.)

(2) Suppose that the relaxation time t_r of a cluster is always proportional to the time until its dissolution (Chapter 29). Suppose the initial distribution of t_r is a power law $f(t_r) \propto t_r^{-\alpha}$, with $\alpha \sim 2$, say. Show that the distribution of $\log t_r$ eventually evolves into a form resembling a normal distribution.

(3) For a tidally limited star cluster, show that the shock-heating time scale (Eq. (12.9)) is of order $M_c R V / (G \sum^2 r_h^3)$.

Appendix A

A Simple N-Body Integrator

N-body codes used in research run to many thousands of lines and hundreds of subroutines. The two main collisional codes/packages are

(1)   *Starlab* See www.manybody.org

(2)   *NBODYx* A series of N-body codes of differing sophistication and adapted to different hardware. See ftp://ftp.ast.cam.ac.uk/pub/sverre/nbody6/

The codes that follow are *much* more modest. We give both Matlab and C codes, but not a FORTRAN code (for which see Binney & Tremaine 1987). Both codes may be found at www.manybody.org/codes.html.

MATLAB

This code is presented mainly because of the integrated graphics capabilities of MATLAB (Fig. A.1). To run the code, enter 'nbody(25,100,10)' in a MATLAB session, assuming that a file containing the following code is in the MATLAB search path. The input data are the number of particles, the number of outputs and the end time, respectively.

The first function initialises, runs at intervals of length dt (collecting data in y), and outputs a final result.

```
function nbody(n,nout,tmax)
cpu = cputime;dt=tmax/nout;t=0;
[h,x,xdot,f,fdot,step,tlast,m,y] = initialise(n);
nsteps=0;
while t<tmax
[x,xdot,f,fdot,step,tlast,m,t,nsteps] = ...
hermite(x,xdot,f,fdot,step,tlast,m,t,t+dt,nsteps);
```

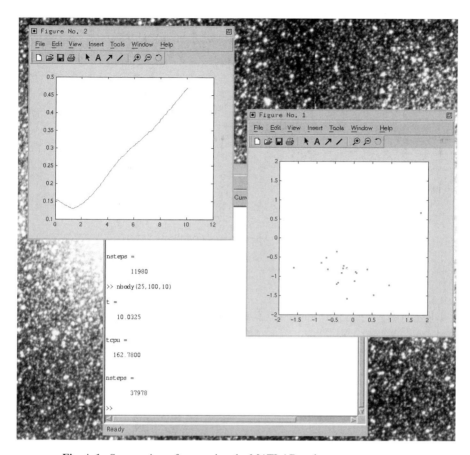

Fig. A.1. Screen view after running the MATLAB code.

```
[y]=runtime_output(h,x,y,t,n);
end
final_output(t,cpu,nsteps,y)
```

The next function initialises various scalars, and chooses n positions and velocities with a uniform distribution inside spheres whose radii are chosen to give *N*-body units approximately. It then sets masses, and times when particles were most recently updated (i.e. 0, initially). Forces, force derivatives and initial time steps are selected (cf. Makino & Aarseth 1992). Finally a plot window for showing a movie is initialised.

```
function [h,x,xdot,f,fdot,step,tlast,m,y] = ...
initialise(n)
y(1)=0;figure(1)
x = uniform(n,6/5);
xdot = uniform(n,sqrt(5/6));
m = ones(1,n)./n;
tlast = zeros(1,n);
for i = 1:n
```

```
[f(1:3,i),fdot(1:3,i)] = ffdot(x,xdot,m,i);
step(i) = 0.01*sqrt(sum(f(1:3,i).^2)/sum(fdot(1:3,i).^2));
end
h=plot(x(1,:),x(2,:),'o');
axis([-2 2 -2 2])
axis square
grid off
set(h,'EraseMode','xor','MarkerSize',2)

function [x] = uniform(n,a)
r = a*rand(1,n).^(1/3);
cos_theta = 2*rand(1,n)-1;
sin_theta = sqrt(1-cos_theta.^2);
phi = 2*pi*rand(1,n);
x(1,1:n) = r.*sin_theta.*cos(phi);
x(2,1:n) = r.*sin_theta.*sin(phi);
x(3,1:n) = r.*cos_theta;
```

Particles are updated in this next function. In the main loop, the next particle for updating is found (i), positions and velocities of all particles are predicted, and force and force derivative on i are computed by a call to ffdot (see below). Higher order Hermite corrections a2 and a3 are computed and the corrections applied. Finally the next time step is estimated and some variables are updated.

```
function [x,xdot,f,fdot,step,tlast,m,t,nsteps] = ...
hermite (x, xdot, f, fdot, step, tlast, m, t, tend, nsteps)
n = size(m,2);
while t < tend
[t,i] = min(step + tlast);
dt = t - tlast;
dts = dt.^2;
dtc = dts.*dt;
dt3 = dt(ones(1,3),:);
dts3 = dts(ones(1,3),:);
dtc3 = dtc(ones(1,3),:);
xtemp = x + xdot.*dt3 + f.*dts3/2 + fdot.*dtc3/6;
xdottemp = xdot + f.*dt3 + fdot.*dts3/2;
[fi,fidot] = ffdot(xtemp,xdottemp,m,i);
a2 = (-6*(f(1:3,i) - fi) - step(i)*(4*fdot(1:3,i) + ...
2*fidot))/step(i)^2;
a3 = (12*(f(1:3,i) - fi) + 6*step(i)*(fdot(1:3,i) + ...
fidot))/step(i)^3;
x(1:3,i) = xtemp(1:3,i) + step(i)^4*a2/24 + ...
step(i)^5*a3/120;
xdot(1:3,i) = xdottemp(1:3,i) + step(i)^3*a2/6 + ...
step(i)^4*a3/24;
f(1:3,i) = fi;
fdot(1:3,i) = fidot;
```

```
modfi = sum(fi.^2);
modfidot = sum(fidot.^2);
moda2 = sum(a2.^2);
moda3 = sum(a3.^2);
step(i) = min(1.2*step(i),sqrt(0.02*(sqrt(modfi*moda2) ...
+ modfidot)/(sqrt(modfidot*moda3) + moda2)));
tlast(i) = t;
nsteps = nsteps + 1;
end

function [fi,fidot] = ffdot(x,xdot,m,i)
n = size(m,2);
j = [1:i-1,i+1:n];
rij = x(:,i)*ones(1,n-1) - x(:,j);
rijdot = xdot(:,i)*ones(1,n-1) - xdot(:,j);
r2 = sum(rij.^2);
rrdot = sum(rij.*rijdot);
r3 = r2.^1.5;
r5 = r3.*r2;
fi = - rij*(m(j)./r3)';
fidot = - rijdot*(m(j)./r3)' + 3*rij*(rrdot.*m(j)./r5)';
```

This customisable function updates the movie and collects data for plotting the rms distance of all particles against time.

```
function [y]=runtime_output(h,x,y,t,n)
set(h,'XData',x(1,:),'YData',x(2,:));
drawnow
m = size(y,1);
y(m+1,1) = sqrt(sum(sum(x(:,:).^2)))/n;
y(m+1,2) = t;

function final_output(t,cpu,nsteps,y)
t
tcpu=cputime-cpu
nsteps
figure(2)
plot(y(2:end,2), y(2:end,1),'-')
```

C

The following code lacks comments, but is structured like the preceding MATLAB code, so that the comments there should be helpful.

The code may be compiled with gcc[1] nbody.c -o nbody -lm, and run with nbody 25 100 10 (say), where the data are the number of stars, the number of outputs and the end time. It runs about six times faster than the MATLAB code. No movie is produced. To plot the accumulated data y, gnuplot (www.gnuplot.info/) could be used.

---

[1] We used version 2.95 on a Sun running Solaris 2.7, for example.

```c
#include <stdio.h>
#include <math.h>
#include <stdlib.h>
#include <time.h>
#include <assert.h>
#define NMAX 1024
#define NDIM 3
#define loop(idx,last) for (idx = 0; idx < last ; idx++)
#define min(x, y) (x<y?x:y)
typedef double real;
static int i, j, k;

void uniform(int n, real a, real temp[NMAX][NDIM]) {
real r, cos_theta, sin_theta, phi;
loop(i, n) {
r = a*pow(drand48(), 1./3.);
cos_theta = 2.*rand()/((real)RAND_MAX+1)-1;
sin_theta = sqrt(1-pow(cos_theta, 2));
phi = 2*M_PI*rand()/((real)RAND_MAX+1);
temp[i][0] = r*sin_theta*cos(phi);
temp[i][1] = r*sin_theta*sin(phi);
temp[i][2] = r*cos_theta;
}
}

void ffdot(real x[NMAX][NDIM], real xdot[NMAX] [NDIM],
   real m[ ], int i, int n, real fi[ ], real fidot [ ]) {
real rij[NDIM], rijdot[NDIM], r2, r3, r5, rrdot;
loop(k, NDIM) { fi[k] = fidot[k] = 0;}
loop(j, n) {
if (j!=i) {
r2 = rrdot = 0;
loop(k, NDIM) {
rij[k] = x[i][k] - x[j][k];
rijdot[k] = xdot[i][k] - xdot[j][k];
r2 += pow(rij[k],2);
rrdot += rij[k]*rijdot[k];
}
r3 = pow(r2,1.5);
r5 = r3*r2;
assert(r3!=0&&r5!=0);
loop(k, NDIM) {
fi[k] = fi[k] - m[j]*rij[k]/r3;
fidot[k] = fidot[k] - m[j]*(rijdot[k]/r3 -
   3*rrdot*rij[k]/r5);
}
}
}
}
```

```
int hermite(real x[NMAX][NDIM], real xdot[NMAX][NDIM],
    real f[NMAX][NDIM], real fdot[NMAX][NDIM],
    real step[ ], real tlast[ ], real m[ ], real *t,
    real tend, int n)
{
real xtemp[NMAX][NDIM], xdottemp[NMAX][NDIM], fi[NDIM],
    fidot [NDIM];
real a2[NDIM], a3[NDIM], modfi, modfidot, moda2, moda3;
real tmin, dt, dts, dtc, dnmtr, temp;
int imin, nsteps;
assert(n<=NMAX);
nsteps=0;
do {
tmin = pow(10, 6);
imin = 0;
loop(i, n) {
if (step[i] + tlast[i] < tmin) {
tmin = step[i] + tlast [i];
imin = i;
}
}
assert(imin>=0);
*t = tmin;
i = imin;
loop(j, n) {
dt = *t - tlast[j];
dts = pow(dt, 2);
dtc = dts*dt;
loop(k, NDIM) {
xtemp[j][k] = x[j][k] + xdot[j][k]*dt + f[j][k]*dts/2
+ fdot[j][k]*dtc/6;
xdottemp[j][k] = xdot[j][k] + f[j][k]*dt +
    fdot[j][k]*dts/2;
}
}
ffdot(xtemp,xdottemp,m,i,n,fi,fidot);
modfi = modfidot = moda2 = moda3 = 0;
assert(step[i]!=0);
loop(k, NDIM) {
a2[k] = (-6*(f[i][k] - fi[k]) - step[i]*(4*fdot[i][k]
+ 2*fidot[k]))/pow(step[i],2);
a3[k] = (12*(f[i][k] - fi[k]) + 6*step[i]*(fdot[i][k]
+ fidot[k]))/pow(step[i],3);
x[i][k] = xtemp[i][k] + pow(step[i],4)*a2[k]/24
+ pow(step[i],5)*a3[k]/120;
xdot[i][k] = xdottemp[i][k] + pow(step[i],3)*a2[k]/6
+ pow(step[i],4)*a3[k]/24;
f[i][k] = fi[k];
```

```
fdot[i][k] = fidot[k];
modfi += pow(fi[k],2);
modfidot += pow(fidot[k],2);
moda2 += pow(a2[k],2);
moda3 += pow(a3[k],2);
}
dnmtr = moda2 + sqrt(modfidot*moda3);
assert(dnmtr!=0);
temp = sqrt(0.02*(sqrt(modfi*moda2) + modfidot)/dnmtr);
step[i] = min(1.2*step[i], temp);
tlast[i] = *t;
nsteps++;
}
while(*t<tend);
return nsteps;
}

void initialise(int n, real x[NMAX][NDIM],
   real xdot[NMAX][NDIM], real f[NMAX][NDIM],
   real fdot[NMAX][NDIM], real step[NMAX],
   real tlast[NMAX], real m[ ]) {
real fi[NDIM],fidot[NDIM], radius, fmod, fdotmod;
radius = 6./5.;
uniform(n,radius,x);
radius = sqrt(5./6.);
uniform(n,radius,xdot);
loop(i, n) { m[i] = 1./(real)n;}
loop(i, n) {
tlast[i] = 0;
ffdot(x,xdot,m,i,n,fi,fidot);
loop(k, NDIM) {
f[i][k] = fi[k];
fdot[i][k] = fidot[k];
}
fmod = fdotmod = 0;
loop(k, NDIM) {
fmod += pow(fi[k],2);
fdotmod += pow(fidot[k],2);
}
step[i] = 0.01*sqrt(fmod/fdotmod);
}
}

void runtime_output(real x[NMAX][NDIM], real t,
   long int n){
real ss=0;
loop (i,n) ss += x[i][0]*x[i][0] + x[i][1]*x[i][1] +
   x[i][2]*x[i][2];
```

```
printf("%lf %lf\n",t,sqrt(ss));
}

void final_output(real t, long int cpu, long int nsteps,
    long int noutp){
printf("\n#%lf, %ld, %ld\n",
time, (clock()-cpu)/CLOCKS_PER_SEC, nsteps);
}

void main(int argc, char **argv) {
real x[NMAX][NDIM], xdot[NMAX][NDIM], f[NMAX][NDIM],
    fdot[NMAX][NDIM];
real step[NMAX], tlast[NMAX], m[NMAX], t, dt;
long int j, i, n, nsteps, noutp, cpu;
cpu = clock();
printf("%ld\n",cpu);
srand(cpu);
assert(argc==4);
n = atoi(argv[1]);
noutp = atoi(argv[2]);
dt = atof(argv[3])/(real)noutp;
t = 0;
initialise(n, x, xdot, f, fdot, step, tlast, m);
nsteps=0;
loop(j, noutp) {
nsteps += hermite(x, xdot, f, fdot, step, tlast, m, &t,
    t+dt, n);
runtime_output(x,t,n);
}
final_output(t,cpu,nsteps,noutp);
}
```

Problems

(1)   Given acceleration a and its derivative at the beginning and end (subscripts 0 and 1, respectively) of a time step of length h, obtain the approximate formulae

$$a_0^{(2)} = \frac{-6(a_0 - a_1) - h(4\dot{a}_0 + 2\dot{a}_1)}{h^2}$$

$$a_0^{(3)} = \frac{12(a_0 - a_1) + 6h(\dot{a}_0 + \dot{a}_1)}{h^3}$$

for the second and third derivatives at the start of the time step. The Hermite integrator first takes the step using terms in a Taylor series up to $\dot{a}_0$, i.e.

$$x_1' = x_0 + hv_0 + \frac{1}{2}h^2 a_0 + \frac{1}{6}h^3 \dot{a}_0$$

$$v_1' = v_0 + ha_0 + \frac{1}{2}h^2 \dot{a}_0,$$

computes a_1 and $\dot{a}_1$, and then adds the terms in $a_0^{(2)}$ and $a_0^{(3)}$, i.e.

$$x_1 = x_1' + \frac{1}{24}h^4 a_0^{(2)} + \frac{1}{120}h^5 a_0^{(3)}$$

$$v_1 = v_1' + \frac{1}{6}h^3 a_0^{(2)} + \frac{1}{24}h^4 a_0^{(3)}.$$

(2)   Convert the code in Binney & Tremaine 1987 to the Hermite integrator.

(3)   Modify the initial conditions of either N-body code to integrate the three-body problem with masses $m_i = 1$ and the following initial conditions:

$$\mathbf{r}_1 = (-0.970\,004\,36, 0.243\,087\,53, 0) = -\mathbf{r}_2,$$

$$\mathbf{v}_1 = (-0.466\,203\,68, -0.432\,365\,73, 0) = \mathbf{v}_2,$$

$$\mathbf{r}_3 = \mathbf{0}, \mathbf{v}_3 = -2\mathbf{v}_1$$

(cf. Fig. 20.7, top left panel).

(4)   At present the output routines use data corresponding to the last update time. Modify so that it uses extrapolated data (as in the Hermite function/routine).

(5)   Perform the same kind of N-body calculation as the simple MATLAB and C codes (with the same initial conditions) using Starlab.

Appendix B

Hints to Solution of Problems

Chapter 1

(1)   See Box 7.2.

(2) $v^2 = GM/r_t$, of course. For N-body units one could convert from c to r_t using Fig. 8.3.

(3)   At distance d the area within a line of sight of solid angle 1 arcsec^2 is $(\pi d/648\,000)^2$.

Chapter 2

(1)   See Chapter 3.

(2)   For the radius of a stellar system, see Eq. (9.2), and footnote 3.

(3)   Choose an arbitrary centre O, and apply Newton's Theorems to shells with centre O.

(4) $\ddot{r} = -(4\pi/3)G\rho r$, which is linear in r, and so collapse is homologous. Solve the equation of motion at the edge of the sphere ($r = a$) using formulae of Box 7.2, in the case $e = 1$. Fig. 2.1: $\pi(6/5)^{3/2}$.

Chapter 3

(1)   Delete all code generating output, and run to (say) $t = 1$. Repeat each run at least once to assess variance. Change N by factor 2.

(2)   Such a code will use the same time step for all masses ('lockstep').

(3)   Even the circular two-body problem in the plane for one period with minimal output will be illuminating. Maple performs many checks on the validity of each arithmetic operation.

Chapter 4

(1)   The collision set has codimension 3, each orbit is one-dimensional. For example, in the two-body problem the condition is $(\mathbf{r}_1 - \mathbf{r}_2) \wedge (\mathbf{v}_1 - \mathbf{v}_2) = \mathbf{0}$.

(2)   First part: $x = (9t^2/2)^{1/3}$, $y = 0$; parabola; straight line.

(3) $c = 0, \pm\sqrt{3}$.

Chapter 5

(2)   For $n \geq 3$, circular motion is unstable (even without friction). For $n = 1$ the speed in circular motion is independent of the radius of the circle (isothermal case, cf. Chapter 8), and for $n < 1$ the speed decreases with decreasing radius.

Chapter 6

(1)   In plane polars in the plane of motion the equations of motion are $\ddot{r} - r\dot{\theta}^2 = -\partial\phi/\partial r$, $r\ddot{\theta} + 2\dot{r}\dot{\theta} = 0$, while $v_r = \dot{r}$ and $v_t = |r\dot{\theta}|$.

(2) $N(E) = \sqrt{2}(\pi GM)^3 f(E)/|E|^{5/2}$ for $E < 0$.

(3)   (a)
$$x\frac{3x + 17x^2 + 22x^3 + 8x^4 + 3\sqrt{-x(1 + x)}\,\arcsin\sqrt{-x}}{-3x - x^2 + 10x^3 + 8x^4 - 3\sqrt{-x(1 + x)}\,\arcsin\sqrt{-x}}$$
(b) $-x$. It is a good idea to check continuity at $x = 0$ and -1.

(4)   Actually the variance of the $1/r^2$ force is infinite, so suppose there is a short-distance cutoff independent of N. The collapse time scale varies with force F as $\sqrt{R/F}$.

Chapter 7

(1)   Eq. (7.4) is true for a circular orbit of radius a around a mass M in spherical symmetry. For both Keplerian and simple harmonic motions the period is independent of eccentricity.

(2)   The approximate differential equation for y is linear.

(3)   This exercise is more challenging than useful. Anyway, from pericentre (r_0) to apocentre (r_1) we have
$$\dot{r} = \sqrt{2(E - \phi) - J^2/r^2},$$
$$r^2\dot{\theta} = J.$$

Hence the change in polar angle is
$$\int_{r_0}^{r_1} \frac{J\,dr}{r^2\sqrt{2(E - \phi) - J^2/r^2}} = \arccos\left(\frac{J}{r_1\sqrt{2(E - \phi_0)}}\right)$$
$$+ J\int_{r_0}^{r_1} \frac{dr}{r^2}\left(\frac{1}{\sqrt{2(E - \phi) - J^2/r^2}} - \frac{1}{\sqrt{2(E - \phi_0) - J^2/r^2}}\right),$$

where $\phi_0 = \phi(r_0)$. To deal with the last integral, plot the integrand for some model potential (e.g. Plummer, Table 8.1) to convince yourself that the limit as $J \to 0$ (and $r_0 \to 0$) poses no problems.

(4) For a given orbit, construct a Keplerian potential $\phi_K = A - B/r$ which agrees with ϕ at r_{min} and r_{max}. Prove that $\phi_K > \phi$ for $r_{min} < r < r_{max}$ using the fact that $M(r)$ is non-decreasing, and then apply the formula in the previous hint. For the other result, fit a harmonic potential.

(5) Use Eq. (7.10) and the transformation $r = b \sinh \theta$, where $\cosh \theta = -\dfrac{GM}{2bE} + \sqrt{\dfrac{J^2}{2b^2 E} + (1 + GM/2bE)^2 \sin \phi}$.

(6) Answer: $\pi(-J + GM/\sqrt{-2E})$. The pedestrian way is to use a trigonometric substitution for $1/r$ and then integrate by parts, leaving an integral of the form $\int_{-\pi/2}^{\pi/2} d\theta/(a + b \sin \theta)$. The smart way is to read Goldstein (1980). (a) I_r remains close to 0, and so the orbit remains nearly circular. Use $\dot{E} = \partial \phi / \partial t$. (b) The orbit is Keplerian and starts at an apse ($\dot{r} = 0$) with a specific velocity.

(7) If so, the centre of the circle must be the midpoint of the values of $\dot{r}$ at peri- and apocentre.

Chapter 8

(1) Answer: $1/(4\pi G j^2 r^2)$, $r/(G j^2)$, $1/(4G j^2 d)$ on a line of sight passing at distance d from the centre.

(2) $\ddot{v} + \dot{v} + 2(\exp(v) - 1) = 0$, which is the equation of motion of a damped oscillator with equilibrium at $v = 0$. (We learned this argument from Donald Lynden-Bell.) Linearised equation: $\ddot{v} + \dot{v} + 2v = 0$.

(3) 1d: find 'energy' integral for $u'' + \exp(u) = 0$. Answer: $-2 \ln \cosh(z/\sqrt{2})$. 2d: transform to $v(t)$ as in Problem 2, and then proceed as in 1d. Answer: $-2 \ln(1 + z^2/8)$. See Ostriker (1964).

(4) In general in spherical symmetry one has $\phi \simeq \phi(0) + 2\pi G \rho(0) r^2 / 3$.

(5) $\rho = 4\pi \int_0^\infty f v^2 dv$. After differentiation, multiply by $1/\sqrt{2(\phi - E')}$ and integrate (with respect to ϕ) from E' to ∞. Reverse the order of integration in the right side, and use the transformation $\phi = E' \cos^2 \theta + E \sin^2 \theta$. Finally, differentiate with respect to E'.

(6) Eq. (8.8) is approximately $f = 2j^2 f_0(E_0 - E)$ if $E < E_0$. Then use the first result of Problem 5 to show that

$$\rho = C(\phi_e - \phi)^{5/2}, \tag{B.1}$$

where C is a constant and e denotes the edge. Next, the potential energy is $W = 2\pi \int \rho \phi r^2 dr = M \phi_e / 2 - 2\pi \int \rho(\phi_e - \phi) r^2 dr$. Use Eq. (B.1), integrate by parts, and use (B.1) again, which gives an integral proportional to $W = -\int GM dM/r$. There is a similar result for any finite polytrope (cf. Eddington 1926b, Sec. 60).

(7) Integrals are performed easily enough in spherical polar coordinates with the corresponding velocity components v_r, v_θ and v_ϕ. Answer: $2^{21/2} G^{1/2} M^{3/2} a^{1/2} / 4851$.

(8) Use radial and transverse velocity components v_r, v_t for all the integrals.

Chapter 9

(1) $T = 6\pi \int pr^2 dr$. Integrate by parts, and use $W = -G \int M(r) dM(r)/r$.

(2) $\sum_i \sum_{j \neq i} m_i m_j \mathbf{r}_i \cdot (\mathbf{r}_i - \mathbf{r}_j)/|\mathbf{r}_i - \mathbf{r}_j|^3 = \sum_j \sum_{i \neq j} m_i m_j \mathbf{r}_j \cdot (\mathbf{r}_j - \mathbf{r}_i)/|\mathbf{r}_j - \mathbf{r}_i|^3 = (1/2) \sum_i \sum_{j \neq i} m_i m_j (\mathbf{r}_i - \mathbf{r}_j) \cdot (\mathbf{r}_i - \mathbf{r}_j)/|\mathbf{r}_i - \mathbf{r}_j|^3$.

(4)   Answers in Table 8.1.

(5)   (a)

$$\int \frac{\partial f}{\partial t} m(\mathbf{v} - \langle \mathbf{v} \rangle)^2 d^3 \mathbf{v} = \frac{\partial}{\partial t} \int f m (\mathbf{v} - \langle \mathbf{v} \rangle)^2 d^3 \mathbf{v}$$

$$+ 2 \int f m (\mathbf{v} - \langle \mathbf{v} \rangle) \cdot \langle \dot{\mathbf{v}} \rangle d^3 \mathbf{v}$$

$$= \frac{\partial}{\partial t} 3\rho\sigma^2 + 0.$$

Similarly, the second term gives

$$\nabla \cdot (\rho \langle (\mathbf{v} - \langle \mathbf{v} \rangle)(\mathbf{v} - \langle \mathbf{v} \rangle)^2 \rangle + 3\rho\sigma^2 \langle \mathbf{v} \rangle) + 2\rho\sigma^2 \nabla \cdot \langle \mathbf{v} \rangle$$

and the last term gives nothing.

(6) $L \sim M\sigma^2/t_r \sim r^2 k \partial \sigma^2/\partial r$.

(7)   See Lightman & Shapiro (1977).

Chapter 10

(1) f is periodic in J with period $2\pi/t$. $t > \pi/(2\epsilon\delta J)$ is sufficient.

(2) $\int C(f) d\tau$ is invariant, by Liouville's Theorem.

(3) $\dot{J} = \epsilon \{\sin(\phi + 2t) + \sin\phi\}$ before averaging. The fact that the width is proportional to $\sqrt{\epsilon}$ is fundamental to the understanding of resonance.

Chapter 11

(1)   For a power law with extreme masses m_{min} and m_{max}, M is proportional to $m_{max}^{2-\alpha} - m_{min}^{2-\alpha}$ if $\alpha \neq 2$. The quoted approximation to the Miller–Scalo mass function (see caption to Fig. 11.1) implies that the fraction of stars with mass less than m is F, where

$$m = \frac{0.19F}{(1 - F)^{0.75} + 0.032(1 - F)^{0.25}}.$$

Show that $\dot{M} = m\dot{m} dF/dm$, where the right side is evaluated at the turnoff mass.

(2)   Set $c^2 = Gm_0 a$, $r \simeq a$ in the right side of the equation of motion.

(3) $(\partial s/\partial t)_E = -\dfrac{(\partial E/\partial t)_s}{(\partial E/\partial s)_t}$.

(5) $f \simeq$ constant, and so $\rho \propto (-\phi)^{3/2}$. Cf. Box 8.1.

Chapter 12

(1)   In Eq. (12.5) $\omega \sim \kappa \sim \sqrt{G\rho_G}$ (cf. Eq. (7.5)).

(3)   Consider the acceleration (due to the Sun) on (i) the sea on the sunlit surface, (ii) the solid Earth, (iii) the sea on the unlit surface.

(4)   Replace t in the trigonometric terms by $t - t_0$. Then $\dot{x} = -a_1\kappa \sin[\kappa(t - t_0)]$ if $\dot{a}_0 +$
       $\dot{a}_1 \cos[\kappa(t - t_0)] + \kappa a_1 \dot{t}_0 \sin[\kappa(t - t_0)] = 0$. Three more conditions come from the
       corresponding treatment of y and the first two equations of motion (12.1). The
       averaging has to be done numerically. See Ross *et al.* (1997).

(5)   See Fukushige & Heggie (2000).

Chapter 13

(1)   See any dynamics book (e.g. Goldstein 1980) for the transformation to rotating
       coordinates.

(2)   Nearest neighbour distance is of order $n^{-1/3}$.

(3)   (i) Rather than looking up general formulae for geodesics in arbitrary metrics, it is
       quicker to derive the equation with bare hands, starting from the Euler–Lagrange
       equation. One obtains

$$\left(f x^{i'}\right)' = -\frac{1}{2f}\frac{\partial V}{\partial x^i},$$

where $f = \sqrt{(E - V)/(x^{i'} x^{i'})}$ and $' = d/ds$. (ii) The space is conformally flat, and
the expressions for Christoffel symbols and the Ricci tensor take fairly simple
forms. See Synge (1964), though he considers only the case of a four-dimensional
space–time. For the last part consider the *two*-dimensional two-body problem.
(iii) The induced metric is $(\rho'^2 + z'^2)dr^2 + \rho^2 d\theta^2$. A global solution seems not to
be possible if $E < 0$.

Chapter 14

(1)   In suitably rotated coordinates the orbit is given by Eq. (7.8), Box 7.2. The first
       formula for ρ/ρ_0 is given by Bisnovatyi-Kogan *et al.* (1979). Eventually it reduces
       to $p^2/(r \sin\theta \sqrt{r^2 \sin^2\theta + 4ra(1 - \cos\theta)})$.

(2)   A reassuring intermediate result is

$$\frac{1}{t}\sum(\delta E_1)^2 = 4G^2 m^2 \int \frac{(\hat{\mathbf{p}} \cdot \mathbf{v}_1)^2}{pV} f(\mathbf{v}_2) d\hat{\mathbf{p}} dp d^3 \mathbf{v}_2,$$

where the integration with respect to $\hat{\mathbf{p}}$ is round a circle orthogonal to $\mathbf{V} = \mathbf{v}_1 - \mathbf{v}_2$.
To perform the integration with respect to $\mathbf{v}_2$ use spherical polar coordinates with
polar axis along $\mathbf{v}_1$, and the generating function of Legendre polynomials.

(3)   Dimensional analysis.

(4)   The intermediate result is

$$\frac{1}{t}\sum(\delta E_1)^2 = 8\pi \frac{G^2 m^2 n_2 \sigma^2 \ln\Lambda}{v_1}$$

$$\times \left(-\sqrt{\frac{2}{\pi}}\frac{v_1}{\sigma} \exp\left(-\frac{v_1^2}{2\sigma^2}\right) + \mathrm{erf}\left(\frac{v_1}{\sqrt{2}\sigma}\right)\right).$$

Your final coefficient should be the same as the one in Eq. (14.12).

(5) Strictly the coefficient should be 0.139.

(8) To exploit the assumption of constant total energy, use the virial theorem and the approximation for the potential energy in Problem 5.

Chapter 15

(1) Simple algebra, using the quaternion commutation relations $i^2 = j^2 = k^2 = -1$, $ij = -ji = k$, $jk = -kj = i$ and $ki = -ik = j$.

(2) As (1).

(3) (ii) Write down the condition that $\lambda + (1 - \lambda)q$ is imaginary.

(4) Some relabelling and further transformation of variables is needed.

(5) See Box 21.1, and Chenciner & Montgomery (2000).

Chapter 16

(1) The integral with respect to E_1 can be performed with the substitution $E_1 = (\phi - E_2)\sin^2\theta$, and the integral with respect to E_2 similarly. The integration over r gives a Beta function. The exact result is

$$\dot{N} = -\frac{2^{37/2}\Gamma(3/4)}{3 \times 5^2 \times 7 \times 11 \times 13^2 \times 17\Gamma(1/4)}\sqrt{\frac{GM}{\pi a^3}}.$$ See Hénon (1960).

(2) Relaxation is dominated by passage through the core, at intervals $\propto r^{3/2}$ (Kepler's Third Law). Treating the effect as a random walk, $\delta E(t) \propto \sqrt{t/r^{3/2}}$. Finally, $E \propto 1/r$. See Lightman & Shapiro (1978, §V.A.2), and also Spitzer (1987, pp. 89ff).

(3) $T = M\langle v^2\rangle/2$, $W = 2\sum m\phi$, $v_e^2 = -2\phi$. Answer: 0.0074.

(4) If $|m_2\phi_c|$ is big enough, then f_1 and f_2 are nearly of Maxwell–Boltzmann form almost everywhere.

Chapter 17

(1) The temperature is $T = -\dfrac{1}{6Nk}\dfrac{(3x-2)W(x)}{x-1}$. The heat capacity changes sign where $dT/dE = 0$ (cf. Binney & Tremaine 1987).

(3) In 3d the virial theorem gives $T = -E$, and so loss of energy increases temperature. What happens in 2d? In using the analytic solution assume that the potential vanishes at the edge radius r_e. Then the work done (per unit length) in constructing the model from material at this radius is $W = \int_0^{r_e} 2GM(r)\ln(r/r_e)dM$. Then plot $\dfrac{GMm}{kT} = \dfrac{z^2}{4(1 + z^2/8)}$ against

$$\frac{E}{GM^2} = \frac{(1 + z^2/8)^2}{z^4}\left(-32\ln(1 + z^2/8) + 8\frac{z^2}{1 + z^2/8}\right)$$

and look for a vertical tangent.

Chapter 18

(1) By Eq. (14.20), $s(E) \propto (E - \phi(0))^3$ at energies E just above the central potential $\phi(0)$.

(2) (a) See Chapter 16 on mass segregation.
(b) The density of black holes varies as r_s^{-3}, where r_s is the core radius of the stars, while the density of stars varies as $r_s^{-\alpha}$ (Eq. (18.1)).
(c) Eq. (14.12). See Kulkarni *et al.* (1993) and Sigurdsson & Hernquist (1993) for more detailed treatments.

(3) Use results of Chapter 13 and Problem 14.5, or see Hut (1997).

Chapter 19

(1) Apply time reversal to the required encounter: in that case a binary, consisting of stars of mass M, is destroyed by a grain of sand of mass m. To see how this can be done, consider a binary of high eccentricity, and let the two stars be at apocentre, their separation being r. Their relative velocity may be made as small as we please, by suitable adjustment of the orbital parameters. The sand grain approaches one star with speed V, and the maximum speed which it can impart to the star is of order mV/M. By choice of V this may exceed the escape speed from the other star, and indeed may be made as large as we please. Now reverse the velocities at the end of the encounter.

(2) $Nm\langle v^2 \rangle / 2 = -E$.

(3) The joint distribution of the two stars is $n^2 (2\pi kT/m)^{-3} \exp(-m(v_1^2 + v_2^2)/(2kT))$. Convert to relative and barycentric coordinates and velocities (cf. Eq. (19.1)), add a factor $\delta(mv^2/4 - Gm^2/r - \varepsilon)$ (cf. Eq. (19.2)) and integrate over all variables except the position of the barycentre. A factor of $1/2$ is needed in order not to count pairs twice.

(4) See Hut (1985). A classical calculation is given by Jeans (1929).

Chapter 20

(1) Kepler's Third Law (cf. Petrosky 1986, who calls the resulting map the 'Kepler' map). Area-preservation: compute the Jacobian. If $\epsilon = -0.1$ and $t_0 = 0$, there is interesting structure near $E_0 = 0.1, 0.24, 0.2344$ and $0.234\,15$. (The number of decimal places indicates the care with which you have to look.) These structures are easier to see if the log of the escape time is plotted against E_0.

(2) Drop a particle parabolically into a Sitnikov machine, and imagine finding a value for the initial phase at which the particle escapes hyperbolically after a number of interactions. Now vary the initial phase.

(3) Adjust the phase of each visit to the binary so that the particle more and more nearly escapes parabolically each time. One gets an impression of what is needed from the case $z(0) = 0$, $\dot{z}(0) = 2.819\,89$ to $t \simeq 1000$, if the binary has $e = 0.6$, $P = 2\pi$ and $a = 1$.

(4) The equation $u'' + (1/3)u' - (2/9)u + 2u(u^2 + 3^{4/3}/4)^{-3/2} = 0$, where u is the scaled z, has equilibria at $u = 0$ and $u = \pm 3^{7/6}/2$. As the energy decreases the

oscillator either escapes or is trapped. In the latter case the amplitude of the z motion increases without limit in general.

Chapter 21

(1) $\dot{\varepsilon} \sim \dfrac{G^2 m^3 n}{V} \displaystyle\int_{Gm/V^2}^{a} \dfrac{dp}{p}$, using notation of Eq. (14.3) and Eq. (19.5).

(3)   Compute $\dot{N}_b R^3 t_r$.

(4)   See Hut (1989).

(5) $\mathbf{h} \cdot \dot{\mathbf{h}} = \mu \mathbf{h} \cdot \mathbf{r} \times \mathbf{f}$, where μ is the reduced mass of the binary. Use Eq. (21.9), and estimate the average over binary motion (though this is zero if the binary is circular). Hence $\delta h \sim Gm_3 \mu a^2 T / R_p^3$, where T is the time scale of the encounter.

(6)   If $t = i(t_0 + s)$ along the cut, where $t_0 = R_p / V$, the exact result is
$$I = \frac{2n}{V^3} \int_0^\infty \frac{\exp(-n[t_0 + s])ds}{(t_0 + s)\sqrt{2t_0 s + s^2}}.$$

(7)   The linearised equations are
$$(-t)^2 \delta\ddot{x} - \frac{4}{3}(-t)\delta\dot{x} = \frac{1}{2}\delta x + \frac{1}{6}\delta y$$
$$(-t)^2 \delta\ddot{y} - \frac{4}{3}(-t)\delta\dot{y} = \frac{1}{6}\delta x + \frac{1}{2}\delta y.$$

Chapter 22

(1)   Seek solutions $\propto \lambda^n$. Consider $|\lambda|$.

(2)   See Hut (1983b).

Chapter 23

(1)   Use Eq. (21.4), and an argument in the discussion of hard binaries in Chapter 19. For the last part see Goodman & Hut (1993) or Retterer (1980).

(2)   Other results of such experiments are described by Giersz & Heggie (1994b).

(3)   See Kulkarni et al. (1993) and Sigurdsson & Hernquist (1993).

Chapter 24

(1)   To help as much as possible, take conditions in the disk: $\rho \sim 0.1 M_\odot \text{ pc}^{-3}$, $\sigma \sim$ 40 km/s, total mass $\sim 6 \times 10^{10} M_\odot$ (adapted from Binney & Tremaine 1987) and $m \sim 1 M_\odot$.

(2)   The number of AU in a parsec is the same as the number of arc seconds in a radian.

(3)   See Klessen & Kroupa (2001) and references therein.

Chapter 25

(1)   The relative motion of the Moon and Earth is governed by the difference in their acceleration.

(2)   Put $e_i(0) = 0$ in Eqs. (25.8) and (25.9), which imply $\cos^2 i = \cos^2 i(0)/(1 - e_i^2)$ and $e^2(2 - 5\sin^2 i \sin^2 \omega) = 0$. Hence $e = 0$ unless $\sin^2 i \geq 2/5$.

(3) $z = (a[\cos E - e] \sin \omega + b \sin E \cos \omega) \sin i$; $\langle f(E) \rangle = \dfrac{1}{2\pi} \displaystyle\int_0^{2\pi} f(E) \times (1 - e \cos E) dE$.

Chapter 26

(2)   In the second case, one component of the wider binary must interact closely with the harder pair.

(3)   In the input string, M is the projectile mass, v is the initial relative velocity of the two binaries, and a and e refer to the projectile (p) and target (t), respectively. The output shows how the particles are grouped, along with position and velocity components, masses and energy control.

Chapter 27

(1)   Use Eq. (27.3) to obtain the estimate of $\dot{M}$. See Problem 8.1, for the mass–radius relation in an isothermal cluster.

(2)   Unless ρ and σ decrease very slowly with increasing r, the fourth equation shows that $L \to 0$ as $r \to \infty$, which is inconsistent with the third equation. Relaxation is too slow in the outer halo of a cluster to remove all the energy generated.

Chapter 28

(1)   Use the fact that $t_{rh} \propto (Nr_h^3)^{1/2}$ (Eq. 14.13). Neglecting variation of the Coulomb logarithm, the unreduced result is approximately

$$\frac{t}{t_{rh}(0)} \simeq \frac{10.8}{0.12k} \left(\frac{N(0)}{2} \right)^k + 18,$$

where $k = 3/0.24 - 1/2$.

(2)   Good values are $c = $ (a) 10^{-3}, (b) 10^{-5}, (c) 10^{-7} and (d) 10^{-8}.

(3)   The expansion of the cluster requires energy generation at a rate of order $Nm\sigma^2/t_{rh}$.

Chapter 29

(1)   Use Eq. (8.9). Equate the estimate to the heat flux at r_h, and use a virial estimate of σ_c, assuming the cluster is nearly isothermal inside r_h.

(2)   For the first step use an estimate like Eq. (12.8), and virial estimates.

(3)   If the initial total mass is large enough, r_h begins by decreasing with M.

Chapter 30

(1)   The energy of the binary is $- GM_\odot^2/(4R_\odot)$.

Chapter 31

(1)   Reinterpret Eqs. (21.1) and (21.4).

(2) $\dfrac{Gm\delta m}{R} = \dfrac{1}{2}\dfrac{m^2}{2m}V^2.$

(4)   Suppose the constant in Eq. (12.6) is $-GM/a$, and develop a simple approximation to $r - a$. Put $\kappa^2 \simeq \omega_z^2 \simeq \omega^2 \simeq V^2/R^2 \simeq 4Gm/R^3$. The quadrupole field of lumps δm is of order $G\delta m r^2/R^3$. For a detailed analysis of *equilibrium* tides see Eggleton *et al.* (1998).

Chapter 32

(1)   See Box 7.2.

(2)   Integrating Eq. (31.1) over all stars counts each collision twice.

Chapter 33

(1)   For a flat rotation curve $\phi \propto \ln r$, whence κ can be calculated for insertion into Eq. (12.5).

(2)   Show that $t_r(t) = t_r(0) - kt$, where k is a constant.

Appendix A

(1)   Use $a_1 = a_0 + h\dot{a}_0 + h^2 a_0^{(2)}/2 + h^3 a_0^{(3)}/6$ and a similar expression for $\dot{a}_1$.

(5)   The command
```
mksphere -n 25|flathermite -D 0.1 -t 1|xstarplot
```
almost does this, but initial velocities are uniform in a cube instead of in a sphere.

References

Aarseth S.J. (1985). Direct methods for *N*-body simulations, in *Multiple Time Scales*, eds. Brackbill J.U. & Cohen B.I. (Academic Press, New York).

Aarseth S.J. (1999). From NBODY1 to NBODY6: the growth of an industry, *PASP* **111,** 1333.

Aarseth S.J., Lecar M. (1975). Computer simulations of stellar systems, *ARA&A* **13,** 1.

Aarseth S.J., Mardling R.A. (2001). The formation and evolution of multiple star systems, in *Evolution of Binary and Multiple Star Systems; A Meeting in Celebration of Peter Eggleton's 60th Birthday*, eds. Podsiadlowski P., Rappaport S., King A.R., D'Antona F., Burder L., ASP Conf. Ser. 229 (ASP, San Francisco).

Aarseth S.J., Zare K. (1974). A regularization of the three-body problem, *Celes. Mech.* **10,** 185.

Aarseth S.J., Hénon M., Wielen R. (1974). A comparison of numerical methods for the study of star cluster dynamics, *A&A* **37,** 183.

Aarseth S.J., Lin D.N.C., Papaloizou J.C.B. (1988). On the collapse and violent relaxation of protoglobular clusters, *ApJ* **324,** 288–310.

Abraham R., Marsden J.E. (1978). *Foundations of Mechanics*, 2nd ed. (Benjamin/Cummings, Reading).

Agekyan T.A., Anosova Z.P., Orlov V.V. (1983). Decay time of triple systems, *Astrophysics* **19,** 66.

Aguilar L., Hut P., Ostriker J.P. (1988). On the evolution of globular cluster systems. I – Present characteristics and rate of destruction in our Galaxy, *ApJ* **335,** 720.

Ahmad A., Cohen L. (1973). A numerical integration scheme for the N-body gravitational problem, *J. Comp. Phys.* **12,** 389.

Airy G.B. (1884). *Gravitation: An Elementary Explanation of the Principal Perturbations in the Solar System* (Macmillan, London).

Alekseev V.M. (1981). Final motions in the three-body problem and symbolic dynamics, *Russ. Math. Surv.* **36,** 181.

Allen F.S., Heggie D.C. (1992). A model gravothermal oscillator, *MNRAS* **257,** 245–56.

Ambartsumian V.A. (1938). On the dynamics of open clusters, *Uch. Zap. LGU* **22**, 19; transl. in *Dynamics of Star Clusters*, eds. Goodman J., Hut P. (D. Reidel, Dordrecht).

Antonov V.A. (1962). Most probable phase distribution in spherical star systems and conditions for its existence, *Vest. Leningrad Univ.* **7**, 135; transl. in *Dynamics of Star Clusters*, eds. Goodman J., Hut P. (D. Reidel, Dordrecht).

Antonov V.A. (1973). On the instability of stationary spherical models with purely radial motion, in *The Dynamics of Galaxies and Star Clusters*, ed. Omarov T.B. (Nauka Kazakh SSR, Alma Ata); transl. in *Structure and Dynamics of Elliptical Galaxies*, ed. de Zeeuw P.T. IAU Symp. 127 (Reidel, Dordrecht).

Applegate J.H., Douglas M.R., Gürsel Y., Hunter P., Seitz C.L., Sussman G.J. (1985). A digital orrery, *IEEE Trans. Comp.* **C-34,** 822.

Arabadjis J.S., Richstone D.O. (1998). The dynamical evolution of dense rotating systems: paper I. Two-body relaxation effects, astro-ph/9810192.

Arnold V.I. (1978a). *Mathematical Methods of Classical Mechanics* (Springer-Verlag, New York).

Arnold V.I. (1978b). *Ordinary Differential Equations* (MIT Press, Cambridge).

Arnold V.I. (1983) *Geometrical Methods in the Theory of Ordinary Differential Equations* (Springer, Berlin).

Arnold V.I., Kozlov V.V., Neishtadt A.I. (1997). *Mathematical Aspects of Classical and Celestial Mechanics*, 2nd edition (Springer-Verlag, Berlin).

Ashman K.M., Zepf S.E. (1998). *Globular Cluster Systems* (Cambridge University Press, Cambridge).

Bahcall J.N., Wolf R.A. (1976). Star distribution around a massive black hole in a globular cluster, *ApJ* **209,** 214.

Bahcall J.N., Hut P., Tremaine S. (1985). Maximum mass of objects that constitute unseen disk material, *ApJ* **290,** 15.

Bailey M.E., Chambers J.E., Hahn G. (1992). Origin of sungrazers – a frequent cometary end-state, *A&A* **257,** 315.

Bailey V.C., Davies M.B. (1999). Red giant collisions in the Galactic Centre, *MNRAS* **308,** 257.

Bailyn C.D. (1996). X-ray binaries in globular clusters, in *The Origins, Evolutions, and Destinies of Binary Stars in Clusters*, eds. Milone E.F. and Mermilliod J.-C. (ASP Conf. Ser. 90, San Francisco).

Barnes J.E. (1985). Dynamical instabilities in spherical stellar systems, in *Dynamics of Star Clusters (IAU Symp. 113)*, eds. Goodman J., Hut P. (Reidel, Dordrecht).

Barnes J.E., Hut P. (1986). A hierarchical $O(N \log N)$ force-calculation algorithm, *Nature* **324,** 446.

Barnes J., Goodman J., Hut P. (1986). Dynamical instabilities of spherical stellar systems. I. Numerical experiments for generalized polytropes, *ApJ* **300,** 112.

Barrow-Green J. (1996). *Poincaré and the Three-Body Problem* (American and London Mathematical Societies, Providence and London).

Baumgardt H. (1998). The initial distribution and evolution of globular cluster systems, *A&A* **330,** 480.

Baumgardt H. (2001). Scaling of N-body calculations, *MNRAS* **325,** 1323.

Becker L. (1920). On capture orbits, *MNRAS* **80,** 590.

Bettwieser E., Sugimoto D. (1984). Post-collapse evolution and gravothermal oscillation of globular clusters, *MNRAS* **208,** 493.

Binney J., Tremaine S. (1987). *Galactic Dynamics* (Princeton University Press, Princeton).

Bisnovatyi-Kogan G.S., Kazhdan Ya.M., Klypin A.A., Lutskii A.E., Shakura N.I. (1979). Accretion onto a rapidly moving gravitational center, *Sov. Astron.* **23,** 201.

Boyd P.T., McMillan S.L.W. (1992). Initial-value space structure in irregular gravitational scattering, *Phys. Rev.* **A46,** 6277.

Breeden J.L., Cohn H.N. (1995). Chaos in core oscillations of globular clusters, *ApJ* **448,** 672.

Breeden J.L., Cohn H.N., Hut P. (1994). The onset of gravothermal oscillations in globular cluster evolution, *ApJ* **421,** 195.

Burdet C.A. (1967). Regularization of the two body problem, *ZAMP* **18,** 434.

Burrau C. (1913). Numerische Berechnung eines Spezialfalles des Dreikörperproblems, *Astron. Nach.* **195,** 113.

Capuzzo-Dolcetta R., Tesseri A. (1999). Globular cluster system erosion and nucleus formation in elliptical galaxies, *MNRAS* **308,** 961.

Casertano S., Hut P. (1985). Core radius and density measurements in N-body experiments: Connections with theoretical and observational definitions, *ApJ* **298,** 80–94.

Cassisi S., Castellani V. (1993). An evolutionary scenario for primeval stellar populations, *ApJS* **88,** 509.

Chandrasekhar S. (1939). *Stellar Structure* (University of Chicago Press, Chicago); also Dover, New York (1957).

Chandrasekhar S. (1942). *Principles of Stellar Dynamics* (University of Chicago Press, Chicago) pp. 55ff.

Chavanis P.-H. (2002). Gravitational instability of finite isothermal spheres, *A&A* **381,** 340.

Chavanis P.-H., Rosier C., Sire C. (2001). Thermodynamics of self-gravitating systems, cond-mat/0107345.

Chenciner A., Montgomery R. (2000). A remarkable periodic solution of the three-body problem in the case of equal masses, *Ann. Math.* **152,** 881.

Chernin A.D., Valtonen M.J. (1998). Intermittent chaos in three-body dynamics, *N. Astron. Rev.* **42,** 41.

Chernoff D.F., Shapiro S.L. (1987). Globular cluster evolution in the Galaxy – a global view, *ApJ* **322,** 113.

Chernoff, D.F., Weinberg, M.D. (1990). Evolution of globular clusters in the galaxy, *ApJ* **351,** 121.

Cipriani P., Pettini M. (2001). Strong chaos in N-body problem and microcanonical thermodynamics of collisionless self-gravitating systems, astro-ph/0102143.

Clarke C.J., Pringle J.E. (1991). The role of discs in the formation of binary and multiple star systems, *MNRAS* **249,** 588.

Clayton D.D. (1983). *Principles of Stellar Evolution and Nucleosynthesis* (University of Chicago Press, Chicago).

Cohen R.S., Spitzer L., Jr., Routly P.McR. (1950). The electrical conductivity of an ionized gas, *Phys. Rev.* **80,** 230.

Cohn H. (1979). Numerical integration of the Fokker–Planck equation and the evolution of star clusters, *ApJ* **234,** 1036.

Cohn H.N. (1985). Direct Fokker–Planck calculations, in *Dynamics of Star Clusters,*

IAU Symp. 113, eds. Goodman J., Hut P. (Reidel, Dordrecht).

Cohn H., Hut P., Wise M. (1989). Gravothermal oscillations after core collapse in globular cluster evolution, *ApJ* **342**, 814–22.

Combes F., Leon S., Meylan G. (1999). *N*-body simulations of globular cluster tides, *A&A* **352**, 149.

Côté P., Fischer P. (1996). Spectroscopic binaries in globular clusters. I. A search for ultra-hard binaries on the main sequence in M4, *AJ* **112**, 565.

Côté P., Pryor C., McClure R.D., Fletcher J.M., Hesser J.E. (1996). Spectroscopic binaries in globular clusters. II. A search for long-period binaries in M22, *AJ* **112**, 574.

Cuperman S., Harten A. (1972). The evolution of a multi-phase space density collisionless one-dimensional stellar system, *A&A* **16**, 13.

Cvitanović P. (1989). *Universality in Chaos*, 2nd edition (Adam Hilger, Bristol).

Danilov V.M. (1997). Numerical experiments simulating the dynamics of open clusters in the Galactic field, *Astron. Rep.* **41**, 163.

David M., Theuns T. (1989). Numerical experiments on radial virial oscillations in *N*-body systems, *MNRAS* **240**, 957.

Davies M.B. (1995). The binary zoo: the calculation of production rates of binaries through 2 + 1 encounters in globular clusters, *MNRAS* **276**, 887.

Davies M.B., Benz W. (1995). A stellar audit: the computation of encounter rates for 47 Tucanae and ω Centauri, *MNRAS* **276**, 876.

Davoust E., Prugniel P. (1990). On the flattening of globular clusters, *A&A* **230**, 67.

Dejonghe H. (1984). The construction of analytical models for globular clusters, *A&A* **133**, 225.

Dejonghe H. (1987a). A completely analytical family of anisotropic Plummer models, *MNRAS* **224**, 13.

Dejonghe H. (1987b). On entropy and stellar systems, *ApJ* **320**, 477–81.

Dejonghe H., Hut P. (1986). Round-off sensitivity in the *N*-body problem, in *The Use of Supercomputers in Stellar Dynamics*, eds. Hut P., McMillan S., Lect. Notes Phys. 267 (Springer-Verlag, Berlin).

de Vega H.J., S'anchez N. (2000). The statistical mechanics of the self-gravitating gas: equation of state and fractal dimension, *Phys. Lett. B* **490**, 180.

Diacu F. (2000). A century-long loop, *Math. Intelligencer* **22**, No. 2, 19.

Diacu F., Holmes P. (1996). *Celestial Encounters* (Princeton University Press, Princeton).

Djorgovski S., Piotto G., Phinney E.S., Chernoff D. (1991). Modification of stellar populations in post-core-collapse globular clusters, *ApJ* **372L**, 41.

Drukier G.A. (1995). Fokker–Planck models of NGC 6397, *ApJS* **100**, 347.

Dull J.D., Cohn H.N., Lugger P.M., Murphy B.W., Seitzer P.O., Callanan P.J., Rutten R.G.M., Charles P.A. (1997). The dynamics of M15: observations of the velocity dispersion profile and Fokker–Planck models, *ApJ* **481**, 267.

Duquennoy A., Mayor M. (1991). Multiplicity among solar-type stars in the solar neighbourhood. II – Distribution of the orbital elements in an unbiased sample, *A&A* **248**, 485.

Eddington A.S. (1926a). Star, in *Encyclopaedia Britannica*, 13th edition (London), Vol. 25, p. 787.

Eddington A.S. (1926b). *The Internal Constitution of the Stars* (Cambridge University Press, Cambridge); also Dover Publications, New York (1959).

Efstathiou G., Davis M., White S.D.M., Frenk C.S. (1985). Numerical techniques for large cosmological N-body simulations, *ApJS* **57**, 241.

Eggleton P., Kiseleva L. (1995). An empirical condition for stability of hierarchical triple systems, *ApJ* **455**, 640.

Eggleton P.P., Tout C.A., Fitchett M.J. (1989). The distribution of visual binaries with two bright components, *ApJ* **347**, 998–1011.

Eggleton P.P., Kiseleva L.G., Hut P. (1998). The equilibrium tide model for tidal friction, *ApJ* **499**, 853.

Einsel C., Spurzem R. (1999). Dynamical evolution of rotating stellar systems – I. Pre-collapse, equal-mass system, *MNRAS* **302**, 81.

Eisenstein D.J., Hut P. (1998). HOP: a new group-finding algorithm for N-body simulations, *ApJ* **498**, 137.

Emden R. (1907). *Gaskugeln* (Teubner, Leipzig).

Engle K. (1999). *N-Body Simulations of Star Clusters* (PhD Thesis, Drexel University).

Euler L. (1776). Nova Methodus Motum Corporum Rigidorum Determinandi, *Novi comment. acad. sc. Petrop.* **20**, 208.

Fabian A.C., Pringle J.E., Rees M.J. (1975). Tidal capture formation of binary systems and X-ray sources in globular clusters, *MNRAS* **172P**, 15.

Fall S.M., Zhang Q. (2001). Dynamical evolution of the mass function of globular star clusters, *ApJ* **561**, 751.

Fanelli D., Merafina M., Ruffo S. (2001). A one-dimensional toy model of globular clusters, *Phys. Rev. E* **63**, 066614-1.

Farouki R.T., Salpeter E.E. (1994). Mass segregation, relaxation, and the Coulomb logarithm in N-body systems (again), *ApJ* **427**, 676.

Ferraro F.R., Fusi Pecci F., Cacciari C., Corsi C., Buonanno R., Fahlman G.G., Richer H.B. (1993). Blue stragglers in the Galactic globular cluster M3: evidence for two populations, *AJ* **106**, 2324.

Feynman R.P. (1965). *Lectures on Physics*, Vol. III (Addison-Wesley, Reading), Chapter 6.

Figer D.F., Kim S.S., Morris M., Serabyn E., Rich R.M., McLean I.S. (1999). Hubble Space Telescope/NICMOS observations of massive stellar clusters near the Galactic Center, *ApJ* **525**, 750.

Fischer P., Welch D.L., Côté P., Mateo M., Madore B.F. (1992). Dynamics of the young LMC cluster NGC 1866, *AJ* **103**, 857.

Fischer P., Pryor C., Murray S., Mateo M., Richtler T. (1998). Mass segregation in young Large Magellanic Cloud clusters. I – NGC 2157, *AJ* **115**, 592.

Fridman A.M., Polyachenko V.L. (1984). *Physics of Gravitating Systems* (Springer Verlag, Berlin), 2 Vols.

Freitag M., Benz W. (2001). A Monte Carlo code to investigate stellar collisions in dense galactic nuclei, in *Stellar Collisions, Mergers and their Consequences*, ed. Shara M., ASP Conf. Ser. 263 (ASP, San Francisco).

Fukushige T. (1995). Gravitational scattering experiments in infinite homogeneous N-body systems, in *Chaos in Gravitational N-body Systems*, eds. Muzzio J.C., Ferraz-Mello S., Henrard J. (Kluwer, Dordrecht).

Fukushige T., Heggie D.C. (2000). The time-scale of escape from star clusters, *MNRAS* **318,** 753.

Fukushige T., Hut P., Makino J. (1999). High-performance special-purpose computers in science, *IEEE Comp. Sci. Eng.* **1,** No. 2, 12.

Fullerton L.W., Hills J.G. (1982). Computer simulations of close encounters between binary and single stars – the effect of the impact velocity and the stellar masses, *AJ* **87,** 175.

Funato Y., Hut P., McMillan S., Makino J. (1996). Time-symmetrized Kustaanheimo–Stiefel regularization, *AJ* **112,** 1697.

Gao B., Goodman J., Cohn H., Murphy B. (1991). Fokker–Planck calculations of star clusters with primordial binaries, *ApJ* **370,** 567.

Garabedian P.R. (1986). *Partial Differential Equations*, 2nd edition (Chelsea, New York).

Gebhardt K., Fischer P. (1995). Nonparametric dynamical analysis of globular clusters: M15, 47 Tuc, NGC 362, and NGC 3201, *AJ* **109,** 209.

Ghez A.M., Morris M., Becklin E.E., Tanner A., Kremenek T. (2000). The accelerations of stars orbiting the Milky Way's central black hole, *Nature* **407,** 349.

Giersz M. (1998). Monte Carlo simulations of star clusters – I. First results, *MNRAS* **298,** 1239.

Giersz M., Heggie D.C. (1994a). Statistics of N-body simulations – I. Equal masses before core collapse, *MNRAS* **268,** 257.

Giersz M., Heggie D.C. (1994b). Statistics of N-body simulations – II. Equal masses after core collapse, *MNRAS* **270,** 298.

Giersz M., Heggie D.C. (1996). Statistics of N-body simulations – III. Unequal masses, *MNRAS* **279,** 1037.

Giersz M., Heggie D.C. (1997). Statistics of N-body simulations – IV. Unequal masses with a tidal field, *MNRAS* **286,** 709–31.

Giersz M., Spurzem R. (1994). Comparing direct N-body integration with anisotropic gaseous models of star clusters, *MNRAS* **269,** 241.

Giersz M., Spurzem R. (2000). A stochastic Monte Carlo approach to model real star cluster evolution – II. Self-consistent models and primordial binaries, *MNRAS* **317,** 581.

Gilliland R.L., Brown T.M., Guhathakurta P., Sarajedini A., Milone E.F., Albrow M.D., Baliber N.R., Bruntt H., Burrows A., Charbonneau D., Choi P., Cochran W.D., Edmonds P.D., Frandsen S., Howell J.H., Lin D.N.C., Marcy G.W., Mayor M., Naef D., Sigurdsson S., Stagg C.R., Vandenberg D.A., Vogt S.S., Williams M.D. (2000). A lack of planets in 47 Tucanae from a Hubble Space Telescope search, *ApJ* **545L,** 47.

Gnedin O.Y., Ostriker J.P. (1997). Destruction of the Galactic globular cluster system, *ApJ* **474,** 223.

Gnedin O.Y., Ostriker J.P. (1999). On the self-consistent response of stellar systems to gravitational shocks, *ApJ* **513,** 626.

Gnedin O.Y., Lee H.M., Ostriker J.P. (1999). Effects of tidal shocks on the evolution of globular clusters, *ApJ* **522,** 935.

Goldstein, H. (1980). *Classical Mechanics*, 2nd edition (Addison Wesley, Reading, MA).

Goodman, J. (1983). Core collapse with strong encounters, *ApJ* **270,** 700–10.

Goodman J. (1984). Homologous evolution of stellar systems after core collapse, *ApJ* **280**, 298.

Goodman J. (1987). On gravothermal oscillations, *ApJ* **313**, 576.

Goodman J. (1989). Late evolution of globular clusters, in *Dynamics of Dense Stellar Systems*, ed. Merritt D. (Cambridge University Press, Cambridge).

Goodman J., Hernquist L. (1991). Hydrodynamics of collisions between binary stars, *ApJ* **378**, 637.

Goodman J., Hut P., eds. (1985). *Dynamics of Star Clusters*, IAU Symp. 113 (Reidel, Dordrecht).

Goodman J., Hut P. (1989). Primordial binaries & globular cluster evolution, *Nature* **339**, 40.

Goodman J., Hut P. (1993). Binary–single-star scattering. V – Steady state binary distribution in a homogeneous static background of single stars, *ApJ* **403**, 271.

Goodman J., Heggie D.C., Hut P. (1993). On the exponential instability of N-body systems, *ApJ* **415**, 715.

Greengard L. (1990). The numerical solution of the N-body problem, *Computers in Phys.* Mar/Apr, 142–52.

Grillmair C.J., Freeman K.C., Irwin M., Quinn P.J. (1995). Globular clusters with tidal tails: deep two-color star counts, *AJ* **109**, 2553.

Grindlay J.E. (1988). X-ray binaries in globular clusters, in *The Harlow–Shapley Symposium on Globular Cluster Systems in Galaxies (Proc. IAU Symp. 126)*, ed. Grindlay J.E., Philip A.G.D. (Kluwer, Dordrecht).

Guhathakurta P., Yanny B., Schneider D.P., Bahcall J.N. (1996). Globular cluster photometry with the Hubble Space Telescope. V. WFPC study of M15's central density cusp, *AJ* **111**, 267.

Gunn J.E., Griffin R.F. (1979). Dynamical studies of globular clusters based on photoelectric radial velocities of individual stars. I – M3, *AJ* **84**, 752–73.

Gurzadyan V.G., Savvidi G.K. (1986). Collective relaxation of stellar systems, *A&A* **160**, 203.

Hachisu I., Nakada Y., Nomoto K., Sugimoto, D. (1978). Gravothermal catastrophe of finite amplitude, *Prog. Theor. Phys.* **60**, 393.

Hamilton W.R. (1853). *Lectures on Quaternions* (Hodges and Smith, Dublin).

Hansen B.M.S. (1999). Cooling models for old white dwarfs, *ApJ* **520**, 680.

Harrington R.S. (1970). Encounter phenomena in triple stars, *AJ* **75**, 1140.

Harris H.C., McClure R.D. (1983). Radial velocities of a random sample of K giant stars and implications concerning multiplicity among giant stars in clusters, *ApJ* **265L**, 77.

Harris W.E. (1996). A catalog of parameters for globular clusters in the Milky Way, *AJ* **112**, 1487.

Hayli A. (1970). More comparisons of numerical integrations in the gravitational N-body problem, *A&A* **7**, 249.

Heggie D.C. (1975). Binary evolution in stellar dynamics, *MNRAS* **173**, 729.

Heggie D.C. (1984). Post-collapse evolution of a gaseous cluster model, *MNRAS* **206**, 179.

Heggie D.C. (1991). Chaos in the N-body problem of stellar dynamics, in *Predictability, Stability, and Chaos in N-Body Dynamical Systems,* ed. Roy A.E. (Plenum Press, New York).

Heggie D.C. (1992). Ecology of globular clusters, *Nature* **359**, 772.

Heggie D.C. (2000). A new outcome of binary–binary scattering, *MNRAS* **318**, 61.

Heggie D.C., Aarseth S.J. (1992). Dynamical effects of primordial binaries in star clusters. I – Equal masses, *MNRAS* **257**, 513.

Heggie D.C., Hut P. (1993). Binary–single-star scattering. IV – Analytic approximations and fitting formulae for cross sections and reaction rates, *ApJS* **85**, 347.

Heggie D.C., Ramamani N. (1989). Evolution of star clusters after core collapse, *MNRAS* **237**, 757.

Heggie D.C., Ramamani N. (1995). Approximate self-consistent models for tidally truncated star clusters, *MNRAS* **272**, 317–22.

Heggie D.C., Sweatman W.L. (1991). Three-body scattering near triple collision or expansion, *MNRAS* **250**, 555.

Heggie D.C., Inagaki S., McMillan S.L.W. (1994). Gravothermal expansion in an *N*-body system, *MNRAS* **271**, 706.

Heggie D.C., Giersz M., Spurzem R., Takahashi K. (1998). Dynamical simulations: methods and comparisons, in *Highlights of Astronomy*, ed. Anderson J., **11B** (Kluwer, Dordrecht).

Hemsendorf M., Merritt D. (2002). Instability of the Gravitational *N*-Body Problem in the Large-*N* Limit, *ApJ*, in press.

Hénon M. (1958). Un calcul amélioré des perturbations des vitesses stellaires, *Ann. d'Astr.* **21**, 186.

Hénon M. (1960). L'évasion des étoiles hors des amas isolés, *Ann. d'Astr.* **23**, 668.

Hénon M. (1961). Sur l'évolution dynamique des amas globulaires, *Ann. d'Astr.* **24**, 369.

Hénon M. (1965). Sur l'évolution dynamique des amas globulaires II. – Amas isolé, *Ann. d'Astr.* **28**, 62.

Hénon M. (1969). Rates of escape from isolated clusters with an arbitrary mass distribution, *A&A* **2**, 151.

Hénon M. (1971). The Monte Carlo method, *Ap. Sp. Sci.* **14**, 151.

Hénon M. (1975). Two recent developments concerning the Monte Carlo method, in *Dynamics of Stellar Systems*, ed. Hayli A. (Proc. IAU Coll. 69, D. Reidel, Dordrecht).

Hénon M. (1976). A complete family of periodic solutions of the planar three-body problem, and their stability, *Celes. Mech.* **13**, 267.

Hénon M. (1982). Vlasov equation, *A&A* **114**, 211.

Hénon M. (1988). Chaotic scattering modelled by an inclined billiard, *Physica D* **33**, 132–56.

Hénon M., Heiles C. (1964). The applicability of the third integral of motion: Some numerical experiments, *AJ* **69**, 73.

Herman J. (1710). Metodo d'investigare l'Orbite de' Pianeti, nell'ipotesi che le forze centrali o pure le gravità degli stessi Pianeti sono in ragione reciproca de' quadrati delle distanze, che i medesimi tengono dal Centro, a cui si dirigono le forze stesse, *Giorn. Lett. d'Italia* **2**, 447.

Hernquist L., Hut P., Kormendy J. (1991). A post-collapse model for the nucleus of M33, *Nature* **354**, 376.

Hernquist L., Hut P., Makino J. (1993). Discreteness noise versus force errors in *N*-body simulations, *ApJ* **402**, L85.

Hill G.W. (1878). Researches in the lunar theory, *Am. J. Math.* **1**, 5, 129 and 245.

Hills J.G. (1975a). Encounters between binary and single stars and their effect on the dynamical evolution of stellar systems, *AJ* **80**, 809.

Hills J.G. (1975b). Effect of binary stars on the dynamical evolution of stellar clusters. II – Analytical evolutionary models, *AJ* **80,** 1075.

Hills J.G. (1980). The effect of mass loss on the dynamical evolution of a stellar system: analytic approximations, *ApJ* **235,** 986–91.

Hills J.G. (1983). The effects of sudden mass loss and a random kick velocity produced in a supernova explosion on the dynamics of a binary star of arbitrary orbital eccentricity – applications to X-ray binaries and to the binary pulsars, *ApJ* **267,** 322.

Hills J.G. (1990). Encounters between single and binary stars – The effect of intruder mass on the maximum impact velocity for which the mean change in binding energy is positive, *AJ* **99,** 979.

Hills J.G., Day C.A. (1976). Stellar collisions in globular clusters, *Astrophys. Lett.* **17,** 87.

Holmberg E. (1941). On the clustering tendencies among the nebulae. II. A study of encounters between laboratory models of stellar systems by a new integration procedure, *ApJ* **94,** 385.

Horwitz G., Katz J. (1978). Steepest descent technique and stellar equilibrium statistical mechanics. III Stability of various ensembles, *ApJ* **222,** 941.

Hurley J., Pols O.R., Tout C.A. (2000). Comprehensive analytic formulae for stellar evolution as a function of mass and metallicity, *MNRAS* **315,** 543.

Hurley J.R., Tout C.A., Aarseth S.J., Pols O. (2001). Direct *N*-body modelling of stellar populations: blue stragglers in M67, *MNRAS* **323,** 630.

Hut P. (1983a). Binary–single-star scattering. II – Analytic approximations for high velocity, *ApJ* **268,** 342.

Hut P. (1983b). The topology of three-body scattering, *AJ* **88,** 1549.

Hut P. (1985). Binary formation and interactions with field stars, in *Dynamics of Star Clusters (Proc. IAU Symp. 113)*, ed. Goodman J., Hut P. (Reidel, Dordrecht).

Hut P. (1989). Fundamental timescales in star cluster evolution, in *Dynamics of Dense Stellar Systems*, ed. D. Merritt (Cambridge University Press), p. 229.

Hut P. (1996). Binary formation and interactions with field stars, in *Dynamical Evolution of Star Clusters, IAU Symp. 174*, eds. Hut P., Makino J. (Kluwer, Dordrecht).

Hut P. (1997). Gravitational thermodynamics, *Complexity* **3**, 38.

Hut P., Bahcall J.N. (1983). Binary–single star scattering. I – Numerical experiments for equal masses, *ApJ* **268,** 319.

Hut P., Djorgovski S. (1992). Rates of collapse and evaporation of globular clusters, *Nature* **359,** 806.

Hut P., Inagaki S. (1985). Globular cluster evolution with finite-size stars – Cross sections and reaction rates, *ApJ* **298,** 502.

Hut P., Makino J., eds. (1996). *Dynamical Evolution of Star Clusters – Confrontation of Theory and Observations*, IAU Symp. 174 (Kluwer, Dordrecht).

Hut P., Makino J. (1999). Astrophysics on the GRAPE family of special purpose computers, *Science* **283,** 501.

Hut P., Paczynski B. (1984). Effects of encounters with field stars on the evolution of low mass semidetached binaries, *ApJ* **284,** 675.

Hut P., Verbunt F. (1983). White dwarfs and neutron stars in globular cluster X-ray sources, *Nature* **301,** 587.

Hut P., Verhulst F. (1981). Explosive mass loss in binary stars – the two time-scale method, *A&A* **101,** 134–7.

Hut P., Makino J., McMillan S. (1988). Modelling the evolution of globular clusters, *Nature* **336,** 31.

Hut P., McMillan S., Goodman J., Mateo M., Phinney E.S., Pryor C., Richer H.B.,

Verbunt F., Weinberg M. (1992a). Binaries in globular clusters, *PASP* **104,** 981.

Hut P., McMillan S., Romani R.W. (1992b). The evolution of a primordial binary population in a globular cluster, *ApJ* **389,** 527.

Hut P., Makino J., McMillan S. (1995). Building a better leapfrog, *ApJ* **443L,** 93.

Iben I., Jr. (1985). The life and times of an intermediate mass star – In isolation/in a close binary, *QJRAS* **26,** 1.

Iben I., Jr., Tutukov A.V. (1997). The lives of stars: from birth to death and beyond (Part 1), *S&T* **94,** 36.

Iben I., Jr., Tutukov A.V. (1998). How close is close?, *S&T* **95,** 47.

Illingworth G., King I.R. (1977). Dynamical models for M15 without a black hole, *ApJ* **218L,** 109.

Inagaki S. (1980). The gravothermal catastrophe of stellar systems, *PASJ* **32,** 213.

Inagaki S. (1984). The effects of binaries on the evolution of globular clusters, *MNRAS* **206,** 149.

Inagaki S., Konishi T. (1993). Dynamical stability of a simple model similar to self-gravitating systems, *PASJ* **45,** 733.

Inagaki S., Wiyanto P. (1984). On equipartition of kinetic energies in two-component star clusters, *PASJ* **36,** 391.

Ipser J.R., Semenzato R. (1983). On the effects of strong encounters in stellar systems. I – A basis for treating anisotropic systems, *ApJ* **271,** 294.

Janes K., ed. (1991) *The Formation and Evolution of Star Clusters*, ASP Conf. Ser. **13** (ASP, San Francisco).

Jeans J.H. (1902). The stability of a spherical nebula, *Phil. Trans. R. Soc. A* **199,** 49.

Jeans, J.H. (1929). *Astronomy and Cosmogony*, 2nd edition (Cambridge University Press, Cambridge); also Dover, New York (1961).

Jeffreys, H. (1929). *The Earth*, 2nd edition (Cambridge University Press, Cambridge).

Jernigan J.G., Porter D.H. (1989). A tree code with logarithmic reduction of force terms, hierarchical regularization of all variables, and explicit accuracy controls, *ApJS* **71,** 871.

Jha S., Torres G., Stefanik R.P., Latham D., Mazeh T. (2000). Studies of multiple stellar systems – III. Modulations of orbital elements in the triple-lined system HD 109648, *MNRAS* **317,** 375.

Johnston H.M., Verbunt F. (1996). The globular cluster population of low-luminosity X-ray sources, *A&A* **312,** 80.

Johnston K.V., Sigurdsson S., Hernquist L. (1999). Measuring mass-loss rates from Galactic satellites, *MNRAS* **302,** 771.

Joshi K.J., Rasio F.A., Portegies Zwart S. (2000). Monte Carlo simulations of globular cluster evolution. I. Method and test calculations, *ApJ* **540,** 969.

Kaliberda V.S., Petrovskaya I.V. (1970). The velocity variation of a star as a purely discontinuous random process. I. Zero mass stars, *Astrofiz.* **6,** 135.

Kalogera V. (1996). Orbital characteristics of binary systems after asymmetric supernova explosions, *ApJ* **471,** 352.

Kaluzny J., Kubiak M., Szymanski M., Udalski A., Krzeminski W., Mateo M., Stanek K.Z. (1998). The optical gravitational lensing experiment. Variable stars in globular clusters. IV. Fields 104A-E in 47 Tucanae, *A&AS* **128,** 19.

Kandrup H.E., Sideris I.V. (2001). Chaos and the continuum limit in the gravitational

N-body problem. I. Integrable potentials, *Phys. Rev. E* **64,** 056209-1.

Katz J.I. (1975). Two kinds of stellar collapse, *Nature* **253,** 698.

Katz J. (1978). On the number of unstable modes of an equilibrium, *MNRAS* **183,** 765.

Katz J., Lynden-Bell D. (1978). The gravothermal instability in two dimensions, *MNRAS* **184,** 709.

Kawai A., Fukushige T., Makino J., Taiji M. (2000). GRAPE-5: a special-purpose computer for N-body simulations, *PASJ* **52,** 659.

Keenan D.W., Innanen K.A. (1975). Numerical investigation of galactic tidal effects on spherical stellar systems, *AJ* **80,** 290.

Kim E., Einsel C., Lee H.-M., Spurzem R., Lee M.G. (2002). Dynamical evolution of rotating stellar systems – II. Post-collapse, equal mass system, *MNRAS*, in press.

Kim S.S., Figer D.F., Lee H.M., Morris M. (2000). N-body simulations of compact young clusters near the Galactic Center, *ApJ* **545,** 301.

King I. (1959). The escape of stars from clusters. IV. The retardation of escaping stars, *AJ* **64,** 351.

King, I.R. (1962). The structure of star clusters. I. An empirical density law, *AJ* **67,** 471–85.

King, I.R. (1966). The structure of star clusters. III. Some simple dynamical models, *AJ* **71,** 64–75.

King I.R., Sosin C., Cool A.M. (1995). Mass segregation in the globular cluster NGC 6397, *ApJ* **452L,** 33.

Kinoshita H., Yoshida H., Nakai H. (1991). Symplectic integrators and their application to dynamical astronomy, *CeMDA* **50,** 59.

Kiseleva L.G., Eggleton P.P., Orlov V.V. (1994). Instability of close triple systems with coplanar initial doubly circular motion, *MNRAS* **270,** 936.

Klessen R.S., Burkert A. (2001). The formation of stellar clusters: Gaussian cloud conditions. II, *ApJ* **549,** 386.

Klessen R., Kroupa P. (2001). The mean surface density of companions in a stellar-dynamical context, *A&A* **372,** 105.

Kochanek C.S. (1992). The dynamical evolution of tidal capture binaries, *ApJ* **385,** 604.

Kontizas M., Hatzidimitriou D., Bellas-Velidis I., Gouliermis D., Kontizas E., Cannon R.D. (1998). Mass segregation in two young clusters in the Large Magellanic Cloud: SL 666 and NGC 2098, *A&A* **336,** 503.

Korycansky D.G., Laughlin G., Adams F.C. (2001). Astronomical engineering: a strategy for modifying planetary orbits, *ApSS* **275,** 349.

Kozai Y. (1962). Secular perturbations of asteroids with high inclination and eccentricity, *AJ* **67,** 591.

Kroupa P. (1995). Inverse dynamical population synthesis and star formation, *MNRAS* **277,** 1491.

Kulkarni S., Anderson S.B. (1996). Pulsars in globular clusters, in *Dynamical Evolution of Star Clusters*, eds. Hut P., Makino J. (Kluwer (IAUS 174), Dordrecht).

Kulkarni S.R., Hut P., McMillan S. (1993). Stellar black holes in globular clusters, *Nature* **364,** 421.

Kumar P., Goodman J. (1996). Nonlinear damping of oscillations in tidal-capture binaries, *ApJ* **466,** 946.

Kundić T., Ostriker J.P. (1995). Tidal-shock relaxation: a reexamination of tidal shocks in stellar systems, *ApJ* **438,** 702–7.

Kustaanheimo P., Stiefel E. (1965). Perturbation theory of Kepler motion based on spinor regularization, *J. Reine Angew. Math.* **218**, 204.

Kuzmin G.G. (1957). The effect of stellar encounters and the evolution of stellar clusters, *Tartu Astr. Obs. Publ.* **33**, 75.

Lada C.J., Margulis M., Dearborn D. (1984). The formation and early dynamical evolution of bound stellar systems, *ApJ* **285**, 141–52.

Lagoute C., Longaretti P.-Y. (1996). Rotating globular clusters. I. Onset of the gravothermal instability, *A&A* **308**, 441.

Lagrange J. (1772). *Essai sur le Problème des Trois Corps*, in *Oeuvres de Lagrange*, **6**, 231, ed. Serret J.-A. (Gauthier-Villars, Paris, 1873).

Lambert D., Kibler M. (1988). An algebraic and geometric approach to non-bijective quadratic transformations, *J. Phys. A: Math. Gen.* **21**, 307.

Lancellotti C., Kiessling M. (2001). Self-similar gravitational collapse in stellar dynamics, *ApJ* **549L**, 93.

Lançon A., Boily C., eds. (2000). *Massive Star Clusters*, ASP Conf. Ser. **211** (ASP, San Francisco).

Landau L.D., Lifshitz E.M. (1977). *Quantum Mechanics* (Pergamon, Oxford).

Larson, R.B. (1970). A method for computing the evolution of star clusters, *MNRAS* **147**, 323–37.

Laskar, J. (1996). Large scale chaos and marginal stability in the solar system, in *Chaos in Gravitational N-Body Systems* eds. Muzzio J.C., Ferraz-Mello S., Henrard J., (Kluwer, Dordrecht); and *CeMDA* **64**, 115.

Lee H.-M. (1987a). Dynamical effects of successive mergers on the evolution of spherical stellar systems, *ApJ* **319**, 801.

Lee H.-M. (1987b). Evolution of globular clusters including a degenerate component, *ApJ* **319**, 772.

Leimkuhler B. (1999). Reversible adaptive regularization: perturbed Kepler motion and classical atomic trajectories, *Phil. Trans. R. Soc. A* **357**, 1101.

Lemaitre G. (1955). Regularization of the three-body problem, *Vistas Astron.* **1**, 207.

Leonard P.J.T. (1996). Blue stragglers in star clusters, in *The Origins, Evolution, and Destinies of Binary Stars in Clusters*, eds. Milone E.F., Mermilliod J.-C., ASP Conf. Ser. 90 (ASP, San Francisco).

Leonard P.J.T., Duncan M.J. (1988). Runaway stars from young star clusters containing initial binaries. I – Equal-mass, equal-energy binaries, *AJ* **96**, 222.

Levi Civita T. (1906). Sur la résolution qualitative du problème des trois corps, *Acta Math.* **30**, 305.

Lichtenberg A.J., Liebermann M.A. (1992). *Regular and Stochastic Motion*, 2nd edition (Springer-Verlag, Berlin).

Lidov M.L. (1963). On the approximate analysis of the evolution of orbits of artificial satellites, in *Problems of the Motion of Artificial Spacecraft* (Izdat. Akad. Nauk. SSSR, Moscow).

Lightman A.P., Shapiro S.L. (1977). The distribution and consumption rate of stars around a massive, collapsed object, *ApJ* **211**, 244.

Lightman A.P., Shapiro S.L. (1978). The dynamical evolution of globular clusters, *Rev. Mod. Phys.* **50**, 437.

Littlewood J.E. (1986). *A Mathematician's Miscellany*, rev. edition (Cambridge University Press, Cambridge).

Liverani C. (2000). Interacting particles, in *Hard Ball Systems and the Lorentz Gas,* ed. Szasz D. (Springer-Verlag, Berlin).

Livio M. (1996). Cataclysmic variables in globular clusters, in *The Origins, Evolution, and Destinies of Binary Stars in Clusters,* eds. Milone E.F., Mermilliod J.-C., ASP Conf. Ser. 90 (ASP, San Francisco).

Lombardi J.C., Jr., Rasio F.A., Shapiro S.L. (1996). Collisions of main-sequence stars and the formation of blue stragglers in globular clusters, *ApJ* **468,** 797.

Louis P.D. (1990). An anisotropic homological model for core collapse in star clusters, *MNRAS* **244,** 478–92.

Louis P.D., Spurzem R. (1991). Anisotropic gaseous models for the evolution of star clusters, *MNRAS* **251,** 408.

Lynden-Bell D. (1967). Statistical mechanics of violent relaxation in stellar systems, *MNRAS* **136,** 101–21.

Lynden-Bell D. (1968). Runaway centres, *Bull. Astron.* **3,** 305.

Lynden-Bell D., Eggleton P.P. (1980). On the consequences of the gravothermal catastrophe, *MNRAS* **191,** 483.

Lynden-Bell D., Kalnajs A.J. (1972). On the generating mechanism of spiral structure, *MNRAS* **157,** 1–30.

Lynden-Bell D., Wood R. (1968). The gravo-thermal catastrophe in isothermal spheres and the onset of red-giant structure for stellar systems, *MNRAS* **138,** 495.

Makino J. (1996). Postcollapse evolution of globular clusters, *ApJ* **471,** 796.

Makino J., Aarseth S.J. (1992). On a Hermite integrator with Ahmad–Cohen scheme for gravitational many-body problems, *PASJ* **44,** 141.

Makino J., Hut P. (1991). On core collapse, *ApJ* **383,** 181.

Makino J., Taiji M. (1998). *Scientific Simulations with Special-Purpose Computers* (Wiley, Chichester).

Makino J., Akiyama K., Sugimoto D. (1990). On the apparent universality of the $r^{1/4}$ law for brightness distribution in galaxies, *PASP* **42,** 205–15.

Mandushev G., Spassova N., Staneva A. (1991). Dynamical masses for Galactic globular clusters, *A&A* **252,** 94–9.

Marchal, C. (1990). *The Three-Body Problem* (Elsevier, Amsterdam). This reference is an unbeatable source of information about the general three-body problem.

Marchant A.B., Shapiro S.L. (1980). Star clusters containing massive, central black holes. III – Evolution calculations, *ApJ* **239,** 685.

Mardling R.A. (1995). The role of chaos in the circularization of tidal capture binaries. I. The chaos boundary; II. Long-time evolution, *ApJ* **450,** 722 and 732.

Mardling R.A. (2001). Stability in the general three-body problem, in *Evolution of Binary and Multiple Star Systems; A Meeting in Celebration of Peter Eggleton's 60th Birthday,* eds. Podsiadlowski P., Rappaport S., King A.R., D'Antona F., Burder L., ASP Conf. Ser. 229 (ASP, San Francisco).

McGehee R. (1974). Triple collision in the collinear three-body problem, *Inv. Math.* **27,** 191.

McMillan S.L.W., Aarseth S.J. (1993). An O($N \log N$) integration scheme for collisional stellar systems, *ApJ* **414,** 200.

McMillan S., Hut P. (1994). Star cluster evolution with primordial binaries. 3: Effect of the galactic tidal field, *ApJ* **427,** 793.

McMillan S.L.W., Hut P. (1996). Binary–single-star scattering. VI. Automatic determination of interaction cross sections, *ApJ* **467,** 348.

McMillan S.L.W., McDermott P.N., Taam R.E. (1987). The formation and evolution of tidal binary systems, *ApJ* **318,** 261.

McVean J.R., Milone E.F., Mateo M., Yan L. (1997). Analyses of the light curves of the eclipsing binaries in the globular cluster M71, *ApJ* **481,** 782.

Matthews R., Gilmore G. (1993). Is Proxima really in orbit about Alpha Cen A/B?, *MNRAS* **261L,** 5.

Merritt D. (1981). Two-component stellar systems in thermal and dynamical equilibrium, *ApJ* **86,** 318.

Merritt D. (1999). Elliptical galaxy dynamics, *PASP* **111,** 129.

Merritt D., Tremblay B. (1994). Nonparametric estimation of density profiles, *AJ* **108,** 514–37.

Merritt D., Meylan G., Mayor M. (1997). The stellar dynamics of omega centauri, *AJ* **114,** 1074.

Meylan G., Heggie D.C. (1997). Internal dynamics of globular clusters, *Astron. Astrophys. Rev.* **8,** 1.

Michie R.W., Bodenheimer P.H. (1963). The dynamics of spherical stellar systems, II, *MNRAS* **126,** 267.

Mihos J.C., Hernquist L. (1996). Gas dynamics and starbursts in major mergers, *ApJ* **464,** 641.

Mikkola S. (1984). Encounters of binaries. II – Unequal energies, *MNRAS* **207,** 115.

Mikkola S. (1997). Numerical treatment of small stellar systems with binaries, *CeMDA* **68,** 87.

Mikkola S., Aarseth S.J. (1993). An implementation of *N*-body chain regularization, *CeMDA* **57,** 439.

Mikkola S., Tanikawa K. (1999). Algorithmic regularization of the few-body problem, *MNRAS* **310,** 745.

Miller R.H. (1964). Irreversibility in small stellar systems, *ApJ* **140,** 250.

Mladenov I.M. (1991). Geometric quantization of the five-dimensional Kepler problem, in *Differential Geometric Methods in Theoretical Physics*, LNP 375 eds. Bartocci C., Bruzzo U., Cianci R. (Springer-Verlag, Berlin), p. 387.

Montgomery R. (2001). A new solution to the three-body problem, *Not. AMS* **48,** 471.

Moore C. (1993). Braids in classical dynamics, *Phys. Rev. Lett.* **70,** 3675.

Moser, J. (1973). *Stable and Random Motions in Dynamical Systems* (Princeton University Press, Princeton).

Moulton F.R. (1914). *An Introduction to Celestial Mechanics* (Macmillan; also Dover, New York (1970)).

Murali C., Weinberg M.D. (1997). Evolution of the Galactic globular cluster system, *MNRAS* **291,** 717.

Murphy B.W., Cohn H.N., Hut P. (1990). Realistic models for evolving globular clusters – II. Post core collapse with a mass spectrum, *MNRAS* **245,** 335.

Muzzio J.C., Vergne M.M., Wachlin F.C., Carpintero D.D. (2001). Stellar motions in galactic satellites, in *Dynamics of Natural and Artificial Celestial Bodies* (Proc. US–European Celestial Mechanics Workshop, Poznan, Poland, July 3–7, 2000), eds. Pretka-Ziomek H., Wnuk E., Seidelmann P.K., Richardson D. (Kluwer, Dordrecht).

Nakada Y. (1978). On the gravothermal catastrophe, *PASJ* **30,** 57.

Nash P.E., Monaghan J.J. (1978). A statistical theory of the disruption of three-body systems. III – Three-dimensional motion, *MNRAS* **184,** 119.

Neutsch W. (1991). Quaternionic regularisation of perturbed Kepler motion, Max-Planck-Inst. Math. Bonn, Preprint MPI/91–7.

Newell B., Dacosta G.S., Norris J. (1976). Evidence for a central massive object in the X-ray cluster M15, *ApJ* **208L,** 55.

Newton Sir I. *Mathematical Principles of Natural Philosophy* and *System of the World*, trans. Motte A. (1729), rev. Cajori F., 1930 (University of California Press, 1962).

Ogorodnikov K.F. (1965). *Dynamics of Stellar Systems* (Pergamon Press, Oxford).

Okumura S.K., Makino J., Ebisuzaki T., Fukushige T., Ito T., Sugimoto D., Hashimoto E., Tomida K., Miyakawa N. (1993). Highly parallelized special-purpose computer, GRAPE-3, *PASJ* **45,** 329.

Ostriker J. (1964). The equilibrium of polytropic and isothermal cylinders, *ApJ* **140,** 1056.

Padmanabhan T. (1990). Statistical mechanics of gravitating systems, *Phys. Rep.* **188,** 285–362.

Palmer P.L. (1994). *Stability of Collisionless Stellar Systems*, Astrophysics and Space Science Library, Vol. 185 (Kluwer, Dordrecht).

Paresce F., De Marchi G. (2000). On the globular cluster initial mass function below $1M_\odot$, *ApJ* **534,** 870.

Peacock J.A. (1999). *Cosmological Physics* (Cambridge University Press, Cambridge).

Peebles P.J.E. (1993). *Principles of Physical Cosmology* (Princeton University Press, Princeton).

Peterson C.J., Reed B.C. (1987). Structural parameters and luminosities of globular clusters, *PASP* **99,** 20.

Petit J.-M, Hénon M. (1986). Satellite encounters, *Icarus* **66,** 536–55.

Petrosky T.Y. (1986). Chaos and cometary clouds in the solar system, *Phys. Lett. A* **117,** 328.

Phinney E.S. (1996). Binaries and pulsars in globular clusters, in *The Origins, Evolution, and Destinies of Binary Stars in Clusters*, eds. Milone E.F., Mermilliod J.-C., ASP Conf. Ser. 90 (ASP, San Francisco).

Phinney E.S., Sigurdsson S. (1991). Ejection of pulsars and binaries to the outskirts of globular clusters, *Nature* **349,** 220.

Plummer H.C. (1911). On the problem of distribution in globular star clusters, *MNRAS* **71,** 460.

Poincaré, H. (1892–9). *Les Méthodes Nouvelles de la Mécanique Céleste* (Gauthier-Villars, Paris).

Pollard, H. (1976). *Celestial Mechanics* (Math. Assoc. Amer., Carus Math. Mon., **18**).

Portegies Zwart S.F. (1996). *Interacting Stars*, PhD Thesis, University of Utrecht.

Portegies Zwart S.F., McMillan S.L.W. (2000). Black hole mergers in the universe, *ApJ* **528,** L17.

Portegies Zwart S.F., Verbunt F. (1996). Population synthesis of high-mass binaries, *A&A* **309,** 179.

Portegies Zwart S.F., Hut P., Verbunt F. (1997). Star cluster ecology. I. A cluster core with encounters between single stars, *A&A* **328,** 130–42.

Portegies Zwart S.F., Makino J., McMillan S.L.W., Hut P. (1999). Star cluster ecology. III. Runaway collisions in young compact star clusters, *A&A* **348,** 117.

Portegies Zwart S.F., Makino J., McMillan S.L.W., Hut P. (2001a). How many young star clusters exist in the Galactic Center?, *ApJ* **546L,** 101.

Portegies Zwart S.F., McMillan S.L.W., Hut P., Makino J. (2001b). Star cluster ecology – IV. Dissection of an open star cluster: photometry, *MNRAS* **321,** 199.

Prata S.W. (1971). Dynamic evolution of rich Galactic star clusters. II, *AJ* **76,** 1029.

Press, W.H., Teukolsky, S.A., Vetterling, W.T. and Flannery B.P. (1992). *Numerical Recipes in FORTRAN*, 2nd edition (Cambridge University Press, Cambridge).

Prialnik D. (2000). *An Introduction to the Theory of Stellar Structure and Evolution* (Cambridge University Press, Cambridge).

Pryor C., Meylan G. (1993). Velocity dispersions for Galactic globular clusters, in *Structure and Dynamics of Globular Clusters*, eds. Djorgovski S., Meylan G. (ASP, San Francisco), Conf. Ser., vol. **50,** p. 357.

Quinlan G., Tremaine S. (1992). On the reliability of gravitational N-body integrations, *MNRAS* **259,** 505.

Rasio F.R., McMillan S., Hut P. (1995). Binary–binary interactions and the formation of the PSR B1620–26 triple system in M4, *ApJL* **438,** L33.

Rauch K.P., Tremaine S. (1996). Resonant relaxation in stellar systems, *New Astron.* **1,** 149–70.

Reichl L.E. (1980). *A Modern Course in Statistical Physics* (University of Texas Press, Austin).

Retterer J.M. (1980). The binding-energy distribution of the binaries in a star cluster, *AJ* **85,** 249.

Rosenbluth M.N., MacDonald W.M., Judd D.L. (1957). Fokker–Planck equation for an inverse-square force, *Phys. Rev.* **107,** 1.

Ross D.J., Mennim A., Heggie D.C. (1997). Escape from a tidally limited star cluster, *MNRAS* **284,** 811.

Rubenstein E.P., Bailyn C.D. (1997). Hubble Space Telescope observations of the post-core-collapse globular cluster NGC 6752. II. A large main-sequence binary population, *ApJ* **474,** 701.

Ruffert M. (1992). Collisions between a white dwarf and a main-sequence star. II – Simulations using multiple-nested refined grids, *A&A* **265,** 82.

Saari D.G. (1990). A visit to the Newtonian N-body problem via elementary complex-variables, *Amer. Math. Mon.* **97,** 105.

Saari D.G., Xia Z. (1995). Off to infinity in finite time, *Not. Amer. Math. Soc.* **42,** 538.

Saffer R.A., ed. (1993). *Blue Stragglers*, Proc. Stars Journal Club Miniworkshop (STScI, Baltimore).

Salpeter, E.E. (1955). The luminosity function and stellar evolution, *ApJ* **121,** 161–7.

Sandage A.R. (1953). The color–magnitude diagram for the globular cluster M3, *AJ* **58,** 61.

Saslaw W.C., De Young D.S. (1971). On equipartition in galactic nuclei and gravitating systems, *ApJ* **170,** 423.

Schwarzschild, B. (2000). Theorists and experimenters seek to learn why gravity is so weak, *Phys. Today* **53,** 9, 22.

Schwarzschild M. (1958). *Structure and Evolution of the Stars* (Princeton University Press, Princeton).

Schwarzschild M., Härm R. (1962). Red giants of Population II. II, *ApJ* **136,** 158.

Seife C. (2000). A slow carousel ride gauges gravity's pull, *Science* **288,** 944.

Shu F.H. (1978). On the statistical mechanics of violent relaxation, *ApJ* **225**, 83.

Siegel C.L., Moser J.K. (1971). *Lectures on Celestial Mechanics* (Springer-Verlag, Berlin).

Siegel M.H., Majewski S.R., Cudworth K.M., Takamiya, M. (2001). A cluster's last stand: the death of Palomar 13, *AJ* **121**, 935.

Sigurdsson S., Hernquist L. (1993). Primordial black holes in globular clusters, *Nature* **364**, 423.

Sigurdsson S., Phinney E.S. (1993). Binary–single star interactions in globular clusters, *ApJ* **415**, 631.

Sills A., Faber J.A., Lombardi J.C., Jr., Rasio F.A., Warren A.R. (2001). Evolution of stellar collision products in globular clusters. II. Off-axis collisions, *ApJ* **548**, 323.

Simonovic N., Grujic P. (1987). Small-energy three-body systems: I Threshold laws for the Coulomb interaction, *J. Phys. B* **20**, 3427.

Smith H., Jr. (1979). The dependence of statistical results from N-body calculations on N, *A&A* **76**, 192.

Sobouti Y. (1985). Linear oscillations of isotropic stellar systems. II – Radial modes of energy-truncated models, *A&A* **147**, 61.

Soker N. (1996). H-function evolution during violent relaxation, *ApJ* **457**, 287–90.

Spitzer L., Jr. (1940). The stability of isolated clusters, *MNRAS* **100**, 396.

Spitzer L., Jr. (1969). Equipartition and the formation of compact nuclei in spherical stellar systems, *ApJ* **158**, L139.

Spitzer L., Jr. (1987). *Dynamical Evolution of Globular Clusters* (Princeton University Press, Princeton).

Spitzer L., Jr., Chevalier R.A. (1973). Random gravitational encounters and the evolution of spherical systems. V. Gravitational shocks, *ApJ* **183**, 565.

Spitzer L., Jr., Härm R. (1958). Evaporation of stars from isolated clusters, *ApJ* **127**, 544.

Spitzer L., Jr., Hart M.H. (1971). Random gravitational encounters and the evolution of spherical systems. I. Method, *ApJ* **164**, 399.

Spitzer L., Jr., Mathieu R.D. (1980). Random gravitational encounters and the evolution of spherical systems. VIII. Clusters with an initial distribution of binaries, *ApJ* **241**, 618.

Spitzer L., Jr., Shapiro S.L. (1972). Random gravitational encounters and the evolution of spherical systems. III. Halo, *ApJ* **172**, 529.

Spitzer L., Jr., Shull J.M. (1975). Random gravitational encounters and the evolution of spherical systems. VI. Plummer's model, *ApJ* **200**, 339.

Spitzer L., Jr., Thuan T.X. (1972). Random gravitational encounters and the evolution of spherical systems. IV. Isolated systems of identical stars, *ApJ* **175**, 31.

Spurzem R. (1999). Direct N-body simulations, *J. Comp. Appl. Math.* **109**, 407–32.

Spurzem R., Baumgardt H. (2001). A parallel implementation of an Aarseth N-body integrator on general and special purpose supercomputers, *MNRAS*, submitted.

Statler T.S., Ostriker J.P., Cohn H.N. (1987). Evolution of N-body systems with tidally captured binaries through the core collapse phase, *ApJ* **316**, 626.

Stiefel E.L., Scheifele G. (1971). *Linear and Regular Celestial Mechanics* (Springer-Verlag, Berlin).

Stodółkiewicz J.S. (1985). Monte-Carlo calculations, in *Dynamics of Star Clusters*, ed. Goodman J., Hut P. (Reidel (IAUS 113), Dordrecht).

Stumpff K. (1962). *Himmelsmechanik*, Band I (VEB Deutscher Verlag der Wissenschaften, Berlin).

Sturrock P.A. (1994). *Plasma Physics* (Cambridge University Press, Cambridge).

Sugimoto D. (1996). A comparative study of globular clusters, in *Dynamical Evolution of Star Clusters*, eds. Hut P., Makino J., IAUS 174 (Kluwer, Dordrecht).

Sugimoto D., Bettwieser E. (1983). Post-collapse evolution of globular clusters, *MNRAS* **204P,** 19.

Sundman K.F. (1913). Mémoire sur le problème des trois corps, *Acta Math.* **36,** 105.

Surdin V.G. (1995). Dynamical properties of globular clusters: primordial or evolutional?, *A&AT* **7,** 147.

Sussman G.J., Wisdom J. (1988). Numerical evidence that the motion of Pluto is chaotic, *Science* **241,** 433.

Sutantyo W. (1975). The formation of globular cluster X-ray sources through neutron star-giant collisions, *A&A* **44,** 227.

Synge J.L. (1964). *Relativity: The General Theory* (North-Holland, Amsterdam).

Szebehely V. (1974). Numerical investigation of a one-parameter family of triple close approaches occurring in stellar systems, *AJ* **79,** 1449.

Szebehely V., Peters C.F. (1967). Complete solution of a general problem of three bodies, *AJ* **72,** 876.

Takahashi K. (1995). Fokker–Planck models of star clusters with anisotropic velocity distributions I. Pre-collapse evolution, *PASJ* **47,** 561.

Takahashi K., Portegies Zwart S.F. (1998). The disruption of globular star clusters in the Galaxy: a comparative analysis between Fokker–Planck and *N*-body models, *ApJ* **503,** 49.

Terlevich E. (1987). Evolution of *N*-body open clusters, *MNRAS* **224,** 193.

Testa V., Ferraro F.R., Chieffi A., Straniero O., Limongi M., Fusi Pecci F. (1999). The Large Magellanic Cloud cluster NGC 1866: new data, new models, new analysis, *AJ* **118,** 2839.

Teuben P.J., Hut P., Levy S., Makino J., McMillan S., Portegies Zwart S., Shara M., Emmart C. (2001). Immersive 4D interactive visualization of large-scale simulations, in *Astronomical Data Analysis Software and Systems X*, eds. Primini F.A., Harnden, Jr., F.R., Payne H.E., ASP Conf. Ser. 238 (ASP, San Francisco).

Theis C., Spurzem R. (1999). On the evolution of shape in *N*-body simulations, *A&A* **341,** 361.

Thompson, W.B. (1962). *An Introduction to Plasma Physics* (Pergamon Press, Oxford).

Thorsett S.E., Arzoumanian Z., Camilo F., Lyne A.G. (1999). The triple pulsar system PSR B1620-26 in M4, *ApJ* **523,** 763.

Timmes F.X., Woosley S.E., Weaver T.A. (1996). The neutron star and black hole initial mass function, *ApJ* **457,** 834.

Tolman, R.C. (1938). *The Principles of Statistical Mechanics* (Clarendon Press, Oxford).

Toomre A., Toomre J. (1972). Galactic bridges and tails, *ApJ* **178,** 623.

Tout C.A., Aarseth S.J., Pols O.R. (1997). Rapid binary star evolution for *N*-body simulations and population synthesis, *MNRAS* **291,** 732.

Tout C.A., Kembhavi A. (1993). Structural changes in stars following tidal capture, in

The Globular Cluster–Galaxy Connection, eds. Smith G.H., Brodie J.P., ASP Conf. Ser. 48 (ASP, San Francisco).

Trager S.C., King I.R., Djorgovski S. (1995). Catalogue of Galactic globular-cluster surface-brightness profiles, *AJ* **109**, 218.

Tremaine S. (1981). Galaxy mergers, in *The Structure and Evolution of Normal Galaxies*, eds. Fall S.M., Lynden-Bell D. (Cambridge University Press, Cambridge).

Tremaine S., Hénon M., Lynden-Bell D. (1986). *H*-functions and mixing in violent relaxation, *MNRAS* **219**, 285–97.

Tremaine S., Weinberg M.D. (1984). Dynamical friction in spherical systems, *MNRAS* **209**, 729.

Tsuchiya T., Gouda N. (2000). Relaxation and Lyapunov time scales in a one-dimensional gravitating sheet system, *Phys. Rev. E* **61**, 948.

van Albada T.S. (1982). Dissipationless galaxy formation and the $r^{1/4}$ law, *MNRAS* **201**, 939–955.

van den Bergh S. (1980). Star clusters as touchstones for theories of galactic evolution – a few examples, in *Star Clusters*, ed. Hesser J.E., IAU Symp. 85 (D. Reidel, Dordrecht).

van Leeuwen F., Le Poole R.S., Reijns R.A., Freeman K.C., de Zeeuw P.T. (2000). A proper motion study of the globular cluster ω Centauri, *A&A* **360**, 472.

Verbunt F., Hut P. (1987). The globular cluster population of X-ray binaries, in *The Origin and Evolution of Neutron Stars*, eds. Helfand D.J., Huang J.-H. IAUS 125 (Reidel, Dordrecht).

Vesperini E. (1998). Evolution of the mass function of the Galactic globular cluster system, *MNRAS* **299**, 1019.

Vivarelli M.D. (1985). The KS-transformation in hypercomplex form and the quantization of the negative-energy orbit manifold of the Kepler problem, *Celes. Mech.* **36**, 349.

Volk O. (1975). Zur Geschichte der Himmelsmechanik: Johannes Kepler, Leonhard Euler und die Regularisierung, Preprint No. 4, Math. Inst., University of Würzburg.

von Hoerner S. (1957). Internal structure of globular clusters, *ApJ* **125**, 451.

von Hoerner S. (2001). How it all started, in *Dynamics of Star Clusters and the Milky Way*, eds. Deiters S., Fuchs B., Just A., Spurzem R., Wielen R. ASP Conf. Ser. 228 (ASP, San Francisco).

Waldvogel J. (1982). Symmetric and regularized coordinates on the plane triple collision manifold, *Celes. Mech.* **28**, 69.

Wannier G.H. (1953). The threshold law for single ionization of atoms or ions by electrons, *Phys. Rev.* **90**, 817.

Webbink R.F. (1988). Kinematics of the Galactic globular cluster system, in *The Harlow–Shapley Symposium on Globular Cluster Systems in Galaxies* IAU Symp. 126 (Kluwer, Dordrecht), p. 49.

Weidemann V. (2000). Revision of the initial-to-final mass relation, *A&A* **363**, 647.

Weinberg M.D. (1994a). Weakly damped modes in star clusters and galaxies, *ApJ* **421**, 481.

Weinberg M.D. (1994b). Adiabatic invariants in stellar dynamics. 1: Basic concepts, *AJ* **108**, 1398–1402.

Whittaker Sir E.T. (1927). *A Treatise on Analytical Dynamics*, 3rd edition (Cambridge University Press, Cambridge).

Wijers R.A.M.J., van Paradijs J. (1991). An upper limit to the number of pulsars in globular clusters, *A&A* **241L,** 37.

Witte M.G., Savonije G.J. (1999). Tidal evolution of eccentric orbits in massive binary systems. A study of resonance locking, *A&A* **350,** 129.

Yoshida H. (1982). A new derivation of the Kustaanheimo–Stiefel variables, *Celes. Mech.* **28,** 239.

Youngkins V.P., Miller B.N. (2000). Gravitational phase transitions in a one-dimensional spherical system, *Phys. Rev. E* **62,** 4583.

Index